HOW TO MAKE
My own RECIPE

김동석 Eric Kim

전 세계 일류 요리사들에게 명성을 인정받고 있는 세계조리사회연맹(WACS) 인증 국제 대회의 한국인 최초 제과제빵 부문 심사 위원. 2022년에는 월드 초콜릿 마스터즈 WCM(World Chocolate Masters) 대회에서 세계 6위, 아시아 1위라는 놀라운 결과를 이뤄냈다. 20여 년간 쌓아온 현장 경험과 지식을 바탕으로 국내와 해외를 오가며 세미나, 강의, 메뉴 컨설팅 등의 다양한 활동을 펼치고 있다. 서울호서직업전문학교 호텔제과제빵 학과에서 12년째 학생들을 가르치며 후학 양성에도 힘쓰고 있다.

수상/ 경력

2022
World Chocolate Masters 아시아 1위

2018
세계조리사회연맹 WACS(World Association of Chefs Societies) 제과 부문 Continental 심사 위원

(사)한국조리사협회 제과 부문 국가 대표 감독

CACAO BARRY Brand Ambassador

CACAO BARRY Collective Lab Head Chef

2016
독일 세계요리올림픽(IKA Culinary Olimpics) Cold Display Dessert 부문 금메달

2014
룩셈부르크 요리월드컵(Expogast Culinary World Cup) 국가 대표

2012
독일 세계요리올림픽(IKA Culinary Olimpics) 국가 대표

HOW TO MAKE *My own* RECIPE

나만의 디저트 레시피를 구상하는 방법

초판 1쇄 인쇄 2024년 05월 01일
초판 1쇄 발행 2024년 05월 17일

지은이 김동석 | **영문 번역** Kim Eunice | **펴낸이** 박윤선 | **발행처** (주)더테이블

기획·편집 박윤선 | **교정** 김영란 | **디자인** 김보라 | **사진·영상** 박성영
영업·마케팅 김남권, 조용훈, 문성빈 | **경영지원** 김효선, 이정민

주소 경기도 부천시 조마루로385번길 122 삼보테크노타워 2002호
홈페이지 www.icoxpublish.com | **쇼핑몰** www.baek2.kr (백두도서쇼핑몰) | **인스타그램** @thetable_book
이메일 thetable_book@naver.com | **전화** 032) 674-5685 | **팩스** 032) 676-5685
등록 2022년 8월 4일 제 386-2022-000050 호 | **ISBN** 979-11-92855-10-3 (14590), 979-11-92855-09-7 (14590) [세트]

Chef Eric's Patisserie Series 2

HOW TO MAKE *My own* RECIPE

나만의 디저트 레시피를 구상하는 방법

김동석 Eric Kim 지음

PROLOGUE

2000년도에 제과제빵 일을 시작하고 벌써 24년이 지났습니다. 윈도우 베이커리부터 프랜차이즈 업장, 대형 제과제빵 공장, 5성급 호텔의 부티크를 거치면서 페이스트리 기술을 배울 수 있는 모든 곳에서 다양한 경험을 쌓았습니다.

처음에는 단순히 업계의 저명한 선배들처럼 기술자가 되고 싶었고, 그 후에는 남들보다 더 잘하고 싶은 욕심이, 그 후에는 '나만의 것을 만들고 싶다'는 원초적인 생각으로 발전했습니다.

온전한 내 것을 만들기란 무척이나 힘들고 어려운 일입니다. 저 또한 기술에 대한 끝없는 갈망으로 수많은 국제 대회에 출전하며 새로운 기술을 익히고 나만의 기술을 찾아가며 노력을 이어나갔습니다. 당시 국내에 아직 들어오지 않은 다양한 서적들과 세계적으로 유명한 셰프들의 연구 자료를 찾아보며 모방과 응용을 거듭했고, 반복된 실패를 통해 다시 배우며 나만의 기술을 만들어나갔습니다.

나만의 방법을 찾고 구체화시키기 위해 노력한 긴 시간의 끝에 터득한 것은 놀랍게도 화려한 테크닉이 아닌, 제과의 기본기와 재료에 대한 깊은 이해였습니다. 기본기를 튼튼하게 다지고, 내가 사용하는 재료를 정확하게 이해하는 것이 그 시작입니다.

이 책은 일반적인 레시피북과는 다른 방향으로 집필했습니다. 단순히 레시피만을 참고하는 책이 아닌, 이해와 응용을 바탕으로 나만의 레시피를 만들 수 있는 방법을 알려드리기 위해 노력했습니다. 레시피를 구상하는 방법부터 재료를 페어링하고 테크닉을 연출하는 방법까지의 모든 과정을 저의 디저트를 통해 여러분들에게 전달하고 싶었고, 여러분들만의 디저트를 만드는 데 도움이 될 수 있는 하나의 가이드북으로 사용되기를 바라며 제가 연구한 많은 테크닉들을 담았습니다. 물론 마법처럼 빠른 방법은 아닙니다. 배워나가는 과정이 때로는 어렵고, 때로는 생각보다 오래 걸릴 수도 있을 것입니다. 하지만 제대로 된 기본을 토대로 꾸준히 공부할 수 있는 좋은 지침서임에는 틀림없을 것이라 자부합니다.

부디 저의 오랜 경험과 노하우가 담긴 이 시리즈를 통해 나만의 레시피를 만드는 것에 목말라했던 이들에게 유용한 가이드북이 되기를 희망합니다.

It's been 24 years since I started baking in 2000. From window bakeries to franchises and large bakeries to boutiques in five-star hotels, I've gained experience everywhere I could learn the art of pastry.

Initially, I simply wanted to be a technical chef like my distinguished seniors in the industry. This desire to do better than others evolved and then into the primal idea of 'building something of my own.'

Creating something entirely one's own is a very demanding and challenging task. With an insatiable appetite for technical skills, I competed in numerous international competitions to learn new techniques and find my own. I studied various books that were not yet available in Korea at the time and researched materials from world-renowned chefs. I imitated and applied them and learned from repeated failures to create my own techniques.

After a long time of trying to find and refine my own methods, what I learned was surprisingly not fancy techniques but a deep understanding of the basics of pastry and ingredients. It all starts with a strong foundation and an accurate understanding of the ingredients you use.

I've written this book differently than an ordinary recipe book - I've tried to give you a way to create your own recipes based on understanding and application rather than simply referencing recipes. I wanted to take you through the entire process, from how I conceived of the recipe to how I paired the ingredients and executed the technique through my desserts. Furthermore, I've included many of the techniques I've researched, hoping they can serve as a guidebook to help you create your own desserts.

It's not a magic bullet, of course - the learning curve will sometimes be strenuous, and sometimes it will take longer than you imagined - but this is a sound guide to get you started on the right foot.

This series will be a helpful guidebook for those thirsty to create their own recipes.

March 2024, Author **김동석 Eric Kim**

HOW TO USE THIS BOOK

이 책을 활용하는 법

이 책은 일반적인 디저트 레시피북과는 다른 방식으로 기획되었습니다. 정해져 있는 레시피가 아닌, 나 스스로 디저트 메뉴를 구상하고자 할 때 활용할 수 있는 가이드북입니다. 메뉴마다의 페어링, 텍스처, 테크닉, 디자인의 포인트를 다양하게 보여주고자 쿠키, 치즈케이크, 피낭시에처럼 홈베이커도 충분히 만들 수 있는 간단한 메뉴부터 비교적 구성 요소가 많은 메뉴들까지 다양하게 담았습니다. (시리즈 1권 『파티스리: 더 베이직』에서 제과에 관한 전반적인 이론을 다뤘다면, 이 책에서는 그 이론들을 활용한 메뉴를 바탕으로 나만의 레시피를 구상하는 방법을 설명합니다.)

책에서 제안하는 디저트 메뉴를 구상하는 순서는 아래와 같습니다.

STEP 1 메뉴를 선택한다.

STEP 2 메인이 되는 맛을 선택한다.

STEP 3 메인이 되는 맛과 어울리는 재료를 페어링한다.

STEP 4 제과에서 기본 구성 요소인 크림, 스펀지(시트), 인서트(충전물)를 선택한다.

STEP 5 필요한 경우 크리스피한 식감을 낼 수 있는 재료를 더하거나, 커버(글레이즈나 피스톨레)를 씌운다.

구성 요소(크림, 스펀지, 인서트, 커버, 크리스피)는 아래와 같이 분류하고 있습니다.

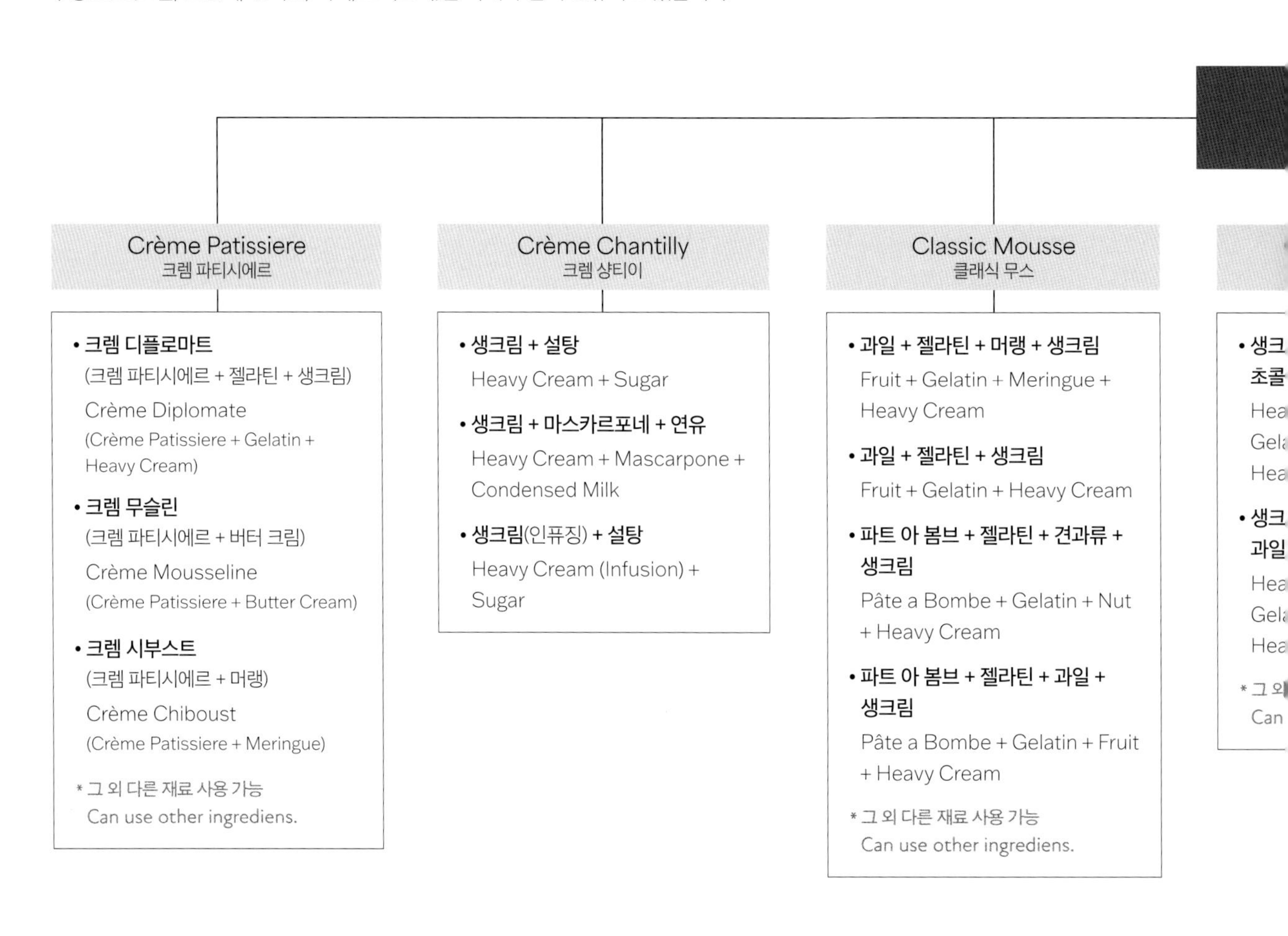

This book is designed to be different than your typical dessert recipe book. It's not a collection of fixed recipes but rather a guidebook you can use to create your own dessert menu. To showcase the different pairings, textures, techniques, and design points of each menu, we've included a variety of dishes that are simple enough for home bakers to make, such as cookies, cheesecake, and financiers, as well as dishes with more components. (Book 1 of the series, *PATISSERIE: THE BASICS*, covered the overall theories of pastry, and this book explains how to use those theories to build your own recipes based on your menu.)

Here's how to plan a dessert menu as suggested in the book.

STEP 1 *Select a menu.*

STEP 2 *Choose a main flavor.*

STEP 3 *Pair the main flavor with the ingredients that go with it.*

STEP 4 *Select the basic building blocks of confectionery: cream, sponge (cake), and insert (filling).*

STEP 5 *Add a crunchy texture or covering (glaze or spray) if needed.*

The components (cream, sponge, insert, cover, and crisp) are categorized as follows;

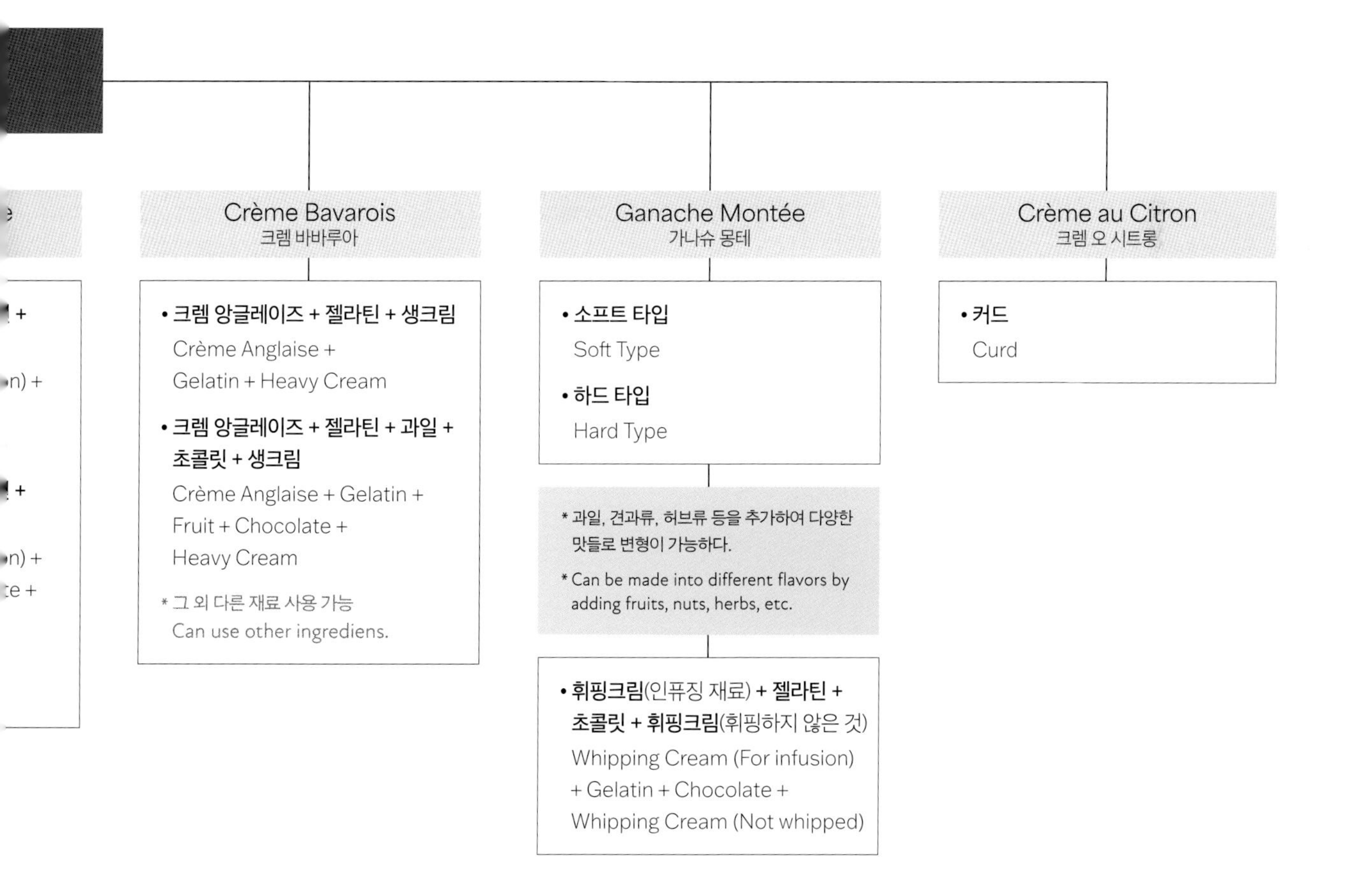

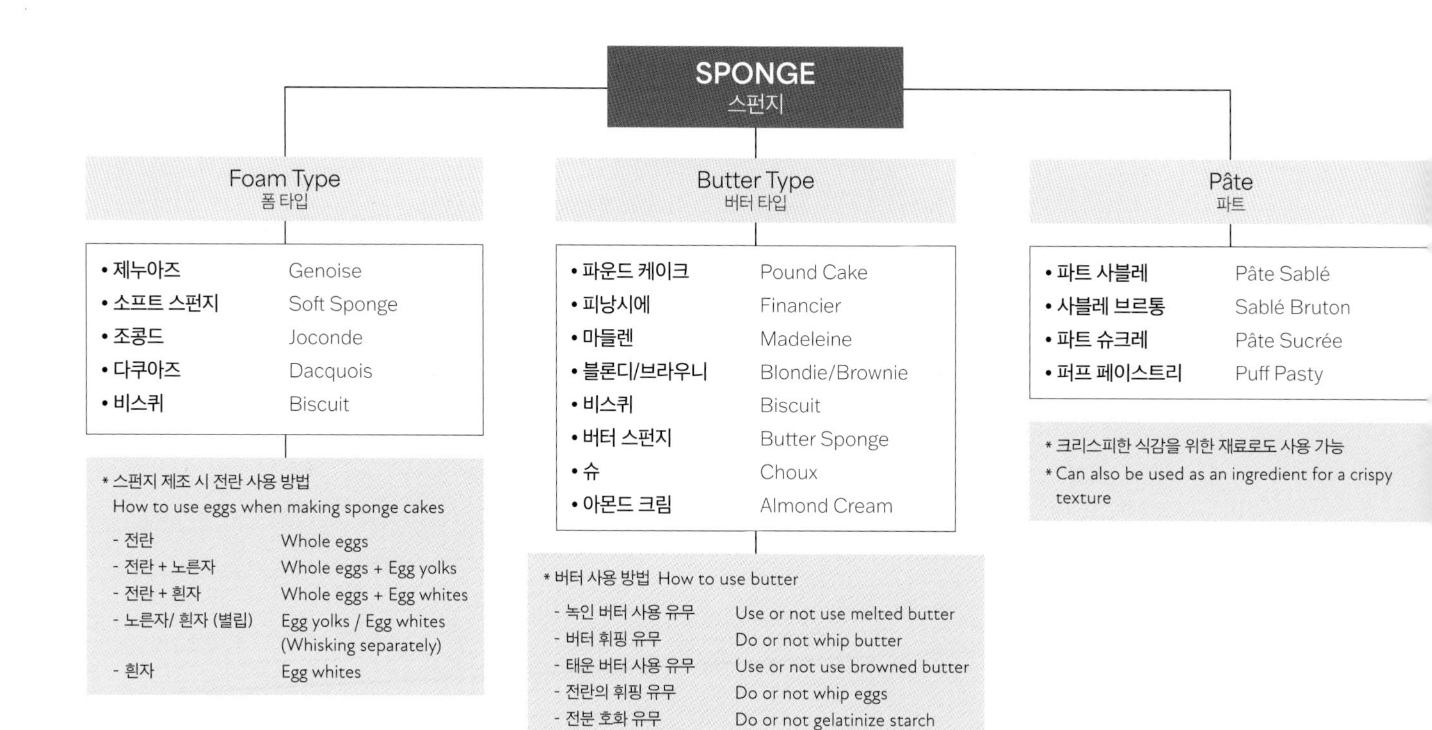

• 쿨리 Coulis
• 콩포트 Comport
• 가나슈 Ganache
• 캐러멜 Caramel
• 젤리 Jelly
• 페이스트 Paste
• 프랄린 Praline
• 콩피 Confit
• 커스터드 Custard
• 가나슈 크레뫼 Ganache Cremeux
• 겔 Gel

• 파에테포요틴 Paillet Feuilletine
• 크럼블 Crumble
• 라이스 크런치 Rice Crunch
• 머랭 Meringue

(그 외 크런치한 식감을 낼 수 있는 모든 것 포함)
(Including anything else that will give you a crunchy texture)

이 책의 '에릭 티라미수 도넛'을 예로 설명해보겠습니다.

Let's use "Eric's Tiramisu Donuts" from this book as an example to illustrate.

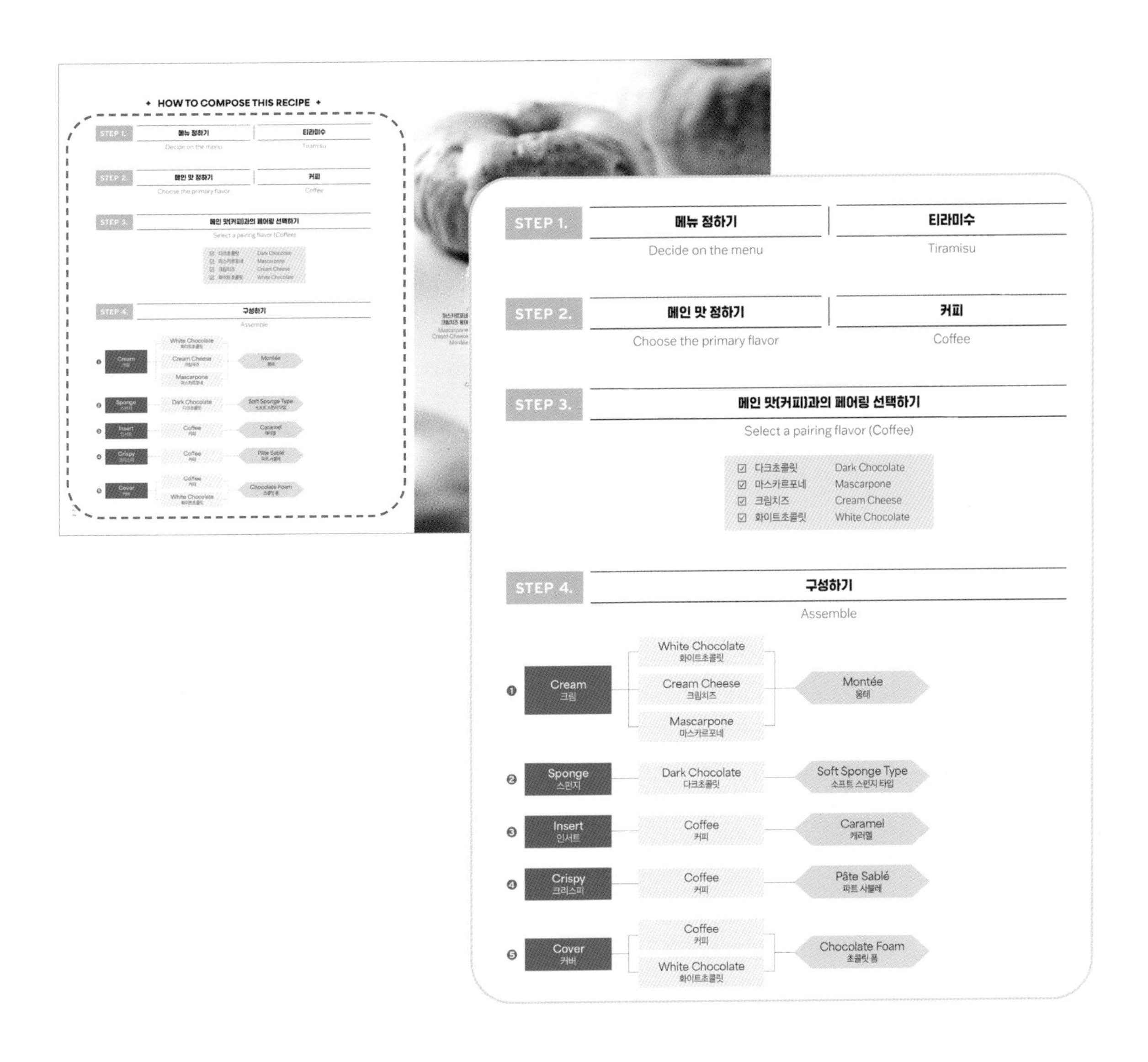

STEP 1. 메뉴를 선택한다.

→ 티라미수

STEP 2. 메인이 되는 맛을 선택한다.

→ 커피

STEP 3. 메인이 되는 맛과 어울리는 재료를 페어링한다.

→ 다크초콜릿, 마스카르포네, 크림치즈, 화이트초콜릿

* 부록 디저트 페어링 북의 '커피' 파트를 참고해 페어링 재료를 선택해도 좋습니다.

STEP 1. Select the menu.

→ Tiramisu

STEP 2. Select the main flavor.

→ Coffee

STEP 3. Pair ingredients that match the main flavor.

→ Dark chocolate, mascarpone, cream cheese, white chocolate

* You can also refer to the 'Coffee' section of the appendix *DESSERT PAIRING BOOK* to help you choose your pairing ingredients.

STEP 4. **제과에서 기본 구성 요소인 크림, 스펀지(시트), 인서트(충전물)를 선택한다.**

→ 크림은 크림치즈와 마스카르포네로 맛을 낸 몽테를 사용했다.

→ 스펀지는 다크초콜릿으로 맛을 낸 소프트 스펀지를 사용했다.

→ 인서트는 커피로 맛을 낸 캐러멜을 사용했다.

STEP 4. **Select the basic components of the pastry: cream, sponge (cake), and insert (filling).**

→ For the cream, we used montée flavored with cream cheese and mascarpone.

→ The sponge is a soft sponge flavored with dark chocolate.

→ The insert is a caramel flavored with coffee.

STEP 5. **필요한 경우 크리스피한 식감을 낼 수 있는 재료를 더하거나, 커버(글레이즈나 피스톨레)를 씌운다.**

→ 크리스피한 식감은 커피로 맛을 낸 파트 사블레를 사용했다.

→ 커버는 커피로 맛을 낸 초콜릿 폼을 사용했다.

STEP 5. **If necessary, add ingredients to create a crunchy texture or coat with a covering (glaze or spray).**

→ For the crispy texture, we used pâte sablé flavored with coffee.

→ For the covering, we used coffee-flavored chocolate foam.

나만의 디저트 개발이 처음이거나, 어려움이 있으신 분들은 책에서 소개하는 순서대로 각각의 디저트가 어떻게 구성되었는지 디저트 페어링 북과 함께 보면서 이해한다면 나만의 디저트를 구상하고 개발하는 데 분명 많은 도움이 될 것입니다.

이 책에 담긴 33가지(베리에이션 메뉴를 포함하면 총 43가지) 메뉴들은 모두 저자의 시그니처 메뉴이거나, 책을 위해 많은 테스트를 거쳐 개발한 메뉴이거나, WCM(월드 초콜릿 마스터즈) 대회에서 높은 점수를 받은 메뉴입니다. 그렇기에 책의 레시피 그대로 만들거나 판매하기에는 물론, 구성 요소나 테크닉 일부를 활용하거나 변형해 사용하기에도, 각 메뉴마다의 구성 요소들을 재조합해 새롭게 만들기에도 더없이 좋을 것입니다.

If you're new to developing your own desserts, or if you're struggling with it, understanding how each dessert is organized in the order presented in the book, along with the *DESSERT PAIRING BOOK*, will definitely help you conceive and develop your own desserts.

All 33 menus in the book (43 including the variations) are either the author's signature menus, heavily tested and developed for the book, or scored highly in the World Chocolate Masters (WCM) competition. Therefore, they're great to make and sell as recipes from the book, to use or adapt some of the components or techniques, or to mix and match components from each menu to create something new.

▶ Chef Eric's Patisserie Series는 총 3권으로 구성되어 있으며 서로 연결되어 있습니다. 1권인 『파티스리: 더 베이직』은 제과에 관한 이론을 설명하며, 2권인 『HOW TO MAKE MY OWN RECIPE』는 1권의 이론을 활용한 레시피를 제안하며 나만의 디저트 레시피를 구상하는 방법을 설명합니다. 1권이 없으신 분들도 무리 없이 보실 수 있도록 공정을 표기하였으나, 과정 사진과 상세 설명이 담긴 1권과 함께 보시는 것을 추천합니다. (3권은 데커레이션에 관한 주제로, 2025년 상반기에 출간될 예정입니다.)

▶ Chef Eric's Patisserie Series consists of three interconnected books. Book 1, *PATISSERIE: THE BASICS*, explains the theory behind patisserie, and Book 2, *HOW TO MAKE MY OWN RECIPE*, shows you how to create your own dessert recipes, with recipe suggestions that utilize the theory from Book 1. The processes are labeled so that you can follow along, even if you don't have the first book. However, we recommend that you read it in conjunction with it, which includes photos and detailed instructions. (Book 3 will be on Decoration and is scheduled to be published in the first half of 2025.)

▶ 아래는 이 책에서 반복적으로 사용되는 주요 구성 요소들입니다. 만드는 공정은 각 페이지마다 표기해두었으나, 아래에 명시된 책과 페이지를 참고하는 것을 추천합니다.

▶ Below are the main components used repeatedly throughout this book. We've labeled each page with the process for making them, but we recommend referring to the books and pages listed below.

PÂTE SABLÉ 파트 사블레 → 1권 Book 1, p.114
SABLÉ BRETON 사블레 브르통 → 1권 Book 1, p.122
PÂTE SUCRÉE 파트 슈크레 → 1권 Book 1, p.132

MIRROR GLAZE 미러 글레이즈 → 1권 Book 1, p.326
PECTIN MIRROR GLAZE 펙틴 미러 글레이즈 → 1권 Book 1, p.336
ROOM TEMPERATURE GLAZE 룸 템퍼라처 글레이즈 → 1권 Book 1, p.344

CHOCOLATE MOUSSE 초콜릿 무스 → 1권 Book 1, p.224

CHOCOLATE POWDER 초콜릿 파우더 → 2권 Book 2, p.188

CRUMBLE 크럼블 → 2권 Book 2, p.388

COULIS 쿨리 → 2권 Book 2, p.422

HAZELNUT CARAMEL 헤이즐넛 캐러멜 → 2권 Book 2, p.446

TOFFEE & TOFFEE POWDER 토피 & 토피파우더 → 2권 Book 2, p.504

▶ 이 책의 레시피에서 사용한 모든 젤라틴은 가루젤라틴(200bloom)과 가루젤라틴 무게 5배의 물을 섞어 굳혀 만든 젤라틴매스입니다.

▶ All gelatin used in the recipes in this book is gelatin mass, made by mixing powdered gelatin (200 bloom) with 5 times the weight of the powdered gelatin in water and letting it set.

▶ 이 책의 배합표에서 갈색으로 표시된 부분은 메인 재료와 페어링되는 재료들입니다.

▶ The ingredients in brown in the recipe table are those that are paired with the main ingredient.

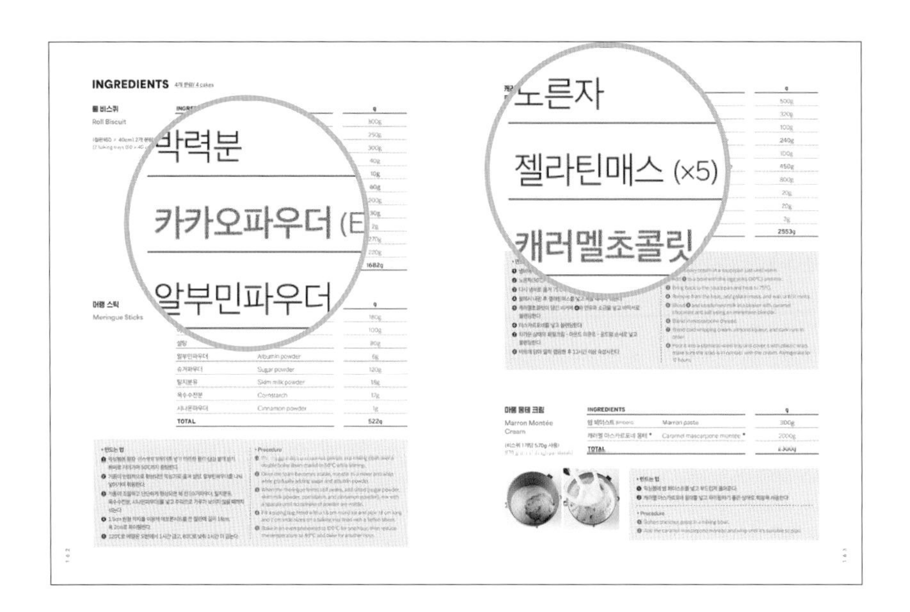

▶ 이 책의 레시피에서 '펙틴(또는 다른 증점제) 반응을 확인한다.'라고 설명된 부분이 있습니다. 이 부분은 주걱으로 들어올렸을 때 물처럼 주르륵 흐르는 것이 아닌, 영상과 같이 천천히 잘리듯 떨어지는 현상을 확인한다는 의미입니다. 점도와 농도로 설명하기 위해 아래의 영상을 첨부하니 참고해주시기 바랍니다.

▶ The recipes says to "check the pectin (or other thickener) reaction/activation." This means that when you lift up the spatula, you're looking for a slow, choppy drip, as in the video, rather than dripping off quickly like water. Please check the video attached below that explains its viscosity and density.

영상으로 확인하기
Check it out in the video

▶ 이 책에서 사용한 몰드는 시판 제품과 자체 제작 몰드 두 가지입니다. 시판 몰드의 경우 제품명을 표기하였고, 자체 제작 몰드의 경우에도 원하는 경우 사용하실 수 있도록 '말랑 홈페이지(smartstore.naver.com/malangstore)'에서 구매하실 수 있도록 하였습니다. 자체 제작 몰드를 사용한 제품은 아래와 같으며, 말랑 홈페이지에서 제품명으로 검색해 몰드를 구입하실 수 있습니다.

▶ This book uses two types of molds: commercial and custom-made. The name of the commercial mold is indicated, and for the custom-made molds, they are available for purchase on the Malang website (smartstore.naver.com/malangstore) so that you can use them if you wish. The products that used custom-made molds are shown below, and you can purchase the molds by searching for the product name on the Malang website.

- BOHEMIAN MANGO 보헤미안 망고 p.172
- MY LOVE RIE 마이 러브 리에 p.206
- CHERRY OATMEAL GATE 체리 오트밀 게이트 p.220
- UNIQUE CUBE PUMPKIN 유니크 큐브 펌킨 p.238
- WHOLE FRUIT FIG 홀 프루트 피그 p.344
- CRISPY BASALT 크리스피 바솔트 p.360
- J'ADORE NUTS 자도르 너츠 p.380
- TINGLING STRAWBERRY CAKE 팅글링 딸기 케이크 p.410

CONTENTS

Chef Eric's SIGNATURE DESSETS

PARING · TEXTURE · TECHNIQUE · DESIGN

21
WHOLE FRUIT FIG
홀 프루트 피그
344

22
CRISPY BASALT
크리스피 바솔트
360

23
FREESTYLE ISPAHAN
프리스타일 이스파한
370

24
J'ADORE NUTS
자도르 너츠
380

25
PEACH BASKET
피치 바스켓
394

26
TINGLING
STRAWBERRY CAKE
팅글링 딸기 케이크
410

Chef Eric's WORLD CHOCOLATE MASTERS (2022)

FOR OUR TOMORROW

Chef Eric's SIGNATURE DESSETS

PARING · TEXTURE ·
TECHNIQUE · DESIGN

1

HYBRID COOKIE

하이브리드 쿠키

PAIRING & TEXTURE
페어링 & 텍스처

- 쫀득하면서도 바삭한 식감의 쿠키 반죽에 여러 가지 재료를 더해 다양한 맛과 식감으로 연출했다.
- 특히 딸기 우유 쿠키와 리얼 다크 체리 쿠키의 경우 토핑으로 기존에 쉽게 볼 수 없었던 동결건조 과일을 사용해 가볍게 바스러지는 식감과 과일 자체의 프레시한 맛을 쿠키에 더했다.
- 토피넛 쿠키 - 초콜릿(다크, 밀크), 땅콩, 땅콩버터를 사용해 메인 재료인 땅콩의 풍미를 끌어올렸다.
- 리얼 다크 체리 쿠키 - 다크초콜릿과 체리는 맛에 있어서 페어링하기 좋은 재료일 뿐만 아니라 함께 배치했을 때 색감도 잘 어울린다. 다크한 초콜릿과 검붉은 체리의 조화가 맛에 있어서도, 시각적으로도 매력적인 제품이다.
- 딸기 우유 쿠키 - 기존 아메리칸 쿠키에서 쉽게 볼 수 없었던 동결건조 과일을 토핑해 먹었을 때 가볍게 바스러지는 식감과 과일 그 자체의 프레시한 맛을 살렸다. 또한 딸기 초콜릿을 사용해 마치 딸기 우유를 먹는 것 같은 기분 좋은 달콤함을 느낄 수 있게 하였다.
- 말차 코코 쿠키 - 말차, 화이트초콜릿, 코코넛을 페어링해 자칫 너무 진하게 느껴질 수 있는 말차의 맛을 부담스럽지 않게 표현했다.

- Different ingredients are added to the chewy and crispy cookie dough to create a variety of flavors and textures.
- Especially for the strawberry milk cookies, I used freeze-dried fruit, which you don't often see as a topping, adding a light, crumbly texture and the fresh taste of the fruit itself to the cookie.
- Toffee Nut Cookie – I used chocolates (dark and milk), peanuts, and peanut butter to bring out the flavor of the main ingredient, peanut.
- Real Dark Cherry Cookie – Dark chocolate and cherries are not only a good pairing in terms of taste, but also in color when placed together. The combination of dark chocolate and black cherries is as visually appealing as it is flavorful.
- Strawberry Milk Cookie – Using freeze-dried fruit as a topping is not a common practice in traditional American cookies. Yet, it creates a light crunchiness and gives a fresh flavor of the fruit itself. Also, the use of strawberry-flavored chocolate creates a pleasant sweetness that feels like drinking strawberry milk.
- Matcha Coco Cookie – The pairing of matcha, white chocolate, and coconut offers an excellent balance to the flavor of matcha, which can often be overpowering.

TECHNIQUE
테크닉

- 베이스가 되는 한 가지 반죽만 있으면 다양한 변형이 가능한 하이브리드 스타일 아메리칸 쿠키.
- 다양한 풍미의 초콜릿을 메인으로 하고 각각의 초콜릿과 어울리는 부재료를 페어링해 기존의 아메리칸 쿠키보다 더 고급스러운 맛과 모양으로 완성했다.

- It is a hybrid-style American cookie you can modify in many ways with just one base dough.
- With the use of different chocolate flavors as the main ingredient and pairing each with matching ingredients, these cookies taste and look more sophisticated than traditional American cookies.

DESIGN
디자인

- 초콜릿을 파우더 형태로 만들어 쿠키에 토핑해 질감과 맛을 업그레이드했다.
- 완성된 쿠키에 토피파우더를 뿌리고 오븐에서 살짝 녹여 광택이 나는 효과를 주어 고급스러움을 강조했다.

- I made chocolate powders, topped the cookies with them, and upgraded the texture and flavor.
- I sprinkled toffee powder on the baked cookies and melted them slightly in the oven to give them a glossy look for a lavish finish.

✦ HOW TO COMPOSE THIS RECIPE ✦

STEP 1.	메뉴 정하기	아메리칸 쿠키
	Decide on the menu	American Cookie

STEP 2 ~ 3. **메인 맛 정하기 & 메인 맛과의 페어링 선택하기**

Choose the primary flavor & Select a pairing flavor

① TOFFEENUT COOKIE
토피넛 쿠키

- 메인 맛 : 땅콩
- 메인 맛과의 페어링
  - 토피
  - 밀크초콜릿
  - 다크초콜릿

- Main flavor: Peanut
- Pairings with:
  - Toffee
  - Milk chocolate
  - Dark chocolate

② STRAWBERRY MILK COOKIE
딸기 우유 쿠키

- 메인 맛 : 딸기
- 메인 맛과의 페어링
  - 동결건조 딸기
  - 딸기초콜릿

- Main flavor: Strawberry
- Pairings with:
  - Freeze-dried strawberries
  - Strawberry-flavored chocolate

③ MATCHA COCO COOKIE
말차 코코 쿠키

- 메인 맛 : 말차
- 메인 맛과의 페어링
  - 코코넛
  - 화이트초콜릿

- Main flavor: Matcha
- Pairings with:
  - Coconut
  - White chocolate

④ REAL DARK CHERRY COOKIE
리얼 다크 체리 쿠키

- 메인 맛 : 체리
- 메인 맛과의 페어링
  - 동결건조 체리
  - 다크초콜릿

- Main flavor: Cherry
- Pairings with:
  - Freeze-dried cherries
  - Dark chocolate

STEP 4.

구성하기

Assemble

Cookie Flavors 쿠키 종류	Main Ingredient 주재료	Dough 반죽	Topping 토핑
TOFFEENUT COOKIE 토피넛 쿠키	Peanut 땅콩	Base Cookie Dough + Dark Chocolate + Milk Chocolate 베이스 쿠키 반죽 + 다크초콜릿 + 밀크초콜릿	Milk Chocolate 밀크초콜릿 Dark Chocolate 다크초콜릿 Peanut Butter 땅콩버터 Toffee Powder 토피파우더 Peanut 땅콩
STRAWBERRY MILK COOKIE 딸기 우유 쿠키	Strawberry 딸기	Base Cookie Dough + Freeze-Dried Strawberries + Strawberry Chocolate 베이스 쿠키 반죽 + 동결건조 딸기 + 딸기초콜릿	Strawberry Chocolate 딸기초콜릿 Freeze-Dried Strawberries 동결건조 딸기
MATCHA COCO COOKIE 말차 코코 쿠키	Matcha 말차	Base Cookie Dough + Matcha + Coconut + White Chocolate 베이스 쿠키 반죽 + 말차 + 코코넛 + 화이트초콜릿	Matcha Crumble 말차 크럼블 Matcha Chocolate Powder 말차 초콜릿 파우더
REAL DARK CHERRY COOKIE 리얼 다크 체리 쿠키	Cherry 체리	Origine Cherry Cookie Dough 오리진 체리 쿠키 반죽	Cacao Crumble 카카오 크럼블 Dark Chocolate 다크초콜릿 Freeze-Dried Cherries 동결건조 체리 Toffee Powder 토피파우더

INGREDIENTS

쿠키 반죽 *

Cookie Dough

- 사용할 반죽의 양을 계산한 후 배합을 조절한다.
- Calculate the amount of dough to use and adjust the recipe.

INGREDIENTS		g
버터 (Bridel)	Butter	260g
설탕	Sugar	140g
황설탕	Brown sugar	160g
소금	Salt	4g
바닐라에센스 (Aroma Piu)	Vanilla essence	4g
중력분	All purpose flour	400g
베이킹소다	Baking soda	4g
베이킹파우더	Baking powder	8g
시나몬파우더	Cinnamon powder	1g
물엿	Corn syrup	100g
달걀	Eggs	120g
TOTAL		**1201g**

① TOFFEENUT COOKIE 토피넛 쿠키

쿠키 반죽

Cookie Dough

8개 분량/ 8 pieces

(쿠키 1개당 반죽 100g 사용)
(100 grams of dough per cookie)

INGREDIENTS		g
BASE 쿠키 반죽 *	Cookie dough *	720g
청크초콜릿 (다크)	Dark chocolate chunks	80g
청크초콜릿 (밀크)	Milk chocolate chunks	50g
TOTAL		**850g**

토핑

Topping

INGREDIENTS (쿠키 1개당 토핑 양, Amount of topping per cookie)		g
다크초콜릿 (Force Noire 50%)	Dark chocolate	6알/ 6 drops
밀크초콜릿 (Ghana 40%)	Milk chocolate	3알/ 3 drops
땅콩버터	Penut butter	8g
땅콩반태	Peanut halves	8알/ 8 pcs
토피파우더 (p.504)	Toffee powder (p.504)	적당량/ QS
TOTAL		**8g**

② STRAWBERRY MILK COOKIE 딸기 우유 쿠키

쿠키 반죽

Cookie Dough

8개 분량/ 8 pieces

(쿠키 1개당 반죽 100g 사용)
(100 grams of dough per cookie)

INGREDIENTS		g
BASE 쿠키 반죽 *	Cookie dough *	720g
동결건조 딸기파우더	Freeze-dried strawberry powder	15g
동결건조 딸기 다이스	Freeze-dried strawberries, diced	30g
다진 딸기초콜릿 (Callebaut)	Strawberry chocolate, chopped	80g
TOTAL		**845g**

토핑

Topping

INGREDIENTS (쿠키 1개당 토핑 양, Amount of topping per cookie)		g
딸기초콜릿 (Callebaut)	Strawberry chocolate	10알/ 10 drops
동결건조 딸기 슬라이스	Freeze-dried strawberries, sliced	적당량/ QS
동결건조 딸기 다이스	Freeze-dried strawberries, diced	1g
TOTAL		**1g**

NOTE.

③ MATCHA COCO COOKIE 말차 코코 쿠키

쿠키 반죽

Cookie Dough

11개 분량/ 11 pieces

(쿠키 1개당 반죽 100g 사용)
(100 grams of dough per cookie)

INGREDIENTS		g
BASE 쿠키 반죽 *	Cookie dough *	950g
말차파우더	Matcha powder	25g
코코넛롱	Shredded dried coconut	15g
물	Water	26g
청크초콜릿 (화이트)	White chocolate chunks	40g
화이트초콜릿 (Zephyr white 34%)	White chocolate	90g
TOTAL		**1146g**

말차 크럼블 *

Matcha Crumble

INGREDIENTS		**g** (37 pieces)	**g** (11 pieces)
버터 (Bridel)	Butter	360g	90g
코코넛슈거	Coconut sugar	200g	50g
설탕	Sugar	100g	25g
소금	Salt	8g	2g
박력분	Cake flour	600g	150g
말차파우더	Matcha powder	60g	15g
옥수수전분	Cornstarch	60g	15g
물	Water	100g	25g
TOTAL		**1488g**	**372g**

*** 만드는 법**

❶ 믹싱볼에 차가운 상태의 버터, 코코넛슈거, 설탕, 소금을 넣고 비터로 부드럽게 풀어준다.

❷ 체 친 박력분, 말차파우더, 옥수수전분을 넣고 물을 흘려가며 믹싱해 한 덩어리로 만든다.

❸ 테프론시트를 깐 철판 위에서 체(간격 약 0.5cm)에 반죽을 내려 크럼블 상태로 만들어 냉동한다.

❹ 냉동시킨 크럼블은 밀폐용기에 담아 냉동 보관하면서 사용하고, 사용하기 전 손으로 가볍게 풀어준다.

*** Procedure**

❶ Soften cold butter, coconut sugar, sugar, and salt in a mixing bowl using a paddle attachment.

❷ Add sifted cake flour, matcha powder, and cornstarch, then drizzle in water; mix until the dough comes together.

❸ Pass the dough through a coarse sieve (about 0.5cm mesh) onto a baking sheet lined with a Teflon sheet to make crumbles and freeze.

❹ Keep the frozen crumbles stored in an airtight container and use as needed. Gently break them apart with your hands before using them.

말차 초콜릿 파우더

Matcha Chocolate Powder

INGREDIENTS		g
화이트초콜릿 (Zephyr white 34%)	White chocolate	300g
말차파우더	Matcha powder	20g
말토덱스트린	Maltodextrin	200g
소금	Salt	2g
바닐라빈	Vanilla bean	1개 분량/ 1 pc
TOTAL		**522g**

*** 만드는 법**

1. 화이트초콜릿을 50℃로 녹이고 35℃로 식혀 볼에 담는다.
2. 말차파우더, 말토덱스트린, 소금, 바닐라빈을 넣고 주걱으로 고르게 섞고 밀착 랩핑한 후 냉장고에 두어 굳힌다.
3. 로보쿱에서 곱게 갈아준 후 밀폐용기에 담아 냉동 보관한다.

*** Procedure**

1. Melt the white chocolate to 50°C, cool it to 35°C, and put it in a bowl.
2. Stir in matcha powder, maltodextrin, salt, and vanilla bean with a spatula. Adhere the plastic wrap to cover and place in a freezer to set.
3. Finely grind using the Robot Coupe and keep it frozen in an airtight container.

토핑

Topping

INGREDIENTS (쿠키 1개당 토핑 양, Amount of topping per cookie)		g
말차 크럼블 *	Matcha crumble *	20g
말차 초콜릿 파우더	Matcha chocolate powder	적당량/ QS
TOTAL		**20g**

④ REAL DARK CHERRY COOKIE 리얼 다크 체리 쿠키

오리진 체리 쿠키 반죽

Origine Cherry Cookie Dough

12개 분량/ 12 pieces

(쿠키 1개당 반죽 100g 사용)
(100 grams of dough per cookie)

INGREDIENTS		g
버터 (Bridel)	Butter	260g
중력분	All purpose flour	350g
카카오파우더 (Cacaobarry, Plain arome)	Cacao powder	50g
베이킹소다	Baking soda	4g
베이킹파우더	Baking powder	8g
시나몬파우더	Cinnamon powder	1g
설탕	Sugar	140g
황설탕	Brown sugar	160g
소금	Salt	4g
바닐라에센스 (Aroma Piu)	Vanilla essence	4g
물엿	Corn syrup	100g
달걀	Eggs	120g
다크초콜릿 (Venezuela 72%)	Dark chocolate	80g
TOTAL		**1281g**

카카오 크럼블 *

Cacao Crumble

INGREDIENTS		g (38 pieces)	g (19 pieces)
버터 (Bridel)	Butter	360g	180g
코코넛슈거	Coconut sugar	200g	100g
설탕	Sugar	100g	50g
소금	Salt	8g	4g
박력분	Cake flour	600g	300g
카카오파우더 (Extra Brute)	Cacao powder	120g	60g
옥수수전분	Cornstarch	60g	30g
물	Water	100g	50g
바닐라에센스 (Aroma Piu)	Vanilla essence	적당량/ QS	적당량/ QS
TOTAL		**1548g**	**774g**

＊만드는 법

❶ 믹싱볼에 차가운 상태의 버터, 코코넛슈거, 설탕, 소금을 넣고 비터로 부드럽게 풀어준다.

❷ 체 친 박력분, 카카오파우더, 옥수수전분을 넣고 물과 바닐라에센스를 흘려가며 믹싱해 한 덩어리로 만든다.

❸ 테프론시트를 깐 철판 위에서 체(간격 약 0.5cm)에 반죽을 내려 크럼블 상태로 만들어 냉동한다.

❹ 냉동시킨 크럼블은 밀폐용기에 담아 냉동 보관하면서 사용하고, 사용하기 전 손으로 가볍게 풀어준다.

＊Procedure

❶ Soften cold butter, coconut sugar, sugar, and salt in a mixing bowl using a paddle attachment.

❷ Add sifted cake flour, cacao powder, and cornstarch, then drizzle in water and vanilla essesnce; mix until the dough comes together.

❸ Pass the dough through a coarse sieve (about 0.5cm mesh) onto a baking sheet lined with a Teflon sheet to make crumbles and freeze.

❹ Store the frozen crumbles in an airtight container and use them as needed. Gently break them apart with your hands before using.

토핑

Topping

INGREDIENTS (쿠키 1개당 토핑 양, Amount of topping per cookie)		g
카카오 크럼블 *	Cacao crumble *	20g
다크초콜릿 (Venezuela 72%)	Dark chocolate	8알/ 8 drops
동결건조 체리 (반태)	Freeze-dried cherries (halves)	3개/ 3 pcs
토피파우더 (p.504)	Toffee powder (p.504)	적당량/ QS
TOTAL		**20g**

COOKIE DOUGH

<베이스> 쿠키 반죽

1 믹싱볼에 포마드 상태의 버터를 넣고 부드럽게 풀어준다.

2 설탕, 황설탕, 소금, 바닐라에센스를 넣고 고르게 섞일 때까지 믹싱한다.

3 체 친 [중력분, 베이킹소다, 베이킹파우더, 시나몬파우더]를 넣고 믹싱한다.

4 **4-1** 사진처럼 가루 재료가 절반 정도 섞이면 물엿과 달걀(30℃)을 넣고 고르게 섞일 때까지 믹싱한다.

TIP. 반죽을 너무 많이 믹싱하지 않도록 주의한다.

5 완성된 반죽은 한 덩어리로 만들어 밀착 랩핑해 냉장고에 보관한다.

TIP. 쿠키 반죽 배합표에 제시된 부재료를 추가해 다양한 쿠키 반죽으로 응용할 수 있다.

1 Beat the softened butter lightly in a mixing bowl.

2 Mix with sugar, brown sugar, salt, and vanilla essence until thoroughly combined.

3 Mix with sifted [all-purpose flour, baking soda, baking powder, and cinnamon powder].

4 When about half the powdered ingredients are mixed in, as shown in picture **4-1**, add corn syrup and eggs (30°C); mix until thoroughly combined.

TIP. Be careful not to overmix the dough.

5 Knead the finished dough into a ball, wrap it tightly, and refrigerate.

TIP. You can use it to make a variety of cookie dough by adding sub-ingredients suggested in the recipe.

1

2-1

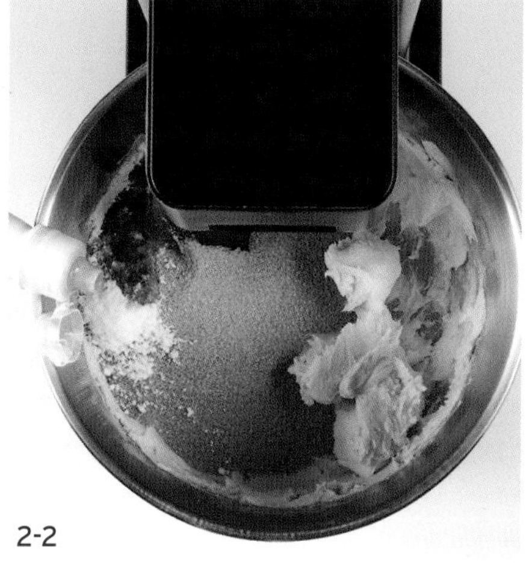

2-2

3

4-1

4-2

4-3

토피넛 쿠키 반죽
Toffeenut
Cookie Dough

말차 코코 쿠키 반죽
Matcha Coco
Cookie Dough

리얼 다크 체리
쿠키 반죽
Real Dark Cherry
Cookie Dough

딸기 우유 쿠키 반죽
Strawberry Milk
Cookie Dough

5

① TOFFEENUT COOKIE
토피넛 쿠키

1 베이스 쿠키 반죽 마지막 공정에서 청크초콜릿(다크, 밀크)을 넣고 섞은 후 한 덩어리로 만들고 밀착 랩핑해 냉장고에서 12시간 정도 숙성시킨다.

2 반죽을 100g씩 분할해 동그랗게 만든 후 테프론시트를 깐 철판에 팬닝한다.

3 180℃로 예열된 오븐에서 8분간 굽다가 잠시 꺼내 세라클 링으로 반죽 가장자리를 돌려가며 최종 크기를 지름 11cm로 만든 후 다시 6분간 굽는다. (총 14~15분)

4 다크초콜릿과 밀크초콜릿을 올린다.

5 땅콩버터를 짤주머니에 담아 쿠키 가장자리에 두 바퀴 원을 그리며 파이핑한다.

6 땅콩반태를 올린다.

7 토피파우더를 뿌린 후 180℃ 오븐에서 약 1분 30초 동안 구워 토피파우더를 녹인다.

8 실온에서 식힌 후 포장한다.

TIP. 포장 시 쿠키 위의 토핑이 망가지지 않도록 제과용 비닐을 덮은 뒤 포장한다.

1 In the last step of the base dough procedure, add chocolate chunks (dark and milk), mix, form into a ball, wrap tightly, and refrigerate for 12 hours.

2 Divide the dough into 100 grams, roll them round, and put on a baking sheet lined with a Teflon sheet.

3 Bake in an oven preheated to 180°C for 8 minutes. Take it out from the oven and roll the edges of the cookies with a round ring to make them 11 cm in diameter. Bake for another 6 minutes. (14~15 minutes in total)

4 Top with milk chocolate and dark chocolate.

5 Put peanut butter in a piping bag and pipe two circles around the edge of the cookies.

6 Arrange peanut halves.

7 Sprinkle toffee powder and bake it at 180°C for 90 seconds to melt the toffee powder.

8 Cool at room temperature before packaging them.

TIP. Place a piece of acetate film on top of the cookies to protect the toppings from damage during packaging.

② STRAWBERRY MILK COOKIE
딸기 우유 쿠키

1. 베이스 쿠키 반죽 마지막 공정에서 동결건조 딸기파우더, 동결건조 딸기 다이스, 다진 딸기초콜릿을 넣고 섞은 후 한 덩어리로 만들고 밀착 랩핑해 냉장고에서 12시간 정도 숙성시킨다.
2. 반죽을 100g씩 분할해 동그랗게 만든 후 테프론시트를 깐 철판에 팬닝한다.
3. 180℃로 예열된 오븐에서 8분간 굽다가 잠시 꺼내 세라클 링으로 반죽 가장자리를 돌려가며 최종 크기를 지름 11cm로 만든 후 다시 10분간 굽는다. (총 18~19분)
4. 딸기초콜릿을 올린다.
5. 동결건조 딸기 슬라이스를 올린 후 동결건조 딸기 다이스를 뿌린다.
6. 실온에서 식힌 후 포장한다.

TIP. 포장 시 쿠키 위의 토핑이 망가지지 않도록 제과용 비닐을 덮은 뒤 포장한다.

1. In the last step of the base dough procedure, add freeze-dried strawberry powder, diced freeze-dried strawberries, and diced strawberry-flavored chocolate, mix, form into a ball, wrap tightly, and refrigerate for 12 hours.
2. Divide the dough into 100 grams, roll them round, and put on a baking sheet lined with a Teflon sheet.
3. Bake in an oven preheated to 180°C for 8 minutes. Take it out from the oven and roll the edges of the cookies with a round ring to make them 11 cm in diameter. Bake for another 10 minutes. (18~19 minutes in total)
4. Top with strawberry-flavored chocolate.
5. Put freeze-dried strawberry slices and sprinkle freeze-dried strawberry powder.
6. Cool at room temperature before packaging them.

TIP. Place a piece of acetate film on top of the cookies to protect the toppings from damage during packaging.

③ MATCHA COCO COOKIE
말차 코코 쿠키

1 베이스 쿠키 반죽 마지막 공정에서 체 친 말차파우더, 코코넛롱, 물을 넣고 섞는다.

2 청크초콜릿, 화이트초콜릿을 넣고 섞은 후 한 덩어리로 만들고 밀착 랩핑해 냉장고에서 12시간 정도 숙성시킨다.

3 반죽을 100g씩 분할해 동그랗게 만든 후 테프론시트를 깐 철판에 팬닝한다.

4 반죽에 말차 크럼블을 소복히 올리고 가볍게 눌러준다.

5 180℃로 예열된 오븐에서 8분간 굽다가 잠시 꺼내 세라클 링으로 반죽 가장자리를 돌려가며 최종 크기를 지름 11cm로 만든 후 다시 10분간 굽는다. (총 18~19분)

6 구워져 나온 쿠키에 말차 초콜릿 파우더를 뿌린다.

7 실온에서 식힌 후 포장한다.

TIP. 포장 시 쿠키 위의 토핑이 망가지지 않도록 제과용 비닐을 덮은 뒤 포장한다.

1 In the last step of the base dough procedure, add sifted matcha powder, shredded coconut, water, and mix.

2 Mix white chocolate chunks and white couverture chocolate to form a ball. Wrap tightly and refrigerate for 12 hours.

3 Divide the dough into 100 grams, roll them around, and put them on a baking sheet lined with a Teflon sheet.

4 Put a heap full of matcha crumble and press gently.

5 Bake in an oven preheated to 180°C for 8 minutes. Take it out from the oven and roll the edges of the cookies with a round ring to make them 11 cm in diameter. Bake for another 10 minutes. (18~19 minutes in total)

6 Sprinkle matcha chocolate powder on the warm cookies.

7 Cool at room temperature before packaging them.

TIP. Place a piece of acetate film on top of the cookies to protect the toppings from damage during packaging.

④ REAL DARK CHERRY COOKIE
리얼 다크 체리 쿠키

1 믹싱볼에 포마드 상태의 버터를 넣고 비터로 부드럽게 풀어준다.

2 체 친 [중력분, 카카오파우더, 베이킹소다, 베이킹파우더, 시나몬파우더], 설탕, 황설탕, 소금, 바닐라에센스를 넣고 섞는다.

3 물엿과 달걀(30℃)을 넣고 섞는다.

4 녹인 다크초콜릿(45~50℃)을 넣고 섞은 후 한 덩어리로 만들고 밀착 랩핑해 냉장고에서 12시간 정도 숙성시킨다.

5 반죽을 100g씩 분할해 동그랗게 만든 후 테프론시트를 깐 철판에 팬닝한다.

6 반죽에 카카오 크럼블을 소복히 올리고 가볍게 눌러준다.

7 180℃로 예열된 오븐에서 8분간 굽다가 잠시 꺼내 세라클 링으로 반죽 가장자리를 돌려가며 최종 크기를 지름 11cm로 만든 후 다시 6분간 굽는다. (총 14~15분)

8 구워져 나온 쿠키에 다크초콜릿과 동결건조 체리(반태)를 올린다.

9 체로 토피파우더를 뿌린 후 180℃ 오븐에서 약 1분 동안 구워 토피파우더를 녹인다.

10 실온에서 식힌 후 포장한다.

TIP. 포장 시 쿠키 위의 토핑이 망가지지 않도록 제과용 비닐을 덮은 뒤 포장한다.

1 Beat the softened butter lightly in a mixing bowl.

2 Mix with sifted [all-purpose flour, cacao powder, baking soda, baking powder, cinnamon powder], sugar, brown sugar, salt, and vanilla essence.

3 Mix with corn syrup and eggs (30°C).

4 Mix with melted dark chocolate (45~50°C), form a ball, wrap tightly, and refrigerate for 12 hours.

5 Divide the dough into 100 grams, roll them round, and put on a baking sheet lined with a Teflon sheet.

6 Put a heap full of cacao crumble and press gently.

7 Bake in an oven preheated to 180°C for 8 minutes. Take it out from the oven and roll the edges of the cookies with a round ring to make them 11 cm in diameter. Bake for another 6 minutes. (14~15 minutes in total)

8 Top with dark chocolate and halved freeze-dried cherries on the warm cookies.

9 Dust toffee powder and bake it at 180°C for 60 seconds to melt the toffee powder.

10 Cool at room temperature before packaging them.

TIP. Place a piece of acetate film on top of the cookies to protect the toppings from damaging during packaging.

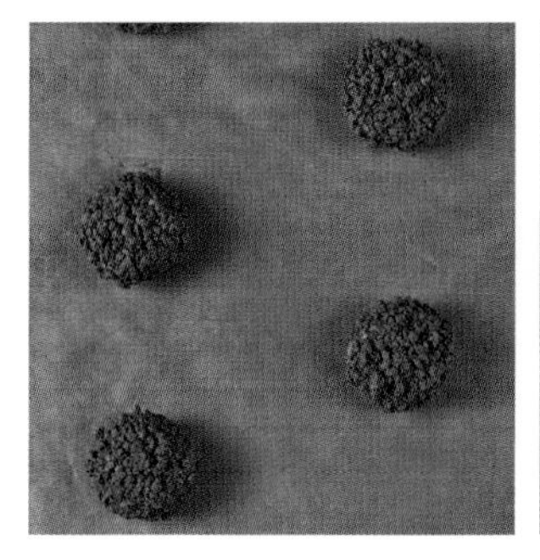

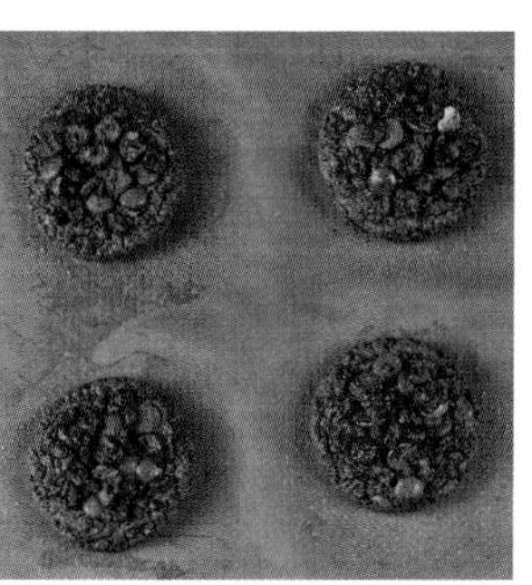

2 OCQUOISE

오쿠아즈

PAIRING & TEXTURE

페어링 & 텍스처

다쿠아즈와 마카롱을 섞어 놓은 듯한 중간 텍스처로, 일반적인 다쿠아즈보다 조금 더 촉촉하고 부드럽게 만든 오렌지 다쿠아즈(오쿠아즈)이다.

The Ocquoise- orange dacquoise - has a texture halfway between dacquoise and macaron, made slightly more moist and soft than the typical dacquoise.

TECHNIQUE

테크닉

알부민파우더 머랭

- 흰자로 머랭을 만드는 일반적인 레시피에서는 퓌레나 과일 등 수분 함량이 높은 신선한 재료를 추가할 수 없어 맛이 한정적이다. 반면 알부민파우더는 건조된 분말 상태이므로 퓌레나 과일 등의 재료를 첨가해 과일 본연의 맛과 향이 느껴지는 제품으로 완성할 수 있다.
- 일반적인 다쿠아즈는 샌딩하는 크림으로 맛에 변화를 준다. 반면 이 다쿠아즈는 머랭을 만들 때 흰자 대신 알부민파우더와 과일 퓌레를 사용해 다쿠아즈 자체에 진한 맛을 연출했다.
- 아몬드파우더와 박력분의 양을 조절해 좀 더 안정적인 제품이 될 수 있도록 완성했다.

ALBUMIN POWDER MERINGUE

- In a typical recipe for meringue made with egg whites, it's challenging to add fresh ingredients with high water content, such as purée or fruit, which limits the flavor. On the other hand, albumin powder is a dried powder in which you can add purée, fruit, and other ingredients to create a product that tastes and smells like the fruit itself.
- A typical dacquoise is flavored with cream filling. As for this dacquoise, I used albumin powder instead of egg whites to make meringue and fruit purée to create a richer flavor.
- The amount of almond powder and cake flour was adjusted to create a more stable product.

✦ HOW TO COMPOSE THIS RECIPE ✦

STEP 1.	메뉴 정하기	다쿠아즈
	Decide on the menu	Dacquoise

STEP 2.	메인 맛 정하기	오렌지
	Choose the primary flavor	Orange

STEP 3. **메인 맛(오렌지)과의 페어링 선택하기**

Select a pairing flavor (Orange)

- ☑ 패션푸르트 — Passion Fruit
- ☑ 생강 — Ginger
- ☑ 크림치즈 — Cream Cheese
- ☑ 바닐라 — Vanilla
- ☑ 꿀 — Honey

STEP 4. **구성하기**

Assemble

오렌지 레몬 다쿠아즈
Orange Lemon
Dacquoise
크림치즈 바닐라 버터 크림
Cream Cheese
Vanilla Butter Cream
만다린 오렌지 쿨리
Mandarin
Orange Coulis

INGREDIENTS

10개 분량/ 10 dacquoises

오렌지 레몬 머랭 *

Orange Lemon Meringue

INGREDIENTS		g
만다린 오렌지 퓌레	Mandarin orange purée	100g
레몬 퓌레	Lemon purée	40g
물	Water	70g
알부민파우더	Albumin powder	30g
설탕	Sugar	200g
옥수수전분	Cornstarch	10g
TOTAL		**450g**

오렌지 레몬 다쿠아즈

Orange Lemon Dacquoise

INGREDIENTS		g
오렌지 레몬 머랭 *	Orange lemon meringue *	250g
설탕	Sugar	25g
오렌지에센스 (Aroma Piu)	Orange essence	5g
아몬드파우더	Almond powder	65g
슈거파우더	Sugar powder	85g
박력분	Cake flour	38g
TOTAL		**468g**

크림치즈 바닐라 버터 크림

Cream Cheese Vanilla Butter Cream

INGREDIENTS		g
버터 (Bridel)	Butter	300g
슈거파우더	Sugar powder	50g
바닐라빈 씨	Vanilla bean seeds	1/4개 분량/ 1/4 pc
연유	Condensed milk	25g
바닐라에센스 (Aroma Piu)	Vanilla essence	5g
크림치즈 (Kiri)	Cream cheese	100g
TOTAL		**480g**

만다린 오렌지 쿨리

Mandarin Orange Coulis

INGREDIENTS		g
만다린 오렌지 퓌레	Mandarin orange purée	160g
패션푸르트 퓌레	Passion fruit purée	90g
꿀	Honey	50g
설탕	Sugar	80g
트레할로스	Trehalose	60g
NH펙틴	NH pectin	13g
생강 퓌레	Ginger purée	5g
오렌지에센스 (Aroma Piu)	Orange essence	3g
TOTAL		**461g**

*** 만드는 법**

❶ 비커에 모든 재료를 담고 바믹서로 블렌딩한다.
❷ 냄비로 옮겨 가열하며 펙틴 반응을 확인한다.
❸ 얼음물에 받쳐 온도를 낮춘다.
❹ 바트에 담아 밀착 랩핑해 냉장고에서 보관한다.

*** Procedure**

❶ Combine all the ingredients in a beaker and mix with an immersion blender.
❷ Cook in a saucepan to activate the pectin.
❸ Let it cool in an ice bath.
❹ Pour it onto a stainless-steel tray and cover it with plastic wrap, making sure the wrap is in contact with the surface of the coulis. Refrigerate.

✤

ORANGE LEMON MERINGUE

오렌지 레몬 머랭

1 비커에 만다린 오렌지 퓌레, 레몬 퓌레, 물을 넣고 50℃로 맞춘다. (전자레인지 사용)

2 미리 섞어둔 [알부민파우더, 설탕, 옥수수전분]을 넣고 바믹서로 블렌딩한다.

3 밀착 랩핑해 냉장고에서 하루 숙성시킨 후 사용한다.

1 Combine mandarin orange purée, lemon purée, and water in a beaker. Use a microwave to heat to 50°C.

2 Add previously mixed [albumin powder, sugar, and cornstarch] and blend with an immersion blender.

3 Adhere the plastic wrap to the preparation to cover, and refrigerate overnight to use.

ORANGE LEMON DACQUOISE

오렌지 레몬 다쿠아즈

1 냉장고에서 숙성시킨 오렌지 레몬 머랭을 믹싱볼에 담아 중탕볼에서 35℃로 맞춘 후 거품이 생기면 설탕을 넣고 휘핑한다.

2 거품이 조밀하고 단단한 상태가 될 때까지 휘핑한다.

TIP. 수분(흰자)이 높은 배합이므로, 이 많은 수분이 빠져나갈 수 있는 공기층이 생길 때까지 꽤 오랫동안 휘핑해야 한다. 휘퍼로 저었을 때 뻑뻑하고 단단한 느낌이 들 때까지 충분히 휘핑한다.

3 볼에 **2**의 2/3와 오렌지에센스를 넣고 주걱으로 가볍게 섞는다.

4 체 친 [아몬드파우더, 슈거파우더, 박력분]을 넣고 섞는다.

5 남은 **2**를 넣고 섞는다.

TIP. 볼에서 섞을 때는 최대한 기포가 죽지 않도록 빠르게 작업하고 곧바로 틀에 파이핑한다. 오래 섞어 기포가 죽고 반죽이 묽어지면 다쿠아즈보다 붓세에 가까운 상태가 된다.

1 Warm the refrigerated orange lemon meringue in a mixing bowl over a double boiler (bain-marie) to 35°C and whip when it becomes frothy.

2 Whip until the foam becomes dense and stiff.

TIP. This meringue has high egg white (water) content, so you will need to whip it for a while to create the air bubbles that allow all this water to escape. Whip until it feels stiff and firm when stirred with the whisk.

3 Add 2/3 of (**2**) and orange essence in a bowl and mix briefly with a spatula.

4 Fold in sifted [almond flour, sugar powder, and cake flour].

5 Fold in the remaining (**2**).

TIP. When mixing in the bowl, work quickly to avoid deflating air bubbles as much as possible and pipe it straight into the molds. If you mix for too long, the air bubbles will deflate, and the batter will become runny, making it more of a bouchée than a dacquoise.

✤

50℃
1
2
3

✤ ✤

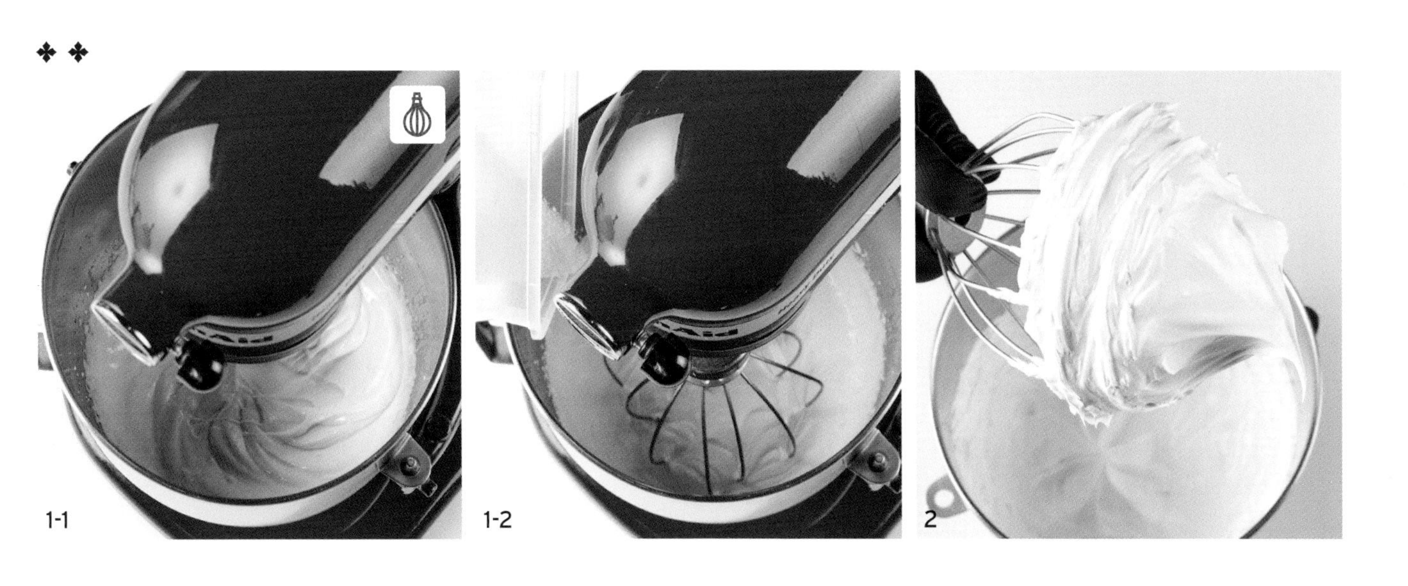
1-1
1-2
2

3
4
5

6 다쿠아즈 틀에 파이핑한 후 슈거파우더를 두 번 뿌린다.

TIP. 다쿠아즈 틀은 안쪽에 물을 묻힌 후 가볍게 털어내 사용한다.
슈거파우더를 뿌린 후 반죽에 흡수되면 다시 한 번 뿌린다.

7 틀을 제거한 후 165~170℃로 예열된 오븐에서 뎀퍼를 100% 열고 10분간 굽는다.

6 Pipe into the dacquoise mold and dust with sugar powder twice.

TIP. Dampen the inside of the mold with water and gently shake it out before using.
After dusting the sugar powder, dust again once the first layer has been absorbed into the batter.

7 Remove the mold and bake in an oven preheated to 165~170°C and damper open at 100% for 10 minutes.

✤ ✤ ✤

CREAM CHEESE VANILLA BUTTER CREAM

크림치즈 바닐라 버터 크림

1 믹싱볼에 포마드 상태의 버터를 넣고 가볍게 풀어준 후 슈거파우더, 바닐라빈 씨를 넣고 믹싱한다.

2 연유, 바닐라에센스를 넣고 믹싱한다.

3 크림치즈를 넣고 충분히 믹싱한다.

4 고르게 섞이면 마무리한다.

1 Beat the softened butter lightly in a mixing bowl, and mix with powdered sugar and vanilla bean seeds.

2 Add condensed milk and vanilla essence to mix.

3 Add cream cheese and mix thoroughly.

4 Finish when evenly combined.

✤ ✤ ✤ ✤

FINISH 마무리

1 다쿠아즈 가장자리에 크림치즈 바닐라 버터 크림을 16g 파이핑한다.

2 가운데에 만다린 오렌지 쿨리를 10g 채운 후 다쿠아즈로 닫아 마무리한다.

1 Pipe 16 grams of cream cheese vanilla buttercream around the edge of the dacquoise.

2 Fill the center with 10 grams of mandarin orange coulis and cover with another dacquoise to finish.

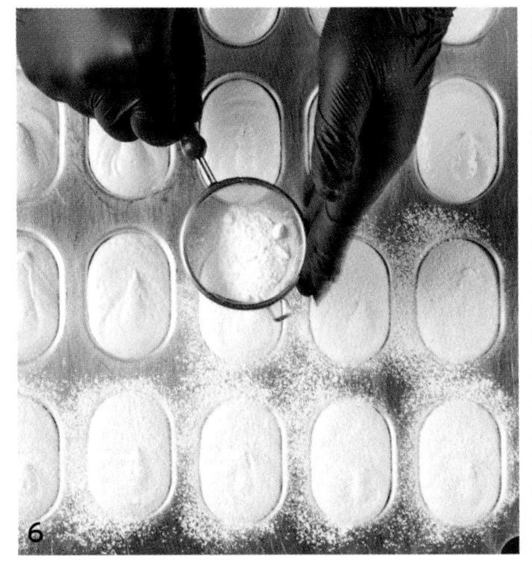
6

7

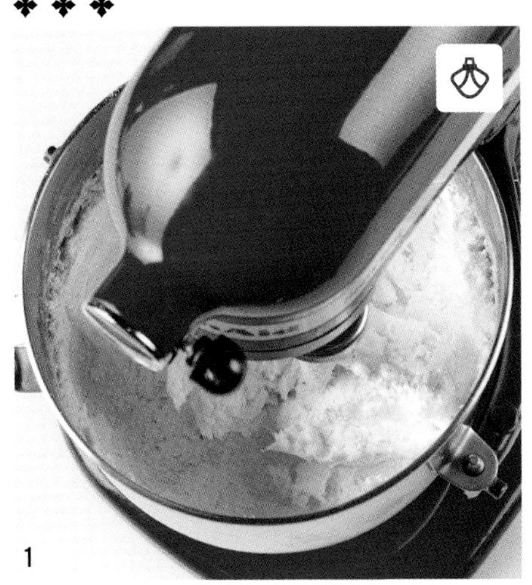
1

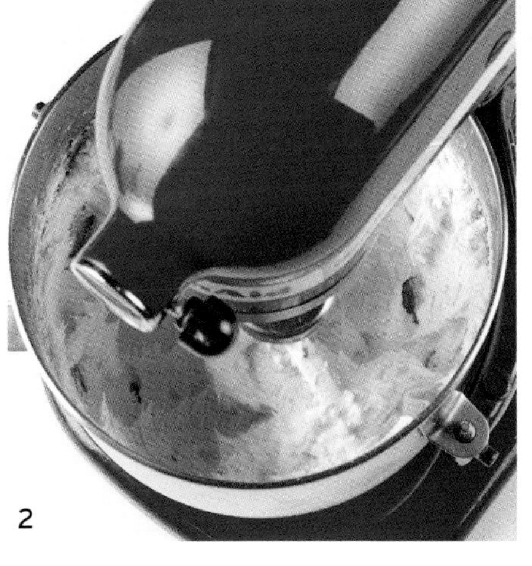
2

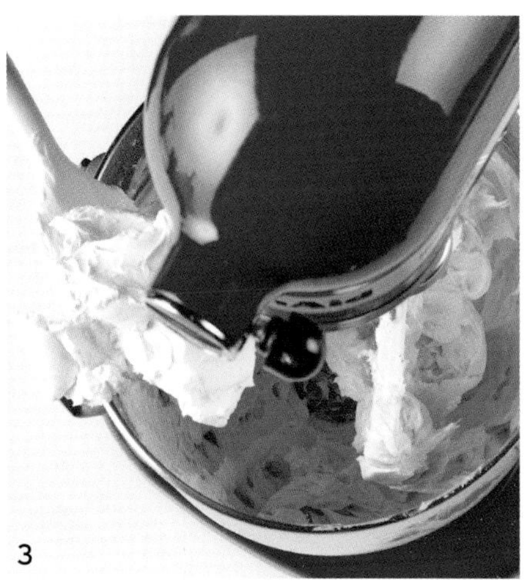
3

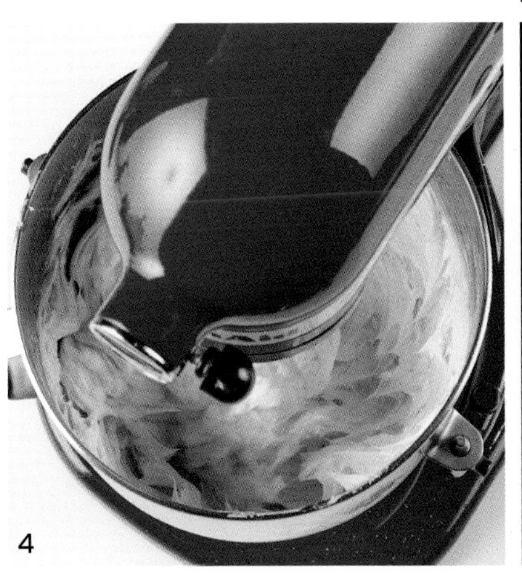
4

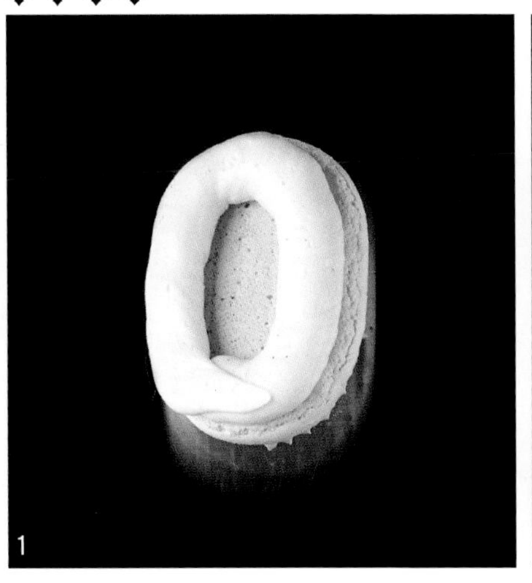
1

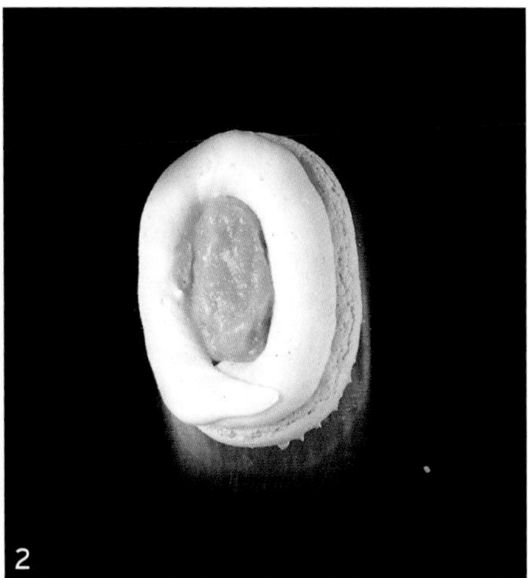
2

3

YUJA INVERTED MACARON

유자 인버트 마카롱

PAIRING & TEXTURE
페어링 & 텍스처

- 일반적인 마카롱은 코크를 기본 맛으로 설정하고(주로 색소로만 변화를 준다.) 샌딩하는 크림이나 충전물로 맛에 변화를 준다. 반면 여기에서 소개하는 마카롱은 머랭을 만들 때 흰자 대신 알부민파우더와 과일 퓌레를 사용해 코크 자체에 진한 과일 맛을 연출했다.
- 샌딩하는 크림은 클래식한 버터 크림에 크림치즈를 더했다.

- Typical macaron coque is made with the same standard flavor (usually only use coloring to give change), and the flavors are added to the cream or filling. On the other hand, the macarons introduced here use albumin powder and fruit purée instead of egg whites to make the meringue, giving a rich fruit flavor in the coque itself.
- The cream used for filling is a classic buttercream with cream cheese.

TECHNIQUE
테크닉

알부민파우더 머랭

- 흰자로 머랭을 만드는 일반적인 레시피에서는 퓌레나 과일 등 수분 함량이 높은 신선한 재료를 추가할 수 없어 맛이 한정적이다. 반면 알부민파우더는 건조된 분말 상태이므로 퓌레나 과일 등의 재료를 첨가해 과일 본연의 맛과 향이 느껴지는 제품으로 완성할 수 있다.
- 일반적인 마카롱 코크는 동일한 레시피를 사용하면서 색소로만 변화를 주고 크림에서 맛을 낸다. 반대로 여기에서 소개하는 알부민파우더 머랭으로 만든 마카롱은 코크에 퓌레나 과일 등의 재료를 첨가하여 코크에서 표현하고자 하는 맛을 진하게 연출하고 크림은 공통 레시피로 사용했다.

다양한 인서트 크림의 맛 + 클래식한 코크의 맛 = 일반 마카롱

다양한 코크의 맛 + 클래식한 바닐라 크림의 맛 = 인버트(반전) 마카롱

ALBUMIN POWDER MERINGUE

- In a classic recipe for meringue made with egg whites, you can't add fresh ingredients with high water content, such as purée or fruit, which limits the flavor. On the other hand, albumin powder is a dry powder, which lets you to add purée, fruit, and other ingredients to create a product that tastes and smells like the fruit itself.
- Traditional macarons use the same recipe for the coque but only change the coloring and add flavor to the cream. However, these macarons are made by adding ingredients such as purée or fruit to meringue using albumin powder to create a richer taste I wanted to show in the coque. The cream is one unified recipe.

Different flavors of insert cream + Classic flavored coque = Typical macaron

Different flavors of coque + Classic flavored vanilla cream = Inverted macaron

✦ HOW TO COMPOSE THIS RECIPE ✦

STEP 1.	메뉴 정하기	마카롱
	Decide on the menu	Macaron

STEP 2.	메인 맛 정하기	유자
	Choose the primary flavor	Yuja

STEP 3. **메인 맛(유자)과의 페어링 선택하기**

Select a pairing flavor (Yuja)

- ☑ 파인애플 Pineapple
- ☑ 크림치즈 Cream Cheese
- ☑ 바닐라 Vanilla

STEP 4. **구성하기**

Assemble

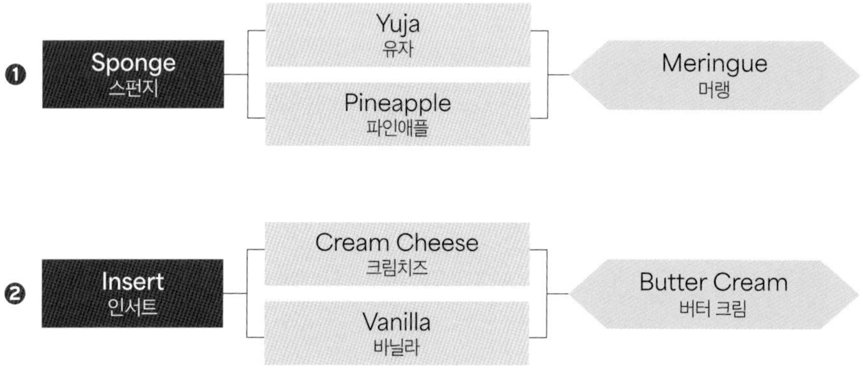

❶ Sponge 스펀지 — Yuja 유자 / Pineapple 파인애플 — Meringue 머랭

❷ Insert 인서트 — Cream Cheese 크림치즈 / Vanilla 바닐라 — Butter Cream 버터 크림

유자 파인애플 코크
Yuja Pineapple Coque
크림치즈 바닐라 버터 크림
Cream Cheese
Vanilla Butter Cream

INGREDIENTS

25개 분량/ 25 macarons

유자 파인애플 머랭 *

Yuja Pineapple Meringue

INGREDIENTS		g
유자 퓌레	Yuja purée	55g
파인애플 퓌레	Pineapple purée	55g
물	Water	70g
알부민파우더	Albumine powder	40g
설탕	Sugar	300g
옥수수전분	Cornstarch	10g
TOTAL		**530g**

유자 파인애플 코크

Yuja Pineapple Coque

INGREDIENTS		g
유자 파인애플 머랭 *	Yuja pineapple meringue *	450g
노란색 색소 (Chefmaster, Lemon yellow)	Yellow food coloring	16방울/ 16 drops
아몬드파우더	Almond powder	150g
슈거파우더	Sugar powder	150g
TOTAL		**750g**

크림치즈 바닐라 버터 크림

Cream Cheese Vanilla Butter Cream

• p.46

INGREDIENTS		g
버터 (Bridel)	Butter	300g
슈거파우더	Sugar powder	50g
바닐라빈 씨	Vanilla bean seeds	1/4개/ 1/4 pc
연유	Condensed milk	25g
바닐라에센스 (Aroma Piu)	Vanilla essence	5g
크림치즈 (Kiri)	Cream cheese	100g
TOTAL		**480g**

NOTE.

✤

YUJA PINEAPPLE MERINGUE

유자 파인애플 머랭

1 비커에 유자 퓌레, 파인애플 퓌레, 물을 넣고 50℃로 맞춘다. (전자레인지 사용)

2 미리 섞어둔 [알부민파우더, 설탕, 옥수수전분]을 넣고 바믹서로 블렌딩한다.

3 밀착 랩핑해 냉장고에서 하루 동안 숙성시킨 후 사용한다.

1 Combine yuja purée, pineapple purée, and water in a beaker. Use a microwave to heat to 50°C.

2 Add previously mixed [albumin powder, sugar, and cornstarch] and blend with an immersion blender.

3 Adhere the plastic wrap to the preparation to cover, and refrigerate overnight to use.

✤ ✤

YUJA PINEAPPLE COQUE

유자 파인애플 코크

1 냉장고에서 숙성시킨 유자 파인애플 머랭을 믹싱볼에 담아 중탕볼에서 35℃로 맞춘 후 거품이 조밀하고 단단한 상태가 될 때까지 휘핑한다.

2 노란색 색소를 넣고 골고루 섞일 때까지 휘핑한다.

3 볼에 체 친 [아몬드파우더, 슈거파우더, **2**의 절반]을 넣고 주걱으로 고르게 섞는다.

4 스크래퍼로 마카로나주 작업을 한다.

5 남은 **2**를 넣고 마카로나주 작업을 한다.

1 Warm the refrigerated yuja pineapple meringue in a mixing bowl over a double boiler (bain-marie) to 35°C and whip until dense and stiff.

2 Add yellow food coloring and whip until evenly incorporated.

3 Add the sifted [almond powder, sugar powder, and half of (**2**)] in a bowl and mix with a spatula.

4 Macaronage with a scraper.

5 Fold in the remaining (**2**) and continue to macaronage.

✤

1
2
3

✤ ✤

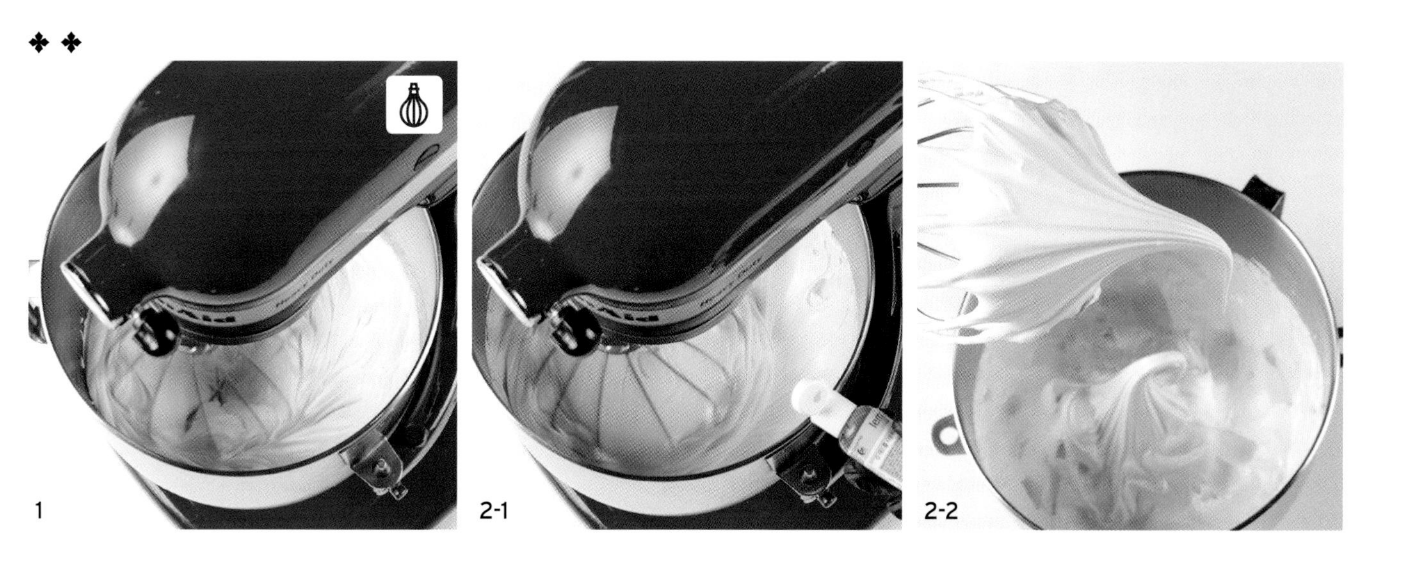

1
2-1
2-2

3
4
5

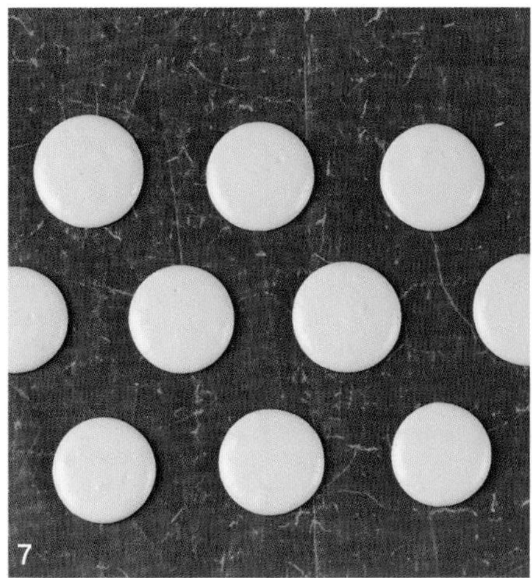

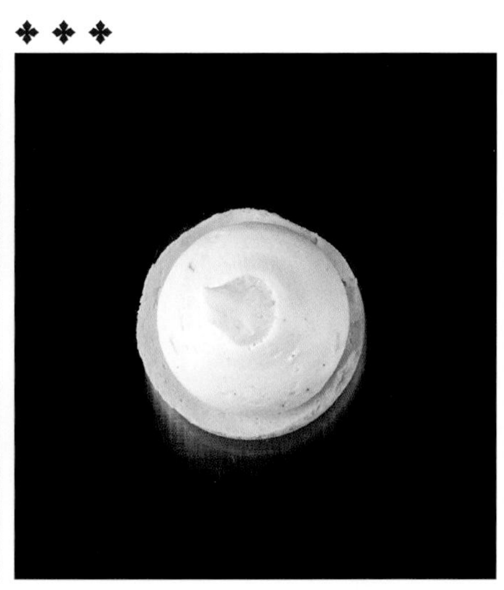

6 테프론시트를 깐 철판에 15g씩 파이핑한다.

7 바닥에 쳐 평평하게 만들고 5분간 실온에 두어 건조시킨 후 140℃로 예열된 오븐에서 댐퍼를 100% 열고 13분간 굽는다.

6 Pipe 15-gram rounds onto a baking sheet lined with a Teflon sheet.

7 Tap the baking tray on the work table to flatten the batter, leave at room temperature for 5 minutes to dry, then bake in an oven preheated to 140°C and damper open at 100% for 13 minutes.

FINISH 마무리

마카롱 코크에 크림치즈 바닐라 버터 크림을 15g씩 파이핑한 후 코크를 덮는다.

Pipe 15 grams of cream cheese vanilla buttercream onto the macaron coque and cover with another coque to finish.

4

CACAO PIE

카카오 파이

PAIRING & TEXTURE

페어링 & 텍스처

- 카카오 스펀지, 동결건조 과일 파우더를 사용해 인공적인 향료가 아닌 내추럴하고 프레시한 맛을 표현하는 것이 가능한 제품이다.
- 이탈리안 머랭 + 젤라틴 + 동결건조 과일 파우더 or 카카오파우더로 다양한 맛의 마시멜로우로 베리에이션할 수 있다.

- Cacao sponge cake and freeze-dried fruit powder are used to make a natural and fresh flavor without artificial flavors.
- Italian meringue + gelatin + freeze-dried fruit powder or cacao powder can be used to make varied marshmallow flavors.

TECHNIQUE
테크닉

카카오 파이 윗면에 초콜릿 파우더를 뿌려 자칫 밋밋해보일 수 있는 제품에 톤온톤으로 질감과 식감을 더했다.

I sprinkled chocolate powder over the Cacao Pie, adding a tone-on-tone texture and mouth feel to the cake that might otherwise look plain.

DESIGN
디자인

마시멜로우를 샌딩한 파이를 초콜릿으로 디핑하는 일반적인 방식의 초콜릿 파이와 다르게 파이를 각각 디핑한 후 마시멜로우를 샌딩해 마카롱이 연상되는 디자인으로 연출했다.

Instead of dipping the marshmallow-filled pies into chocolate, which is the typical method for making chocolate pies, I dipped each pie shell and then filled it with marshmallows to create a design reminiscent of macarons.

✦ HOW TO COMPOSE THIS RECIPE ✦

STEP 1.	STEP 2.	STEP 3.
메뉴 정하기 Decide on the menu	**메인 맛 정하기** Choose the primary flavor	
Types 종류	Main Ingredients 주재료	Dough 반죽
① MI-AMÈRE CACAO PIE 미아메르 카카오 파이	Dark Chocolate 다크초콜릿	Dark Chocolate + Cacao Powder + Sour Cream + Heavy Cream 다크초콜릿 + 카카오파우더 + 사워크림 + 생크림
② CARAMEL CACAO PIE 캐러멜 카카오 파이	Caramel Chocolate 캐러멜초콜릿	Caramel Chocolate + Cacao Powder + Plain Yogurt 캐러멜초콜릿 + 카카오파우더 + 플레인 요거트
③ DOUBLE BERRY CACAO PIE 더블 베리 카카오 파이	Strawberry, Raspberry 딸기, 라즈베리	Strawberry Chocolate + Freeze-Dried Strawberry Powder + Sour Cream + Heavy Cream 딸기초콜릿 + 동결건조 딸기파우더 + 사워크림 + 생크림
④ MATCHA & RED BEAN CACAO PIE 말차 팥 카카오 파이	Matcha 말차	White Chocolate + Matcha Powder + Sour Cream 화이트초콜릿 + 말차파우더 + 사워크림

구성하기
Assemble

Insert 인서트	Glaze 글레이즈	Chocolate Powder 초콜릿 파우더
Montée : Mlik Chocolate + Dark Chocolate 몽테 : 밀크초콜릿 + 다크초콜릿	Dark Chocolate + Cacao butter + Vegetable Oil 다크초콜릿 + 카카오버터 + 식용유	Dark Chocolate Powder 다크초콜릿 파우더
Cream : Caramel 크림 : 캐러멜	Caramel Chocolate + Cacao butter + Vegetable Oil 캐러멜초콜릿 + 카카오버터 + 식용유	Caramel Chocolate Powder 캐러멜초콜릿 파우더
Coulis : Strawberry Purée + Raspberry Purée 쿨리 : 딸기 퓌레 + 라즈베리 퓌레	Strawberry Chocolate + Cacao Butter + Vegetable Oil + Freeze-Dried Raspberry Powder 딸기초콜릿 + 카카오버터 + 식용유 + 동결건조 라즈베리파우더	Strawberry Chocolate Powder (Strawberry Chocolate + Freeze-Dried Raspberry Powder) 딸기초콜릿 파우더 (딸기초콜릿 + 동결건조 라즈베리파우더)
Montée : White Chocolate Sweetened Red Beans 몽테: 화이트초콜릿 + 통팥 앙금	White Chocolate + Cacao Butter + Vegetable Oil + Matcha Powder 화이트초콜릿 + 카카오버터 + 식용유 + 말차파우더	Matcha Chocolate Powder (White Chocolate + Matcha Powder) 말차초콜릿 파우더 (화이트초콜릿 + 말차파우더)

DOUBLE BERRY CACAO PIE
더블 베리 카카오 파이

MATCHA & RED BEAN CACAO PIE
말차 팥 카카오 파이

MI-AMÈRE CACAO PIE
미아메르 카카오 파이

CARAMEL CACAO PIE
캐러멜 카카오 파이

INGREDIENTS

① MI-AMÈRE CACAO PIE 미아메르 카카오 파이

28개 분량/ 28 pies

미아메르 도우 Mi-Amère Batter

INGREDIENTS		g
달걀	Eggs	60g
설탕	Sugar	150g
사워크림	Sour cream	100g
생크림	Heavy cream	50g
바닐라에센스 (Aroma Piu)	Vanilla essence	5g
우유	Milk	90g
다크초콜릿 (Mi-Amère 58%)	Dark chocolate	50g
버터 (Bridel)	Butter	80g
박력분	Cake flour	190g
카카오파우더 (Extra Brute)	Cacao powder	70g
옥수수전분	Cornstarch	10g
베이킹파우더	Baking powder	2g
베이킹소다	Baking soda	2g
TOTAL		**859g**

카카오 마시멜로우 Cacao Marshmallow

INGREDIENTS		g
흰자	Egg whites	90g
설탕A	Sugar A	15g
알부민파우더	Albumin powder	3g
물	Water	70g
설탕B	Sugar B	180g
물엿	Corn syrup	100g
카카오파우더 (Legere 1%)	Cacao powder	15g
젤라틴매스 (×5)	Gelatin mass (×5)	70g
바닐라에센스 (Aroma Piu)	Vanilla essence	5g
TOTAL		**548g**

미아메르 몽테

Mi-Amère Montée

INGREDIENTS		g
휘핑크림A	Whipping cream A	150g
젤라틴매스 (×5)	Gelatin mass (×5)	30g
바닐라빈	Vanilla bean	1.5개 분량/ 1.5 pcs
다크초콜릿 (Mi-Amère 58%)	Dark chocolate	120g
밀크초콜릿 (Alunga 41%)	Milk chocolate	90g
소금	Salt	0.5g
휘핑크림B	Whipping cream B	400g
TOTAL		**790.5g**

* 만드는 법

❶ 냄비에 휘핑크림A, 젤라틴매스, 바닐라빈 껍질을 넣고 가열한다.

❷ 다크초콜릿, 밀크초콜릿, 소금이 든 비커에 ❶을 넣고 바믹서로 블렌딩한다.

❸ 휘핑크림B, 바닐라빈 씨를 넣고 블렌딩한다.

❹ 바트에 담아 밀착 랩핑한 후 냉장고에서 약 12시간 숙성시킨다.

❺ 사용할 때는 비터로 부드럽게 풀어 사용한다.

* Procedure

❶ In a saucepan, heat whipping cream A, gelatin mass, and vanilla bean pod (without seeds).

❷ Pour ❶ into a beaker with dark chocolate, milk chocolate, and salt. Combine with an immersion blender.

❸ Add whipping cream B and vanilla bean seeds; continue to blend.

❹ Pour into a stainless-steel tray, cover with plastic wrap, making sure the wrap is in contact with the cream, and refrigerate for 12 hours.

❺ Soften with the paddle attachment to use.

룸 템퍼라처 글레이즈

Room Temperature Glaze

INGREDIENTS		g
다크초콜릿 (Mi-Amère 58%)	Dark chocolate	500g
카카오버터	Cacao butter	50g
식용유	Vegetable oil	100g
TOTAL		**650g**

* 만드는 법

❶ 비커에 녹인 다크초콜릿(45~50℃), 50℃ 이하로 녹인 카카오버터, 식용유를 담고 바믹서로 블렌딩한다.

❷ 30℃로 맞춰 사용한다.

* Procedure

❶ In a beaker, blend melted dark chocolate (45~50°C), cacao butter melted to below 50°C, and vegetable oil with an immersion blender.

❷ Use at 30°C.

다크초콜릿 파우더

Dark Chocolate Powder

INGREDIENTS		g
다크초콜릿 (Mi-Amère 58%)	Dark chocolate	300g
말토덱스트린	Maltodextrin	300g
소금	Salt	2g
바닐라빈 씨	Vanilla bean seeds	1개 분량/ 1 pc
TOTAL		**602g**

* 만드는 법

❶ 다크초콜릿을 50℃로 녹이고 35℃로 식혀 볼에 담는다.

❷ 말토덱스트린, 소금, 바닐라빈 씨를 넣고 주걱으로 고르게 섞은 후 냉장고에 두어 굳힌다.

❸ 로보쿱에서 곱게 갈아준 후 밀폐용기에 담아 냉동 보관한다.

* Procedure

❶ Melt the dark chocolate to 50°C. Cool to 35°C and pour into a bowl.

❷ Stir in maltodextrin, salt, and vanilla bean seeds. Let it set in a refrigerator.

❸ Finely grind in the Robot Coupe. Keep frozen in an air-tight container.

✤

MI-AMÈRE BATTER

미아메르 도우

1 볼에 달걀(30℃), 설탕, 사워크림(30℃), 생크림(30℃), 바닐라에센스를 넣고 휘퍼로 섞는다.

2 가나슈, 녹인 버터(45℃)를 넣고 섞는다.

TIP. 가나슈는 우유와 녹인 다크초콜릿(40℃)을 비커에 넣고 바믹서로 블렌딩해 사용한다.

3 체 친 [박력분, 카카오파우더, 옥수수전분, 베이킹파우더, 베이킹소다]를 넣고 섞는다.

4 지름 1cm 원형 깍지를 끼운 짤주머니에 담아 15g씩 팬닝한다.

5 바닥에 쳐 평평하게 만든 후 윗불 180℃ 아랫불 140℃에서 18분간 굽고, 도우가 식으면 냉장고에 보관한다.

TIP. 지름은 약 5cm가 적당하다.

6 30℃로 온도를 맞춘 룸 템퍼라처 글레이즈를 코팅한다.

7 윗면이 될 도우에는 다크초콜릿 파우더를 뿌린다.

8 실리콘패드로 옮겨 굳힌다.

1 Stir eggs (30°C), sugar, sour cream (30°C), heavy cream (30°C), and vanilla essence with a whisk.

2 Mix with ganache and melted butter (45°C).

TIP. Blend milk and melted dark chocolate (40°C) with an immersion blender to make ganache.

3 Mix with sifted [cornstarch, cake flour, cacao powder, baking powder, and baking soda].

4 Put the mixture in a piping bag with a 1 cm round tip and pipe 15 grams onto a baking tray.

5 Tap the tray several times to flatten the batter. Bake for 18 minutes at 180°C on top and 140°C on the bottom. Refrigerate once cooled.

TIP. The diameter should be about 5 cm.

6 Coat with the room temperature glaze at 30°C.

7 Dust with dark chocolate powder on the pieces that will be the top.

8 Transfer them on a silicon mat to set.

1

2

3-1

3-2

4

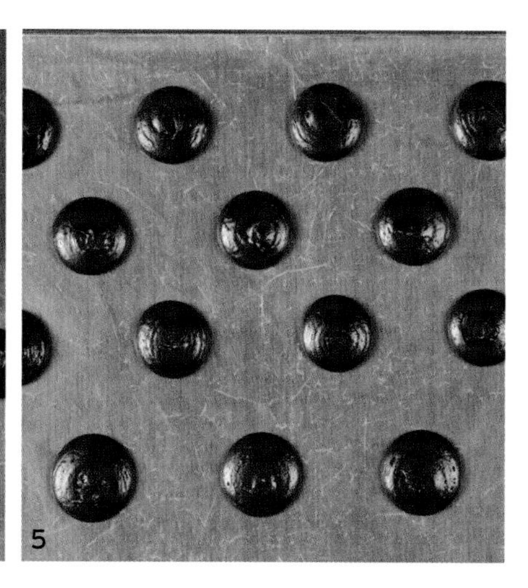
5

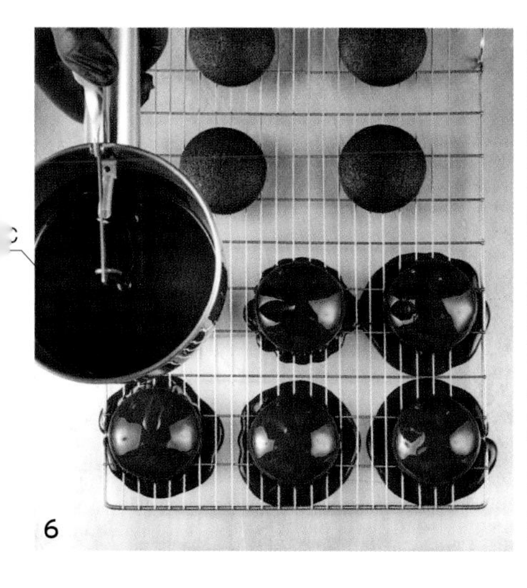
6

7

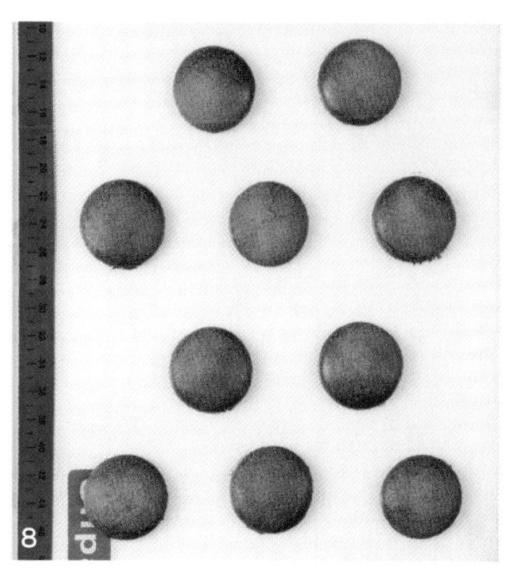
8

✤ ✤

CACAO MARSHMALLOW

카카오 마시멜로우

1 믹싱볼에 흰자(30℃)를 넣고 휘핑하면서 설탕A와 알부민파우더를 흘려 넣어가며 거품이 형성될 정도(약 50%)로 휘핑한다.

TIP. 동시에 냄비에 물, 설탕B, 물엿을 넣고 118℃로 가열해 시럽을 만든다.

2 118℃의 시럽을 흘려 넣어가면서 휘핑한다.

3 카카오파우더를 조금씩 나눠 넣어가면서 휘핑한다.

TIP. 먼저 넣은 카카오파우더가 다 섞이면 남은 카카오파우더를 더 넣고 섞는다.
여기에서 사용한 카카오파우더(Legere 1%)는 코코아버터 함량이 낮아 일반적인 카카오파우더를 사용할 때보다 볼륨이 좋게 완성된다.

4 녹인 젤라틴매스를 넣고 휘핑한다.

5 바닐라에센스를 넣고 휘핑한다.

6 주걱으로 들어올렸을 때 쉽게 흘러내리지 않는 상태(최종 온도 약 31℃)로 마무리한다.

1 Whip egg whites (30°C) in a mixing bowl while adding sugar A and albumin powder until foamy peaks form (about 50%).

TIP. At the same time, heat water, sugar B, and corn syrup in a saucepan to 118°C to make syrup.

2 Slowly drizzle in the syrup cooked to 118°C while whipping.

3 Add cacao powder a little bit at a time while whipping.

TIP. When the first addition of powder is evenly mixed in, add the next.
The cacao powder I used here (Legere 1%) has a lower cacao butter content, which gives it a better volume than regular cacao powder.

4 Add melted gelatin and whip.

5 Add vanilla essence and continue to whip.

6 Finish when it drips off slowly when lifted with a spatula (final temperature: approximately 31°C).

FINISH

마무리

1 미아메르 도우에 카카오 마시멜로우 9g을 파이핑하고, 정중앙에 미아메르 몽테 3g을 파이핑한다.

2 다크초콜릿 파우더를 뿌린 미아메르 도우를 올려 마무리한다.

1 Pipe 9 grams of cacao marshmallow on a Mi-Amère pie shell and pipe 3 grams of Mi-Amère montée in the center.

2 Top with the pie shell dusted with dark chocolate powder to finish.

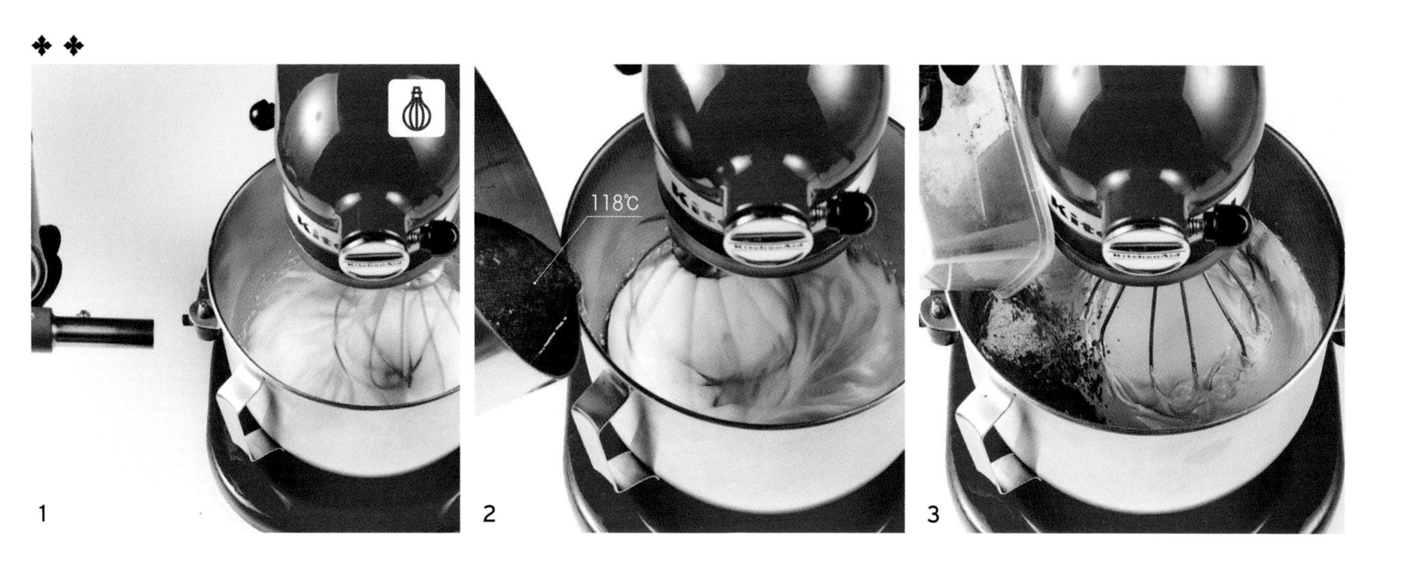
✤ ✤
118℃
1
2
3

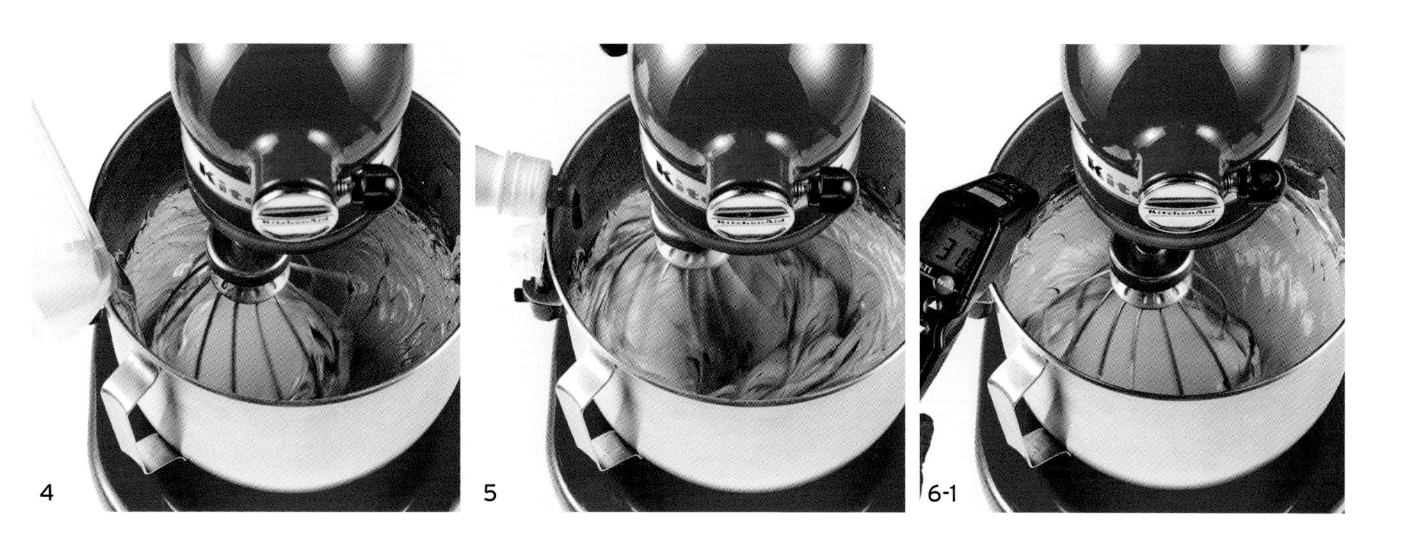
4
5
6-1

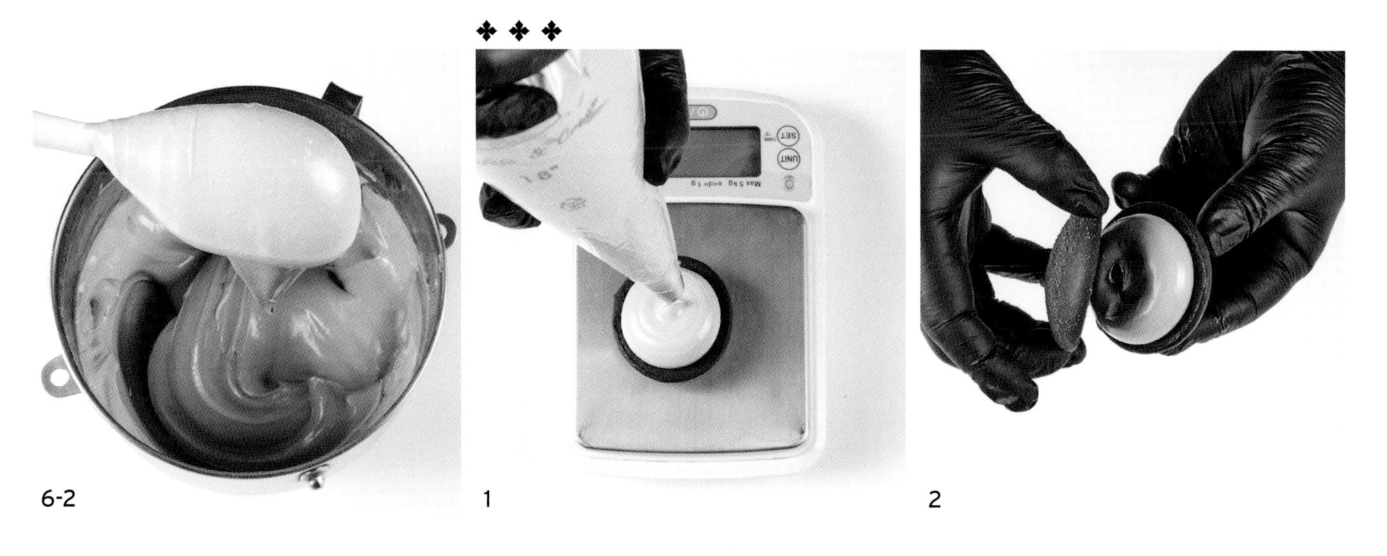
✤ ✤ ✤
SET
UNIT
6-2
1
2

② CARAMEL CACAO PIE 캐러멜 카카오 파이

28개 분량/ 28 pies

제피르 도우 Zephyr Batter

INGREDIENTS		g
달걀	Eggs	55g
설탕	Sugar	150g
플레인요거트	Plain yogurt	150g
바닐라에센스 (Aroma Piu)	Vanilla essence	4g
우유	Milk	90g
캐러멜초콜릿 (Zephyr Caramel 35%)	Caramel chocolate	50g
버터 (Bridel)	Butter	75g
옥수수전분	Cornstarch	10g
박력분	Cake flour	190g
카카오파우더 (CACAOBARRY, Rouge Ultime)	Cacao powder	70g
베이킹파우더	Baking powder	1g
베이킹소다	Baking soda	3g
TOTAL		**848g**

*** 만드는 법**

❶ 볼에 달걀(30℃), 설탕, 플레인요거트(30℃), 바닐라에센스를 넣고 휘퍼로 섞는다.

❷ 가나슈, 녹인 버터(45℃)를 비커에 넣고 바믹서로 블렌딩해 사용한다.
- 가나슈는 우유와 녹인 캐러멜초콜릿(40℃)을 블렌딩해 만들어 사용한다.

❸ 체 친 [옥수수전분, 박력분, 카카오파우더, 베이킹파우더, 베이킹소다]를 넣고 섞는다.

❹ 지름 1cm 원형 깍지를 끼운 짤주머니에 담아 15g씩 팬닝한다.

❺ 바닥에 쳐 평평하게 만든 후 윗불 180℃ 아랫불 140℃에서 18분간 굽고, 도우가 식으면 냉장고에 보관한다.
- 지름은 약 5cm가 적당하다.

❻ 30℃로 온도를 맞춘 룸 템퍼라처 글레이즈를 코팅한다.

❼ 윗면이 될 도우에는 캐러멜초콜릿 파우더를 뿌린다.

❽ 실리콘패드로 옮겨 굳힌다.

*** Procedure**

❶ Stir eggs (30°C), sugar, plain yogurt (30°C), and vanilla essence with a whisk.

❷ Blend ganache and melted butter (45°C) in a beaker with an immersion blender.
- To make the ganache, blend milk and melted caramel chocolate (40°C).

❸ Mix with sifted [cornstarch, cake flour, cacao powder, baking powder, and baking soda].

❹ Put in a piping bag with a 1 cm round tip and pipe 15 grams on a baking tray.

❺ Tap the tray several times to flatten the batter. Bake for 18 minutes at 180°C on top and 140°C on the bottom. Refrigerate once cooled.
- The diameter should be about 5 cm.

❻ Coat with the room temperature glaze at 30°C.

❼ Dust with caramel chocolate powder on the pieces that will be the top.

❽ Transfer them on a silicon mat to set.

바닐라 마시멜로우

Vanilla Marshmallow

INGREDIENTS		g
흰자	Egg whites	90g
설탕A	Sugar A	15g
알부민파우더	Albumin powder	3g
물	Water	70g
설탕B	Sugar B	180g
물엿	Corn syrup	100g
젤라틴매스 (×5)	Gelatin mass (×5)	70g
바닐라에센스 (Aroma Piu)	Vanilla essence	5g
TOTAL		**533g**

*** 만드는 법**

❶ 믹싱볼에 흰자를 넣고 휘핑하면서 설탕A와 알부민 파우더를 흘려 넣어가며 거품이 형성될 정도(약 50%)로 휘핑한다. (최종 온도 약 30℃)

- 동시에 냄비에 물, 설탕B, 물엿을 넣고 118℃로 가열해 시럽을 만든다.

❷ 118℃의 시럽을 흘려 넣어가면서 휘핑한다.

❸ 녹인 젤라틴매스를 넣고 휘핑한다.

❹ 바닐라에센스를 넣고 휘핑한다.

❺ 주걱으로 들어올렸을 때 쉽게 흘러내리지 않는 상태(최종 온도 약 31℃)로 마무리한다.

*** Procedure**

❶ Whip egg whites in a mixing bowl while adding sugar A and albumin powder until foamy peaks form (about 50%). (final temperature: approximately 30°C)

- At the same time, heat water, sugar B, and corn syrup in a saucepan to 118°C to make syrup.

❷ Slowly drizzle in the syrup cooked to 118°C while whipping.

❸ Add melted gelatin and whip.

❹ Add vanilla essence and continue to whip.

❺ Finish when it drips off slowly when lifted with a spatula (final temperature: approximately 31°C).

캐러멜 크림

Caramel Cream

INGREDIENTS		g
설탕	Sugar	130g
생크림	Heavy cream	70g
정향	Anise	1/2개/ 1/2 pc
바닐라빈 껍질	Vanilla bean pod	1개 분량/ 1 pc
버터A (Bridel)	Butter A	25g
우유	Milk	25g
노른자	Egg yolks	70g
소금	Salt	1g
버터B (Bridel)	Butter B	90g
TOTAL		**411g**

*** 만드는 법**

❶ 냄비에 설탕을 넣고 캐러멜화시킨다.
- 캐러멜화를 진행시키는 정도는 제품에 따라, 개인의 취향에 따라 선택할 수 있다. 캐러멜 색이 연할수록 단맛이 강하고, 진할수록 캐러멜 특유의 쌉싸래한 맛이 강하다.

❷ 다른 냄비에 생크림, 정향, 바닐라빈 껍질을 넣고 한 번 끓어오를 때까지 가열한다.
❸ ❶에 ❷를 체에 걸러가며 부어준 후 섞는다.
❹ 불에서 내린 후 버터A를 넣고 녹을 때까지 섞는다.
❺ 미리 섞어둔 우유와 노른자(30℃)에 ❹를 체에 걸러가며 부어준 후 섞는다.
❻ 다시 냄비로 옮겨 85℃까지 가열하고 체에 거른 후 비커로 옮겨 소금을 넣고 바믹서로 블렌딩한다.
❼ 50℃로 쿨링한 후 버터B를 넣고 바믹서로 블렌딩한다.
❽ 바트에 담아 밀착 랩핑해 냉장고에 보관한다.
❾ 사용할 때는 미리 꺼내두어 실온 상태(파이핑하기 좋은 상태)로 사용한다.

*** Procedure**

❶ Caramelize sugar in a pot.
- You can control the caramelization stage depending on your preference. The lighter the color, the sweeter it is; darker the color, more bitter the flavor.

❷ In a different pot, heat cream, anise, and vanilla bean pod until it boils.
❸ Sieve ❷ into ❶ to mix.
❹ Remove from heat and stir in butter A until it melts.
❺ Sieve ❹ into pre-mixed milk and egg yolks (30°C) and mix.
❻ Pour back into the pot and cook to 85°C. Sieve into a beaker and combine with salt using an immersion blender.
❼ Cool to 50°C, add butter B, and blend with an immersion blender.
❽ Pour into a stainless-steel tray, cover with plastic wrap, making sure the wrap is in contact with the cream, and refrigerate.
❾ Bring to room temperature (suitable for piping) ahead of time to use.

캐러멜초콜릿 파우더

Caramel Chocolate Powder

INGREDIENTS		g
캐러멜초콜릿 (Zephyr Caramel 35%)	Caramel chocolate	300g
말토덱스트린	Maltodextrin	200g
소금	Salt	3g
바닐라빈 씨	Vanilla bean seeds	1개 분량/ 1 pc
TOTAL		**503g**

∗ 만드는 법

❶ 캐러멜초콜릿을 50℃로 녹이고 35℃로 식혀 볼에 담는다.

❷ 말토덱스트린, 소금, 바닐라빈 씨를 넣고 주걱으로 고르게 섞은 후 냉장고에 두어 굳힌다.

❸ 로보쿱에서 곱게 갈아준 후 밀폐용기에 담아 냉동 보관한다.

∗ Procedure

❶ Melt caramel chocolate to 50°C. Cool to 35°C and pour into a bowl.

❷ Mix with maltodextrin, salt, and vanilla bean seeds with a spatula. Refrigerate to set.

❸ Finely grind in the Robot Coupe. Keep frozen in an air-tight container.

룸 템퍼라처 글레이즈

Room Temperature Glaze

INGREDIENTS		g
캐러멜초콜릿 (Zephyr Caramel 35%)	Caramel chocolate	500g
카카오버터	Cacao butter	50g
식용유	Vegetable oil	100g
TOTAL		**650g**

∗ 만드는 법

❶ 비커에 녹인 캐러멜초콜릿(45~50℃), 50℃ 이하로 녹인 카카오버터, 식용유를 담고 바믹서로 블렌딩한다.

❷ 30℃로 맞춰 사용한다.

∗ Procedure

❶ In a beaker, blend melted caramel chocolate (45~50°C), cacao butter melted to below 50°C, and vegetable oil with an immersion blender.

❷ Use at 30°C.

∗ 마무리

❶ 룸 템퍼라처 글레이즈(30℃)를 코팅한 제피르 도우에 바닐라 마시멜로우 9g을 파이핑하고, 정중앙에 캐러멜 크림 6g을 파이핑한다.

❷ 캐러멜초콜릿 파우더를 뿌린 제피르 도우를 올려 마무리한다.

∗ Finish

❶ Pipe 9 grams of vanilla marshmallow on a pie shell coated with the room temperature glaze (30°C) and pipe 6 grams of caramel cream in the center.

❷ Top with the pie shell dusted with caramel chocolate powder to finish.

③ DOUBLE BERRY CACAO PIE 더블 베리 카카오 파이

26개 분량/ 26 pies

스트로베리 도우 Strawberry Batter

INGREDIENTS		g
달걀	Eggs	55g
설탕	Sugar	150g
사워크림	Sour cream	120g
바닐라에센스 (Aroma Piu)	Vanilla essence	4g
우유	Milk	30g
생크림	Heavy cream	50g
딸기초콜릿 (Callebaut)	Strawberry chocolate	60g
버터 (Bridel)	Butter	75g
옥수수전분	Cornstarch	10g
박력분	Cake flour	230g
동결건조 딸기파우더	Freeze-dried strawberry powder	20g
베이킹파우더	Baking powder	1g
베이킹소다	Baking soda	3g
TOTAL		**808g**

- 동결건조 딸기파우더는 동결건조 딸기를 곱게 갈아 사용한다.
- Freeze-dried strawberry powder: Finely grind freeze-dried strawberries.

*** 만드는 법**

❶ 볼에 달걀(30℃), 설탕, 사워크림(30℃), 바닐라에센스를 넣고 휘퍼로 섞는다.
❷ 가나슈, 녹인 버터(45℃)를 비커에 넣고 바믹서로 블렌딩해 사용한다.
- 가나슈는 우유, 생크림, 녹인 딸기초콜릿(40℃)을 블렌딩해 만들어 사용한다.

❸ 체 친 [옥수수전분, 박력분, 딸기파우더, 베이킹파우더, 베이킹소다]를 넣고 섞는다.
❹ 지름 1cm 원형 깍지를 끼운 짤주머니에 담아 15g씩 팬닝한다.
❺ 바닥에 쳐 평평하게 만든 후 윗불 180℃ 아랫불 140℃에서 18분간 굽고, 도우가 식으면 냉장고에 보관한다.
- 지름은 약 5cm가 적당하다.

❻ 30℃로 온도를 맞춘 룸 템퍼라처 글레이즈를 코팅한다.
❼ 윗면이 될 도우에는 딸기 초콜릿 파우더를 뿌린다.
❽ 실리콘패드로 옮겨 굳힌다.

*** Procedure**

❶ Stir eggs (30°C), sugar, sour cream (30°C), and vanilla essence with a whisk.
❷ Blend ganache and melted butter (45°C) in a beaker with an immersion blender.
- To make the ganache, blend milk, heavy cream, and melted strawberry chocolate (40°C).

❸ Mix with sifted [cornstarch, cake flour, strawberry powder, baking powder, and baking soda].
❹ Put in a piping bag with a 1 cm round tip and pipe 15 grams on a baking tray.
❺ Tap the tray several times to flatten the batter. Bake for 18 minutes at 180°C on top and 140°C on the bottom. Refrigerate once cooled.
- The diameter should be about 5 cm.

❻ Coat with the room temperature glaze at 30°C.
❼ Dust with strawberry chocolate powder on the pieces that will be the top.
❽ Transfer them on a silicon mat to set.

더블 베리 마시멜로우

Double Berry Marshmallow

INGREDIENTS		g
흰자	Egg whites	90g
설탕A	Sugar A	15g
알부민파우더	Albumin powder	3g
물	Water	70g
설탕B	Sugar B	180g
물엿	Corn syrup	100g
동결건조 라즈베리파우더	Freeze-dried raspberry powder	10g
동결건조 딸기파우더	Freeze-dried strawberry powder	10g
젤라틴매스 (×5)	Gelatin mass (×5)	70g
바닐라에센스 (Aroma Piu)	Vanilla essence	5g
TOTAL		**553g**

- 동결건조 딸기파우더는 동결건조 딸기를 곱게 갈아 사용한다.
- 동결건조 라즈베리파우더는 동결건조 라즈베리를 곱게 갈아 사용한다.
- Freeze-dried strawberry powder: Finely grind freeze-dried strawberries.
- Freeze-dried raspberry powder: Finely grind freeze-dried raspberries.

*** 만드는 법**

❶ 믹싱볼에 흰자(30℃)를 넣고 휘핑하면서 설탕A와 알부민파우더를 흘려 넣어가며 거품이 형성될 정도(약 50%)로 휘핑한다.
- 동시에 냄비에 물, 설탕B, 물엿을 넣고 118℃로 가열해 시럽을 만든다.

❷ 118℃의 시럽을 흘려 넣어가면서 휘핑한다.

❸ 동결건조 라즈베리파우더와 딸기파우더를 조금씩 나눠 넣어가면서 휘핑한다.
- 먼저 넣은 가루 재료가 다 섞이면 남은 가루 재료를 더 넣고 섞는다.

❹ 녹인 젤라틴매스를 넣고 휘핑한다.

❺ 바닐라에센스를 넣고 휘핑한다.

❻ 주걱으로 들어올렸을 때 쉽게 흘러내리지 않는 상태(최종 온도 약 31℃)로 마무리한다.

*** Procedure**

❶ Whip egg whites (30°C) in a mixing bowl while adding sugar A and albumin powder until foamy peaks form (about 50%).
- At the same time, heat water, sugar B, and corn syrup in a saucepan to 118°C to make syrup.

❷ Slowly drizzle in the syrup cooked to118°C while whipping.

❸ Add the fruit powders a little bit at a time while whipping.
- When the first addition of fruit powder is evenly mixed in, add the next addition of powder.

❹ Add melted gelatin and whip.

❺ Add vanilla essence and continue to whip.

❻ Finish when it drips off slowly when lifted with a spatula (final temperature: approximately 31°C).

더블 베리 쿨리

Double Berry Coulis

INGREDIENTS		g
딸기 퓌레	Strawberry purée	140g
라즈베리 퓌레	Raspberry purée	140g
설탕	Sugar	70g
NH펙틴	NH pectin	8g
물엿	Corn syrup	46g
TOTAL		**404g**

*** 만드는 법**

❶ 냄비에 모든 재료를 넣고 가열한다.

❷ 펙틴 반응을 확인한 후 바트에 담아 밀착 랩핑해 냉장 보관한다.

❸ 사용할 때는 바믹서로 블렌딩해 사용한다.

*** Procedure**

❶ Heat all the ingredients in a pot.

❷ After the pectin activates, pour into a stainless-steel tray, cover with plastic wrap, making sure the wrap is in contact with the coulis, and refrigerate.

❸ Blend with an immersion blender to use.

딸기 초콜릿 파우더

Strawberry Chocolate Powder

INGREDIENTS		g
딸기초콜릿 (Callebaut)	Strawberry chocolate	400g
동결건조 라즈베리파우더	Freeze-dried raspberry powder	30g
말토덱스트린	Maltodextrin	290g
소금	Salt	2g
바닐라빈 씨	Vanilla bean seeds	1g
TOTAL		**723g**

*** 만드는 법**

❶ 딸기초콜릿을 50℃로 녹이고 35℃로 식혀 볼에 담는다.

❷ 동결건조 라즈베리파우더, 말토덱스트린, 소금, 바닐라빈 씨를 넣고 주걱으로 고르게 섞은 후 냉장고에 두어 굳힌다.

❸ 로보쿱에서 곱게 갈아준 후 밀폐용기에 담아 냉동 보관한다.

*** Procedure**

❶ Melt the strawberry chocolate to 50°C. Cool to 35°C and pour into a bowl.

❷ Stir in freeze-dried raspberry powder, maltodextrin, salt, and vanilla bean seeds. Let it set in a refrigerator.

❸ Finely grind in the Robot Coupe. Keep frozen in an air-tight container.

룸 템퍼라처 글레이즈

Room Temperature Glaze

INGREDIENTS		g
딸기초콜릿 (Callebaut)	Strawberry chocolate	500g
카카오버터	Cacao butter	50g
식용유	Vegetable oil	100g
동결건조 라즈베리파우더	Freeze-dried raspberry powder	6g
TOTAL		**656g**

*** 만드는 법**

❶ 비커에 녹인 딸기초콜릿, 50℃ 이하로 녹인 카카오버터, 식용유, 동결건조 라즈베리파우더를 담고 바믹서로 블렌딩한다.

❷ 30℃로 맞춰 사용한다.

*** Procedure**

❶ In a beaker, blend melted strawberry chocolate, cacao butter melted to below 50°C, freeze-dried raspberry powder, and vegetable oil with an immersion blender.

❷ Use at 30°C.

*** 마무리**

❶ 룸 템퍼라처 글레이즈(30℃)를 코팅한 스트로베리 도우에 더블 베리 마시멜로우 9g을 파이핑하고, 정중앙에 더블 베리 쿨리 6g을 파이핑한다.

❷ 딸기 초콜릿 파우더를 뿌린 스트로베리 도우를 올려 마무리한다.

*** Finish**

❶ Pipe 9 grams of double berry marshmallow on a pie shell coated with the room temperature glaze (30°C) and pipe 6 grams of double berry coulis in the center.

❷ Top with the pie shell dusted with strawberry chocolate powder to finish.

④ MATCHA & RED BEAN CACAO PIE 말차 팥 카카오 파이

28개 분량/ 28 pies

말차 도우 Matcha Batter

INGREDIENTS		g
달걀	Eggs	55g
설탕	Sugar	150g
사워크림	Sour cream	150g
바닐라에센스 (Aroma Piu)	Vanilla essence	4g
우유	Milk	80g
화이트초콜릿 (Zephyr white 34%)	White chocolate	50g
버터 (Bridel)	Butter	75g
옥수수전분	Cornstarch	10g
박력분	Cake flour	250g
말차파우더	Matcha powder	40g
베이킹파우더	Baking powder	1g
베이킹소다	Baking soda	3g
TOTAL		**868g**

*** 만드는 법**

❶ 볼에 달걀(30℃), 설탕, 사워크림(30℃), 바닐라에센스를 넣고 휘퍼로 섞는다.

❷ 가나슈, 녹인 버터(45℃)를 비커에 넣고 바믹서로 블렌딩해 사용한다.
- 가나슈는 우유와 녹인 화이트초콜릿(40℃)을 블렌딩해 만들어 사용한다.

❸ 체 친 [옥수수전분, 박력분, 말차파우더, 베이킹파우더, 베이킹소다]를 넣고 섞는다.

❹ 지름 1cm 원형 깍지를 끼운 짤주머니에 담아 15g씩 팬닝한다.

❺ 바닥에 쳐 평평하게 만든 후 윗불 180℃ 아랫불 140℃에서 18분간 굽고, 도우가 식으면 냉장고에 보관한다.
- 지름은 약 5cm가 적당하다.

❻ 30℃로 온도를 맞춘 룸 템퍼라처 글레이즈를 코팅한다.

❼ 윗면이 될 도우에는 말차 초콜릿 파우더를 뿌린다.

❽ 실리콘패드로 옮겨 굳힌다.

*** Procedure**

❶ Stir eggs (30°C), sugar, sour cream (30°C), and vanilla essence with a whisk.

❷ Blend ganache and melted butter (45°C) in a beaker with an immersion blender.
- To make the ganache, blend milk and melted white chocolate (40°C).

❸ Mix with sifted [cornstarch, cake flour, matcha powder, baking powder, and baking soda].

❹ Put in a piping bag with a 1 cm round tip and pipe 15 grams on a baking tray.

❺ Tap the tray several times to flatten the batter. Bake for 18 minutes at 180°C on top and 140°C on the bottom. Refrigerate once cooled.
- The diameter should be about 5 cm.

❻ Coat with the room temperature glaze at 30°C.

❼ Dust with matcha chocolate powder on the pieces that will be the top.

❽ Transfer them on a silicon mat to set.

말차 마시멜로우

Matcha Marshmallow

INGREDIENTS		g
흰자	Egg whites	90g
설탕A	Sugar A	15g
알부민파우더	Albumin powder	3g
물	Water	70g
설탕B	Sugar B	180g
물엿	Corn syrup	100g
말차파우더	Matcha powder	15g
젤라틴매스 (×5)	Gelatin mass (×5)	70g
바닐라에센스 (Aroma Piu)	Vanilla essence	5g
TOTAL		**548g**

* 만드는 법

❶ 믹싱볼에 흰자(30℃)를 넣고 휘핑하면서 설탕A와 알부민파우더를 흘려 넣어가며 거품이 형성될 정도(약 50%)로 휘핑한다.
- 동시에 냄비에 물, 설탕B, 물엿을 넣고 118℃로 가열해 시럽을 만든다.

❷ 118℃의 시럽을 흘려 넣어가면서 휘핑한다.

❸ 말차파우더를 조금씩 나눠 넣어가면서 휘핑한다.
- 먼저 넣은 말차파우더가 다 섞이면 남은 말차파우더를 더 넣고 섞는다.

❹ 녹인 젤라틴매스를 넣고 휘핑한다.

❺ 바닐라에센스를 넣고 휘핑한다.

❻ 주걱으로 들어올렸을 때 쉽게 흘러내리지 않는 상태(최종 온도 약 31℃)로 마무리한다.

* Procedure

❶ Whip egg whites (30°C) in a mixing bowl while adding sugar A and albumin powder until foamy peaks form (about 50%).
- At the same time, heat water, sugar B, and corn syrup in a saucepan to 118°C to make syrup.

❷ Slowly drizzle in the syrup cooked to118°C while whipping.

❸ Add matcha powder a little bit at a time while whipping.
- When the first addition of powder is evenly mixed in, add the next addition.

❹ Add melted gelatin and whip.

❺ Add vanilla essence and continue to whip.

❻ Finish when it drips off slowly when lifted with a spatula (final temperature: approximately 31°C).

화이트 몽테 *

White Montée

INGREDIENTS		g
휘핑크림A	Whipping cream A	140g
젤라틴매스 (×5)	Gelatin mass (×5)	30g
연유	Condensed milk	60g
화이트초콜릿 (Zephyr white 34%)	White chocolate	110g
휘핑크림B	Whipping cream B	320g
바닐라에센스 (Aroma Piu)	Vanilla essence	6g
골드럼 (PAN RUM)	Gold rum	6g
TOTAL		**672g**

* 만드는 법

❶ 냄비에 휘핑크림A , 젤라틴매스를 넣고 젤라틴매스가 녹을 때까지 가열한 후(약 60℃) 연유를 넣는다.

❷ 화이트초콜릿이 담긴 비커에 ❶을 넣고 바믹서로 블렌딩한다.

❸ 휘핑크림B를 넣어가며 블렌딩한다.

❹ 바닐라에센스와 골드럼을 넣어가며 블렌딩한다.

❺ 바트에 담아 밀착 랩핑한 후 냉장고에서 약 12시간 숙성시킨다.

* Procedure

❶ Heat whipping cream A and gelatin mass until the gelatin melts (about 60°C) and add condensed milk.

❷ Pour ❶ into a beaker with white chocolate and mix with an immersion blender.

❸ Blend while drizzling in whipping cream B.

❹ Continue to blend while adding vanilla essence and the rum.

❺ Pour into a stainless-steel tray, cover with plastic wrap, making sure the wrap is in contact with the cream, and refrigerate for about 12 hours.

팥 크림

Red Bean Cream

INGREDIENTS		g
화이트 몽테 *	White montée *	200g
팥 앙금 (대두식품, M48)	Red bean paste (Daedoo Foods, M48)	150g
TOTAL		**350g**

*** 만드는 법**

믹싱볼에 화이트 몽테와 팥 앙금을 넣고 비터로 고르게 섞어 사용한다.

*** Procedure**

Combine white montée and red bean paste in a mixing bowl with a paddle.

말차 초콜릿 파우더

Matcha Chocolate Powder

INGREDIENTS		g
화이트초콜릿 (Zephyr white 34%)	White chocolate	300g
말차파우더	Matcha powder	20g
말토덱스트린	Maltodextrin	200g
소금	Salt	2g
바닐라빈 씨	Vanilla bean seeds	1개 분량/ 1 pc
TOTAL		**522g**

*** 만드는 법**

❶ 화이트초콜릿을 50℃로 녹이고 35℃로 식혀 볼에 담는다.

❷ 말차파우더, 말토덱스트린, 소금, 바닐라빈 씨를 넣고 주걱으로 고르게 섞은 후 냉장고에 두어 굳힌다.

❸ 로보쿱에서 곱게 갈아준 후 밀폐용기에 담아 냉동 보관한다.

*** Procedure**

❶ Melt white chocolate to 50°C. Cool to 35°C and pour into a bowl.

❷ Stir in matcha powder, maltodextrin, salt, and vanilla bean seeds. Let it set in a refrigerator.

❸ Finely grind in the Robot Coupe. Keep frozen in an air-tight container.

말차 글레이즈

Matcha Glaze

INGREDIENTS		g
화이트초콜릿 (Zephyr white 34%)	White chocolate	500g
카카오버터	Cacao butter	50g
식용유	Vegetable oil	100g
말차파우더	Matcha powder	20g
TOTAL		**670g**

*** 만드는 법**

❶ 비커에 녹인 화이트초콜릿(45~50℃), 50℃ 이하로 녹인 카카오버터, 식용유, 말차파우더를 담고 바믹서로 블렌딩한다.

❷ 30℃로 맞춰 사용한다.

*** Procedure**

❶ In a beaker, blend melted white chocolate (45~50°C), cacao butter melted to below 50°C, matcha powder, and vegetable oil with an immersion blender.

❷ Use at 30°C.

*** 마무리**

❶ 룸 템퍼라처 글레이즈(30℃)를 코팅한 말차 도우에 말차 마시멜로우 9g을 파이핑하고, 정중앙에 팥 크림 7g을 파이핑한다.

❷ 말차 초콜릿 파우더를 뿌린 말차 도우를 올려 마무리한다.

*** Finish**

❶ Pipe 9 grams of matcha marshmallow on a pie shell coated with the room temperature glaze (30°C) and pipe 7 grams of red bean cream in the center.

❷ Top with the pie shell dusted with matcha chocolate powder to finish.

5

APPLE GORGONZOLA TRAVEL CAKE

애플 고르곤졸라 트래블 케이크

PAIRING & TEXTURE
페어링 & 텍스처

- 캐러멜라이즈 사과를 케이크에 넣어 씹히는 식감을 더했다.
- 케이크 위에 크럼블을 올려 구워 바삭한 식감을 더해 촉촉한 케이크와 대비되는 식감을 표현했다.

- I added caramelized apples to the cake to give it a chewing texture.
- The crumble is baked on top of the cake to add a crunchy texture that contrasts with the moist cake.

TECHNIQUE
테크닉

올인원법

- 모든 재료를 한 번에 섞어 만드는 올인원 방식으로 만들었다.
- 버터를 크림화하지 않고 녹여 사용하는 레시피이므로, 버터 대신 식물성 오일을 사용해 비건 레시피로 변형할 수 있다.

ALL-IN-ONE METHOD

- It's an all-in-one method that mixes all the ingredients in one go.
- Since the recipe uses melted butter instead of creaming it, you can make it vegan using vegetable oil instead of butter.

DESIGN
디자인

- 일반적인 직사각형 파운드 틀 대신 보트 형태의 실리콘몰드를 사용해 개성 있는 디자인으로 연출했다.
- 레몬 초콜릿 파우더를 뿌려 마무리해 시각적으로 좀 더 고급스러운 이미지로 연출했다.

- Instead of the usual rectangular pound cake pans, I used boat-shaped silicone molds to create a unique design.
- I dusted the cake with lemon chocolate powder to give it a more glamorous look.

✦ HOW TO COMPOSE THIS RECIPE ✦

STEP 1.	메뉴 정하기 Decide on the menu	파운드 케이크 Pound Cake
STEP 2.	메인 맛 정하기 Choose the primary flavor	사과 Apple

STEP 3. 메인 맛(사과)과의 페어링 선택하기

Select a pairing flavor (Apple)

- ☑ 코코넛 Coconut
- ☑ 고르곤졸라 Gorgonzola
- ☑ 캐러멜초콜릿 Caramel Chocolate
- ☑ 시나몬 Cinnamon

STEP 4. 구성하기

Assemble

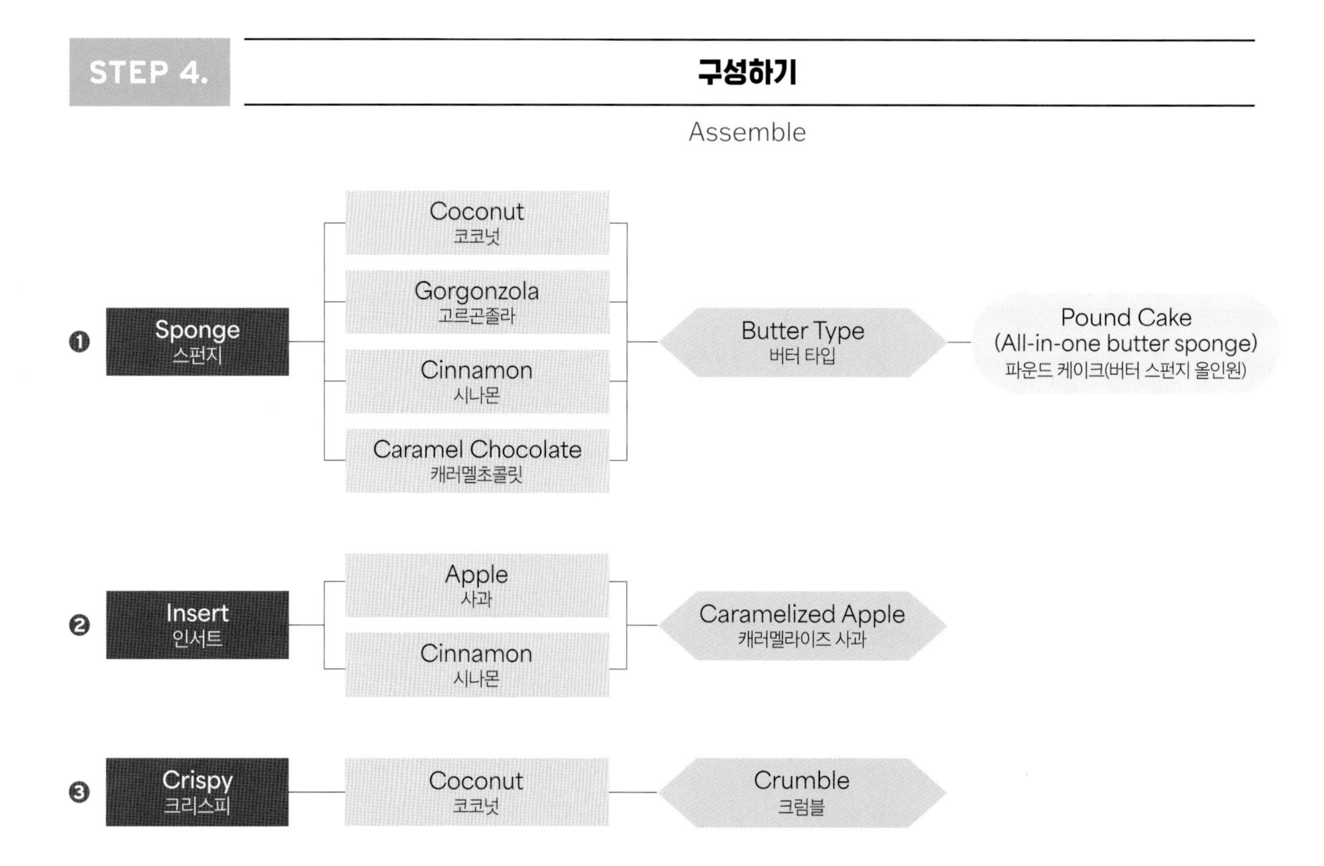

레몬초콜릿 파우더
Lemon Chocolate Powder
코코넛 크럼블
Coconut Crumble
사과 케이크
Apple cake
캐러멜라이즈 사과
Caramelized Apple

INGREDIENTS 6개 분량/ 6 cakes

캐러멜라이즈 사과 *

Caramelized Apple

INGREDIENTS		g
깍둑썬 사과 (사방 1.5cm)	Apple, diced into 1.5 cm cubes	900g
설탕	Sugar	240g
레몬즙	Lemon juice	70g
바닐라빈 껍질	Vanilla bean pod	1개 분량/ 1 pc
시나몬파우더	Cinnamon powder	1g
칼바도스 (Dijon Calvados)	Calvados	15g
TOTAL		**1226g**

사과 케이크

Apple Cake

INGREDIENTS		g
설탕	Sugar	350g
코코넛롱	Shredded dried coconut	80g
버터 (Bridel)	Butter	390g
캐러멜초콜릿 (Zephyr Caramel 35%)	Caramel chocolate	140g
고르곤졸라	Gorgonzola cheese	200g
바닐라에센스 (Aroma Piu)	Vanilla essence	10g
소금	Salt	5g
달걀	Eggs	225g
생크림	Heavy cream	150g
중력분	All-purpose flour	240g
박력분	Cake flour	150g
베이킹파우더	Baking powder	6g
베이킹소다	Baking soda	2g
시나몬파우더	Cinnamon powder	7g
캐러멜라이즈 사과 *	Caramelized apple *	300g
TOTAL		**2255g**

코코넛 크럼블

Coconut Crumble

INGREDIENTS		g
버터 (Bridel)	Butter	180g
코코넛슈거	Coconut sugar	100g
설탕	Sugar	50g
소금	Salt	4g
박력분	Cake flour	300g
코코넛롱	Shredded dried coconut	60g
옥수수전분	Cornstarch	30g
물	Water	40g
TOTAL		**764g**

*** 만드는 법**

❶ 믹싱볼에 차가운 상태의 버터, 코코넛슈거, 설탕, 소금을 넣고 비터로 부드럽게 풀어준다.

❷ 체 친 박력분, 코코넛롱, 옥수수전분을 넣고 물을 흘려가며 믹싱해 반죽 상태로 만든다.

❸ 테프론시트를 깐 철판 위에서 체(간격 약 0.5cm)에 반죽을 내려 크럼블 상태로 만들어 냉동한다.

❹ 냉동시킨 크럼블은 밀폐용기에 담아 냉동 보관하면서 사용하고, 사용하기 전 손으로 가볍게 풀어준다.

*** Procedure**

❶ Cream together the cold butter, coconut sugar, sugar, and salt in a mixing bowl with the paddle.

❷ Add sifted flour, shredded coconut, and cornstarch and mix while drizzling in water until the mixture comes together.

❸ Pass the dough through a coarse sieve (about 0.5cm mesh) onto a baking sheet lined with a Teflon sheet to make crumbles and freeze.

❹ Keep the frozen crumbles stored in an airtight container and use as needed. Gently break them apart with your hands before using them.

✤

CARAMELIZED APPLE

캐러멜라이즈 사과

1 냄비에 사방 1.5cm 크기로 깍둑썬 사과, 설탕, 레몬즙, 바닐라빈 껍질을 넣고 가열하면서 졸인다.

TIP. 하루 전날 레몬즙에 바닐라빈 껍질을 넣어 냉침한다. 완성 후 사용할 때는 바닐라빈 껍질을 제거한다.

2 사과에서 나온 수분이 완전히 날아가면 시나몬파우더, 칼바도스를 넣고 섞어 마무리한다.

TIP. 바트에 담고 밀착 랩핑해 냉장고에 보관한다.

1 In a saucepan, combine apples, diced into 1.5 cm cubes, sugar, lemon juice, and a vanilla bean pod (without seeds), and heat to simmer.

TIP. Soak the vanilla bean in the lemon juice to cold-infuse the day before. Remove the vanilla bean when ready to use.

2 Once all the moisture from the apple evaporates, stir in cinnamon powder and Calvados to finish.

TIP. Pour into a tray, cover with plastic wrap, make sure it is in contact with the apples, and refrigerate.

✤ ✤

APPLE CAKE

사과 케이크

1 써머믹서에 설탕, 코코넛롱을 넣고 곱게 갈아준다.

TIP. 코코넛롱은 오븐에서 살짝 로스팅해 사용한다.

2 녹인 버터(45℃), 녹인 캐러멜초콜릿(45~50℃), 고르곤졸라, 바닐라에센스, 소금을 넣고 갈아준다.

TIP. 분리가 난 것처럼 보이지만 수분이 적기 때문이니 신경쓰지 않아도 된다.

3 달걀(30℃), 생크림(30℃)을 넣고 매끈한 상태가 될 때까지 갈아준다.

4 볼에 옮긴 후 체 친 [중력분, 박력분, 베이킹파우더, 베이킹소다, 시나몬파우더]를 넣고 섞는다.

5 캐러멜라이즈 사과를 넣고 섞는다.

1 Add sugar and shredded coconut in a ThermoMixer and grind to a fine powder.

TIP. Lightly roast the shredded coconut before use.

2 Blend with melted butter (45°C), melted caramel chocolate (45~50°C), gorgonzola, vanilla essence, and salt.

TIP. Because it has little moisture, it may look as if it separates. But it's normal, so it's okay.

3 Add eggs (30°C) and heavy cream (30°C) and blend until smooth.

4 Transfer to a bowl and mix in sifted [all-purpose flour, cake flour, baking powder, baking soda, and cinnamon powder].

5 Add the caramelized apples and mix.

✤

1

2

✤ ✤

1-1

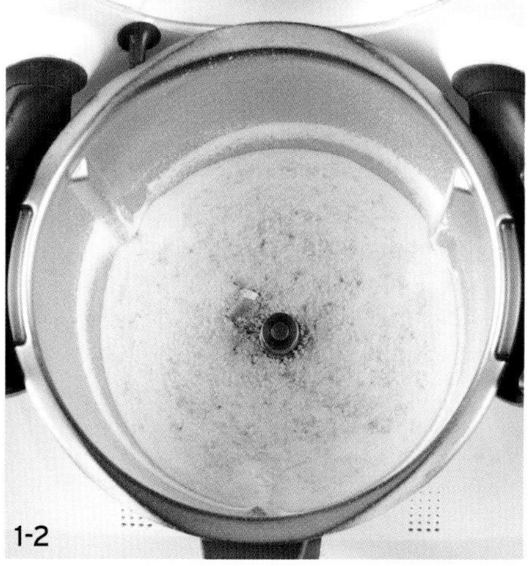
1-2

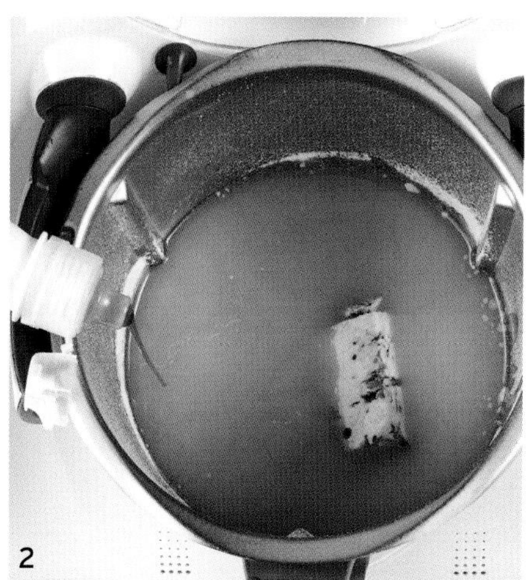
2

3

4

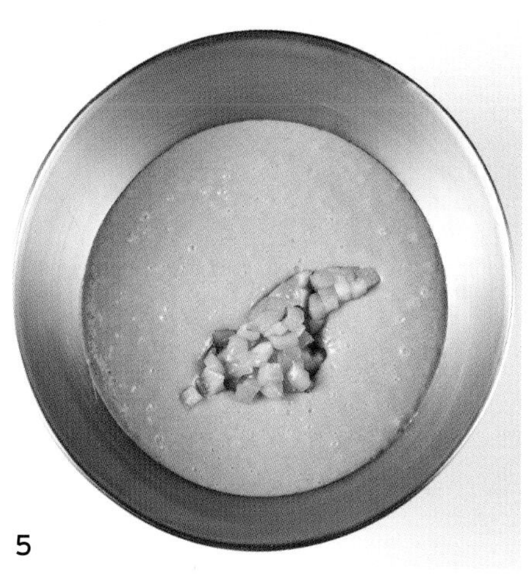
5

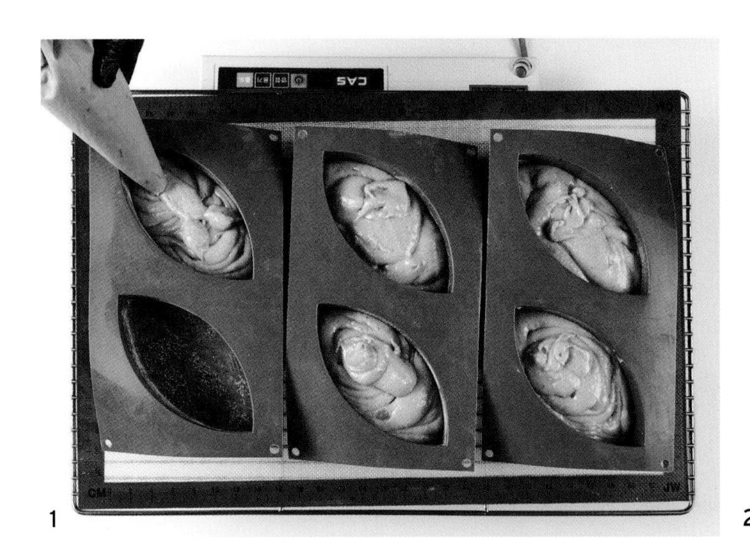

1

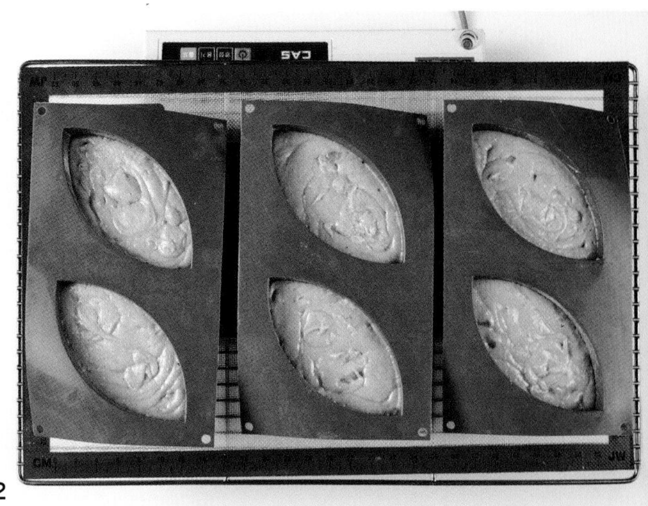

2

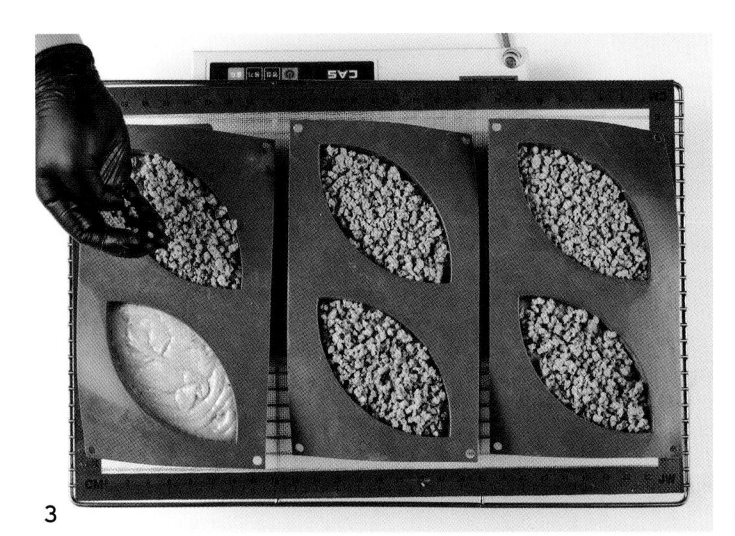

3

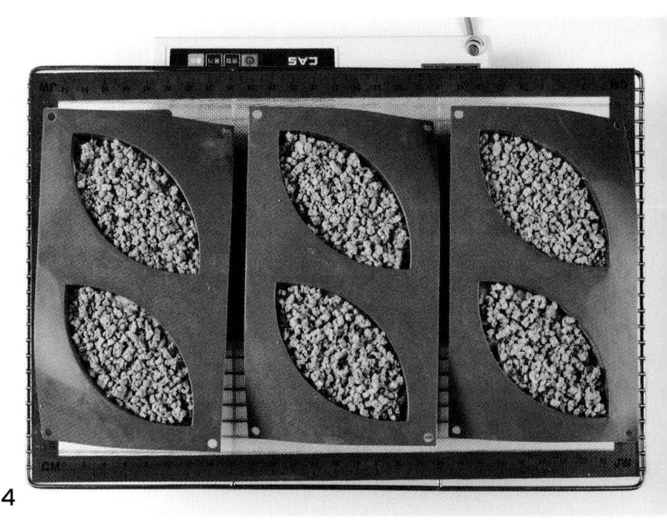

4

✤ ✤ ✤

FINISH 마무리

1 버터를 바르고 밀가루를 뿌린 후 털어낸 몰드(실리코마트 SF218)에 330g씩 채운다.

TIP. 크림법으로 만드는 파운드케이크류는 일반적으로 틀에서 굽지만 이 레시피는 올인원법으로 만들어 실리콘 몰드에 구워도 잘 나온다.

2 바닥에 쳐 평평하게 만든다.

3 코코넛 크럼블을 55g씩 올린다.

4 170℃로 예열된 오븐에서 45분간 구운 후, 레몬초콜릿 파우더(p.188)를 체로 뿌려 마무리한다.

1 Butter and flour the molds (Silikomart SF218) and shake off the excess flour. Fill them with 330 g of batter.

TIP. Pound cakes made with the creaming method generally requires a metal baking pan to bake, but the silicon mold also works well because this recipe is made with all-in-one method.

2 Tap on the work table to flatten.

3 Top with 55 grams of coconut crumble.

4 Bake in a preheated oven at 170°C for 45 minutes and dust with lemon chocolate powder (p.188) to finish.

NOTE.

6

EASY BAKED CHEESECAKE

이지 베이크 치즈 케이크

PAIRING & TEXTURE

페어링 & 텍스처

개인적으로 치즈의 맛이 진하며 묵직하고 꾸덕한 스타일의 치즈 케이크를 좋아해 개발하게 된 나의 시그니처 메뉴이다.

This is my signature menu I developed because I personally like heavy and thick cheesecakes with rich flavors.

TECHNIQUE

테크닉

- 기존 치즈 케이크가 고온의 오븐에서 장시간 굽는 방식이라면, 이 치즈 케이크는 오븐에 들어가기 전 반죽을 익히기 때문에 오븐에서 단시간에 구움색을 얻어낼 수 있다.
- 크렘 파티시에르의 마지막 가열 공정 후 공기의 온도 차에 의해 표면에 막이 생기는 현상에서 착안한 제품이다.

- While traditional cheesecakes are baked in a hot oven for a long time, this cheesecake gets cooked before being put in the oven, allowing it to achieve the toasted color in a short time.
- This cake was inspired by the phenomenon that a film forms on the surface of a crème patissière after the final cooking process due to the temperature difference in the air.

✦ HOW TO COMPOSE THIS RECIPE ✦

STEP 1.

메뉴 정하기	베이크 치즈 케이크
Decide on the menu	Baked Cheesecake

STEP 2.

메인 맛 정하기	크림치즈
Choose the primary flavor	Cream Cheese

STEP 3.

메인 맛(크림치즈)과의 페어링 선택하기

Select a pairing flavor (Cream Cheese)

- ☑ 로투스 — Lotus Biscoff Cookies
- ☑ 다이제스티브 — Digestive Biscuit
- ☑ 레몬즙 — Lemon Juice

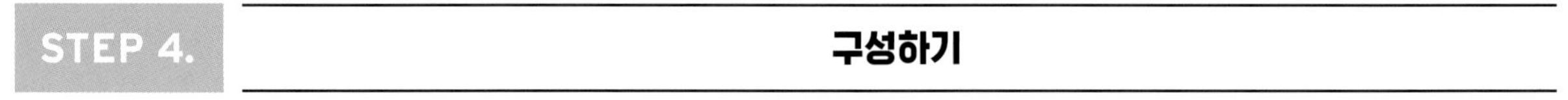

STEP 4.

구성하기

Assemble

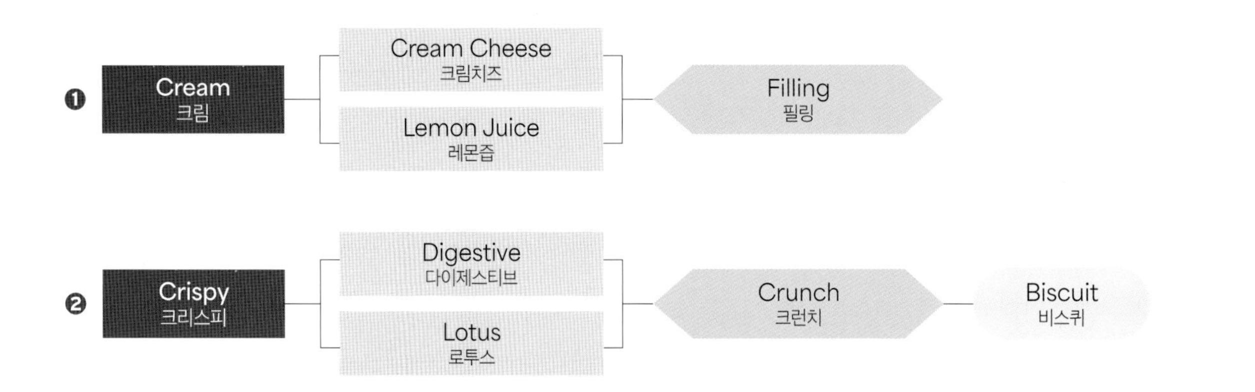

비스퀴
Biscuit
크림치즈 필링
Cream Cheese Filling

INGREDIENTS 4개 분량/ 4 cheesecakes

비스퀴
Biscuit

INGREDIENTS		g
로투스 (시판 과자)	Lotus Biscoff (Store-bought)	210g
다이제스티브 (시판 과자)	Digestive biscuits (Store-bought)	210g
버터 (Bridel)	Butter	164g
TOTAL		**584g**

*** 만드는 법**

❶ 푸드프로세서에 로투스, 다이제스티브를 넣고 갈아준다.
❷ 녹인 버터(45℃)를 넣고 섞는다.

*** Procedure**

❶ Grind Lotus Biscoff and Digestive biscuits in a food processor.
❷ Add the butter melted to 45°C.

치즈 케이크
Cheesecake

INGREDIENTS		g
크림치즈 (Kiri)	Cream cheese	1700g
버터 (Bridel)	Butter	100g
생크림	Heavy cream	850g
우유	Milk	140g
설탕	Sugar	450g
옥수수전분	Cornstarch	110g
바닐라빈 씨	Vanilla bean seeds	2개 분량/ 2 pcs
달걀	Eggs	400g
노른자	Egg yolks	100g
바닐라에센스 (Aroma Piu)	Vanilla essence	20g
레몬에센스 (Aroma Piu)	Lemon essence	2g
레몬즙	Lemon juice	60g
소금	Salt	6g
TOTAL		**3938g**

NOTE.

✤

CHEESECAKE

치즈 케이크

1 로보쿱에 크림치즈(45℃), 포마드 상태의 버터, 생크림과 우유(50℃), 미리 섞어둔 [설탕, 옥수수전분, 소금, 바닐라빈 씨]를 넣고 블렌딩한다.

TIP. 로보쿱이 없는 경우 푸드프로세서로도 작업이 가능하다.

2 달걀과 노른자(30℃) - 바닐라에센스와 레몬에센스 순서로 넣고 블렌딩한 후 체에 거르고 냉장고에서 12시간 동안 숙성시킨다.

TIP. 숙성시키지 않고 바로 사용해도 크게 지장은 없지만 약간의 식감 차이는 생긴다. 숙성시키는 동안 반죽 속 지방, 당분, 수분이 안정되므로 숙성시키지 않은 것과 비교했을 때 더 크리미한 질감으로 완성된다.

3 냉장고에서 숙성시킨 **2**를 냄비로 옮긴다.

4 70℃가 될 때까지 주걱으로 저어가며 가열한다.

TIP. 70℃ 이하로 가열하면 거품이 많이 생기고, 70℃ 이상으로 가열하면 표면의 막이 너무 빨리 생겨 작업성이 떨어진다.

5 70℃가 되면 불에서 내려 잔열이 있는 상태에서 좀 더 저어주어 온도를 일정하게 맞춘다.

6 다시 비커에 옮긴 다음 레몬즙을 넣고 블렌딩해 크림치즈 필링을 완성한다.

TIP. 최대한 거품이 형성되지 않도록 주의한다. 거품이 형성되면 표면과 내상이 매끈하게 완성되지 않는다.

7 지름 15cm, 높이 5cm 세라클 링 바닥을 알루미늄 호일로 감싼 후 바닥과 옆면에 테프론시트를 두른다. 비스퀴를 100g씩 넣고 평평하게 만들어준다.

8 크림치즈 필링을 760g씩 팬닝한다.

1 Combine cream cheese (45°C), softened butter, heavy cream and milk (50°C), and previously mixed [sugar, cornstarch, salt and vanilla bean seeds] in the Robot Coupe.

TIP. You can also use a food processor.

2 Blend in the eggs, egg yolks (30°C), vanilla and lemon essence in order. Strain and refrigerate for 12 hours.

TIP. You can use the mixture immediately without resting it, but it will have a slightly different texture. During resting, the fat, sugar, and moisture in the batter stabilize, resulting in a creamier texture than without resting.

3 Pour the refrigerated (**2**) into a pot.

4 Cook until it reaches 70°C while stirring.

TIP. Cooking it below 70°C will make it foamy, and if it's above 70°C, it will form a film on the surface too quickly, making it less workable.

5 When it reaches 70°C, remove from the heat and stir some more while it's still hot to equalize the temperature.

6 Pour it back into the beaker and blend with lemon juice to finish.

TIP. Try to avoid bubbles as much as possible. Bubbles will result in an uneven surface and center.

7 Cover the bottom of the ring mold (Ø 15 cm, H 5 cm) with aluminum foil, then line with Teflon sheets on the sides and the bottom. Put 100 grams of biscuit and flatten.

8 Fill it with 760 grams of the cream cheese filling.

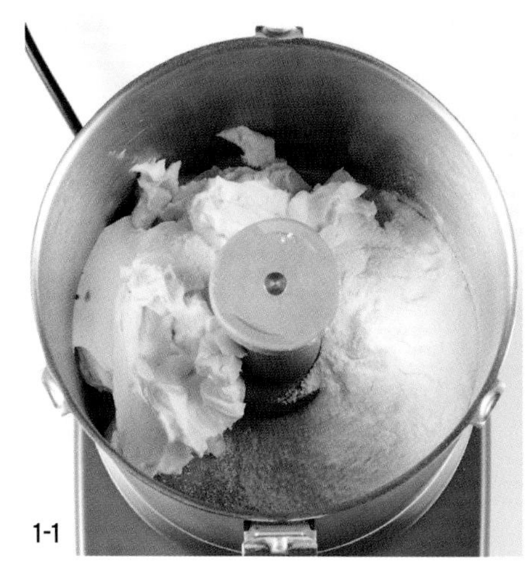
1-1

1-2

2

3

70℃
4

5

6

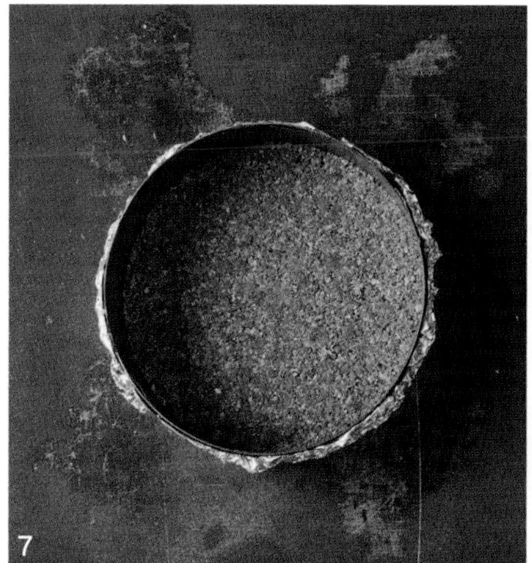
7

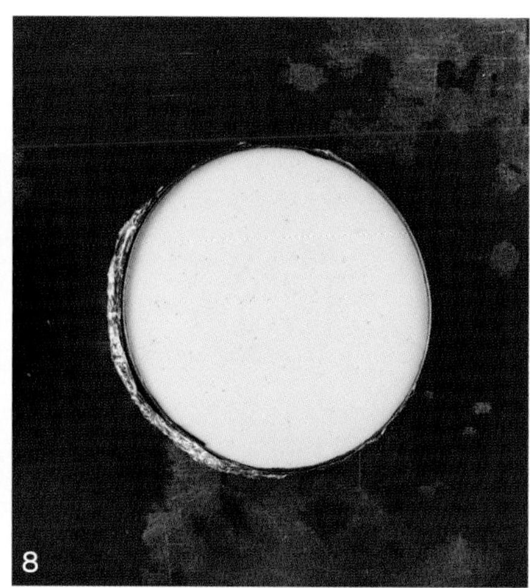
8

9 반죽 표면이 마를 때까지 기다린다.

10 200℃로 예열된 오븐에서 25분간 굽는다.

TIP. 박텔 컨벡션 오븐 기준 뎀퍼를 100%로 열고 바람 세기 2로 맞춰 굽는다.
구워져 나온 치즈 케이크는 형태 유지를 위해 철판째 냉장 보관한 후 완전히 식으면 옆면에 토치로 살짝 열을 가해 세라클 링을 분리한다.

9 Wait until the surface of the batter is dry.

10 Bake in an oven preheated to 200°C for 25 minutes.

TIP. When using the Bechtel convention oven, open the damper to 100% and set the fan speed at 2.
Once baked, refrigerate the cheesecake as is on the baking tray to retain the shape. When it's completely cooled, heat the sides with a blow torch and remove the ring mold.

팬닝 후 약 5분 정도 실온에 두면 반죽 표면의 광택이 없어지면서 얇은 막이 생긴다. (사진은 막을 보여주기 위해 핀셋으로 막을 일부러 벗겨낸 것이다.) 필링은 냄비에서 이미 익혔으므로, 오븐에서는 구움색을 낼 정도로만 짧게 구워주면 되는데, 이때 반죽 표면에 생긴 얇은 막이 빠르게 구움색을 내게 도와준다.

After letting the batter sit at room temperature for about 5 minutes in the mold, it loses its shine and develops a thin film on the surface. (In the photo, the film was intentionally removed with tweezers to demonstrate.) Since the filling is already cooked, it only needs to be baked for a short time to develop a toasty color, and the thin film on the surface will help it to brown quickly.

7 BLONDIE STICK CAKES

블론디 스틱 케이크

PAIRING & TEXTURE
페어링 & 텍스처

- 구움과자를 케이크 형태로 복잡하지 않게 만들어본 제품.
- 크림을 올려 케이크 형태로 만들지 않고 블론디 그 자체만으로도 맛있게 먹을 수 있는 레시피로, 구워져 나온 블론디를 작게 잘라 구움과자로 판매해도 손색없는 제품이다.
- 여기에서는 맛의 임팩트가 강한 블론디에 글레이즈를 씌우고 베이직한 크림을 더해 케이크의 맛과 텍스처를 완성했다.

- A travel cake made into a cake-with-cream style in a simple method
- You can enjoy this recipe on its own without adding cream. This blondie can be cut into small sizes and sold as small travel cakes.
- I created the taste and texture of the cake by covering the blondie, which has a strong taste, with a glaze and using basic cream.

TECHNIQUE
테크닉

- 실온 상태보다 쇼케이스 또는 냉장(5~10℃) 온도에서 맛있게 먹을 수 있는 텍스처로 만들기 위해 지방, 당 등의 양을 조절해 완성한 레시피이다.
- 스틱 형태로 작게 만들거나, 파운드 케이크 형태로 크게 만들거나, 크림을 이용해 케이크 형태로 만들어 선물용으로도 판매하기에 좋은 제품이다.

- The amount of fat, sugar, etc., has been adjusted to adapt a texture that is more palatable at the showcase or refrigerated temperature (5~10°C) than at room temperature.
- It is a great gift product, whether as a stick or large cake, such as a pound cake or a cream-frosted cake.

DESIGN
디자인

- 직사각형 바 형태의 디자인에 얇은 초콜릿 장식을 올려 심플하지만 개성 있는 느낌으로 표현했다.
- 초콜릿 장식물의 반짝이는 면과 반짝이지 않는 면을 번갈아가며 장식해 질감의 대비를 느낄 수 있게 했다.

- I designed it on a rectangular bar shape and decorated it with thin chocolate strips to give a simple but unique ambiance.
- The alternated placement of glossy and matte sides of chocolate strips creates a contrast in texture.

✦ HOW TO COMPOSE THIS RECIPE ✦

STEP 1.	STEP 2.	STEP 3.
메뉴 정하기 Decide on the menu	**메인 맛 정하기** Choose the primary flavor	
Types 종류	Main Ingredient 주재료	Pairing 페어링
ALUNGA EARL GREY BLONDIE STICK CAKE 알룬가 얼그레이 블론디 스틱 케이크	Milk Chocolate 밀크초콜릿	Milk Chocolate + Earl Grey + Diced Dried Orange 밀크초콜릿 + 얼그레이 + 건조 오렌지 다이스
MATCHA BLONDIE STICK CAKE 말차 블론디 스틱 케이크	Matcha 말차	White Chocolate + Matcha Powder + Green Raisins + Dried Cranberries 화이트초콜릿 + 말차파우더 + 건청포도 + 건크랜베리
STRAWBERRY BLONDIE STICK CAKE 스트로베리 블론디 스틱 케이크	Strawberry 딸기	Strawberry Chocolate + Freeze-Dried Strawberry Powder + Freeze-Dried Raspberry Powder 딸기초콜릿 + 동결건조 딸기파우더 + 동결건조 라즈베리파우더
CARAMEL BLONDIE STICK CAKE 캐러멜 블론디 스틱 케이크	Caramel 캐러멜	Caramel Chocolate 캐러멜초콜릿
DARK BLONDIE STICK CAKE 다크 블론디 스틱 케이크	Dark Chocolate 다크초콜릿	Dark Chocolate 다크초콜릿

구성하기
Assemble

Topping 토핑	Montée 몽테	Room Temperature Glaze 룸 템퍼라처 글레이즈
Almond Cream + Earl Grey Crumble 아몬드 크림 + 얼그레이 크럼블	Milk Chocolate + Earl Grey 밀크초콜릿 + 얼그레이	Milk Chocolate + Vegetable Oil + Almond Slices 밀크초콜릿 + 식용유 + 다진 아몬드 슬라이스
Matcha Crumble 말차 크럼블	White Chocolate + Matcha Powder 화이트초콜릿 + 말차파우더	White Chocolate + Vegetable Oil + Matcha Powder + Chopped Pistachios 화이트초콜릿 + 식용유 + 말차파우더 + 다진 피스타치오
Strawberry Jam + Matcha Crumble 딸기잼 + 말차 크럼블	Ruby Chocolate + Freeze-Dried Strawberry Powder + Freeze-Dried Raspberry Powder 루비초콜릿 + 동결건조 딸기파우더 + 동결건조 라즈베리파우더	Strawberry Chocolate + Ruby Chocolate + Vegetable Oil + Chopped Raspberry Crispies 딸기초콜릿 + 루비초콜릿 + 식용유 + 다진 코팅 라즈베리 크리스피
Hazelnut Caramel + can Paste + Caramelized can + Toffee + Fleur de Sel 헤이즐넛 캐러멜 + 피칸 페이스트 + 캐러멜라이즈 피칸 + 토피 + 꽃소금	Caramel Chocolate + Pecan Paste 캐러멜초콜릿 + 피칸 페이스트	Caramel Chocolate + Vegetable Oil + Caramelized hazelnuts, chopped 캐러멜초콜릿 + 식용유 + 다진 캐러멜라이즈 헤이즐넛
Hazelnut Caramel + Hazelnut Paste + ark Chunk Chocolate + Cacao Crumble 헤이즐넛 캐러멜 + 헤이즐넛 페이스트 + 다크 청크 초콜릿 + 카카오 크럼블	Dark Chocolate + Almond Paste 다크초콜릿 + 아몬드 페이스트	Dark Chocolate + Cacao Butter + Vegetable Oil + Chopped Caramelized Hazelnuts 다크초콜릿 + 카카오버터 + 식용유 + 다진 아몬드 슬라이스

INGREDIENTS

① ALUNGA EARL GREY BLONDIE STICK CAKE
알룬가 얼그레이 블론디 스틱 케이크

40개 분량/ 40 stick cakes

반죽

Batter

(철판(42 × 32cm) 1개 분량)
(1 baking tray (42 × 32 cm))

INGREDIENTS		g
달걀	Eggs	290g
얼그레이 티백 잎 (Twinings)	Earl Grey tea bag leaves	27g
설탕	Sugar	320g
전화당	Inverted sugar	50g
밀크초콜릿 (Alunga 41%)	Milk chocolate	290g
버터 (Bridel)	Butter	330g
오렌지에센스 (Aroma Piu)	Orange essence	5g
건조 오렌지 다이스	Diced dried orange	120g
박력분	Cake flour	150g
아몬드파우더	Almond powder	70g
베이킹파우더	Baking powder	2g
크렘 다망드 *	Crème d'amande *	220g
얼그레이 크럼블 *	Earl Grey crumble *	230g
TOTAL		2104g

크렘 다망드 *

Crème d'Amande

• p.200

INGREDIENTS		g
통아몬드파우더	Ground whole almonds	62g
설탕	Sugar	125g
소금	Salt	1.5g
아몬드파우더 (시판)	Almond powder (Store-bought)	62g
박력분	Cake flour	12g
버터 (Bridel)	Butter	125g
달걀	Eggs	125g
골드럼 (PAN RUM)	Gold rum	12g
다진 아몬드	Chopped almonds	62g
TOTAL		**586.5g**

- 통아몬드파우더는 통아몬드를 직접 갈아 사용했다.
- Ground whole almonds are freshly ground before use.

얼그레이 크럼블 *

Earl Grey Crumble

INGREDIENTS		g
버터 (Bridel)	Butter	90g
코코넛슈거	Coconut sugar	50g
설탕	Sugar	25g
소금	Salt	2g
박력분	Cake flour	150g
얼그레이 티백 잎 (Twinings)	Earl Grey tea bag leaves	3g
옥수수전분	Cornstarch	15g
물	Water	20g
TOTAL		**355g**

*** 만드는 법**

❶ 믹싱볼에 차가운 상태의 버터, 코코넛슈거, 설탕, 소금을 넣고 비터로 부드럽게 풀어준다.

❷ 체 친 [박력분, 얼그레이, 옥수수전분]을 넣고 물을 흘려가며 믹싱해 반죽 상태로 만든다.

❸ 테프론시트를 깐 철판 위에서 체(간격 약 0.5cm)에 반죽을 내려 크럼블 상태로 만들어 냉동한다.

❹ 냉동시킨 크럼블은 밀폐용기에 담아 냉동 보관하면서 사용하고, 사용하기 전 손으로 가볍게 풀어준다.

*** Procedure**

❶ Soften cold butter, coconut sugar, sugar, and salt using a paddle attachment.

❷ Mix with sifted [cake flour, tea leaves, and cornstarch], while drizzling the water until it forms a dough.

❸ Pass the dough through a coarse sieve (about 0.5cm mesh) onto a baking sheet lined with a Teflon sheet to make crumbles and freeze.

❹ Store the frozen crumbles in an airtight container and use as needed. Gently break them apart with your hands before using them.

룸 템퍼라처 글레이즈

Room Temperature Glaze

INGREDIENTS		g
밀크초콜릿 (Alunga 41%)	Milk chocolate	1000g
식용유	Vegetable oil	200g
다진 아몬드 슬라이스	Almond slices, chopped	40g
TOTAL		**1240g**

- 아몬드 슬라이스는 로스팅 후 다져 사용한다.
- Chop the roasted almond slices to use.

*** 만드는 법**

❶ 비커에 녹인 밀크초콜릿(45~50℃), 식용유를 담고 바믹서로 블렌딩한다.
❷ 다진 아몬드 슬라이스를 넣고 블렌딩한다.
❸ 30℃로 맞춰 사용한다.

*** Procedure**

❶ In a beaker, blend melted milk chocolate (45~50°C) and vegetable oil with an immersion blender.
❷ Blend with chopped almond slices.
❸ Use at 30°C.

알룬가 얼그레이 몽테 크림

Alunga Earl Grey Montée Cream

INGREDIENTS		g
얼그레이 티백 잎 (Twinings)	Earl Grey tea bag leaves	25g
물	Water	100g
휘핑크림A	Whipping cream A	400g
우유	Milk	120g
전화당	Inverted sugar	84g
젤라틴매스 (×5)	Gelatin mass (×5)	68g
밀크초콜릿 (Alunga 41%)	Milk chocolate	400g
소금	Salt	1g
휘핑크림B	Whipping cream B	840g
TOTAL		**2038g**

NOTE.

✤

BATTER

반죽

1 믹싱볼에 달걀, 얼그레이 티백 잎, 설탕, 전화당을 넣고 따뜻한 물이 담긴 볼에 받쳐 휘퍼로 저어가며 45℃까지 온도를 높인다.

TIP. 거품이 올라오지 않게 주의하면서 작업한다.

2 녹인 밀크초콜릿(45~50℃)과 녹인 버터(45℃)를 붓고 바믹서로 블렌딩한다.

3 오렌지에센스를 넣고 블렌딩한다.

4 건조 오렌지 다이스를 넣고 휘퍼로 섞는다.

5 체 친 [박력분, 아몬드파우더, 베이킹파우더]를 넣고 날가루가 보이지 않을 때까지 휘퍼로 섞는다.

6 테프론시트를 깐 철판(42 × 32cm)에 반죽을 붓고 스패출러로 평평하게 정리한다.

7 크렘 다망드 220g을 얇게 파이핑한다.

8 얼그레이 크럼블 230g을 고르게 뿌린다.

9 160℃로 예열된 오븐에서 뎀퍼를 100% 열고 40분간 굽고 식힌 후 12 × 2.5cm 크기로 잘라 냉동 보관한다.

TIP. 유산지를 올린 식힘망 위에 뒤집어 빼 윗면이 눌리지 않게 한다.
블론디 그 자체만으로도 구움과자로 판매해도 손색없을 정도로 맛이 뛰어나다.

1 Add eggs, tea leaves from Earl Grey tea bags, sugar, and inverted sugar in a mixing bowl. Place it over hot water and whisk until it reaches 45°C.

TIP. Be careful not to make bubbles.

2 Blend with melted milk chocolate (45~50°C) and melted butter (45°C) using an immersion blender.

3 Blend with orange essence.

4 Add diced dried oranges and stir with a whisk.

5 Mix sifted [cake flour, almond powder, and baking powder] with a whisk until the powder ingredients are no longer visible.

6 Spread onto a baking tray (42 × 32 cm) lined with a Teflon sheet using an offset spatula.

7 Thinly pipe 230 grams of crème d'amande.

8 Evenly sprinkle 230 grams of Earl Grey crumble.

9 Bake in an oven preheated to 160°C and damper open at 100% for 40 minutes. Cool, cut to 12 × 2.5 cm, and freeze.

TIP. Place it upside down on a cooling rack lined with parchment paper and remove the tray so the top doesn't get pressed.
The blondies are so delicious that they could be sold on their own as the travel cake.

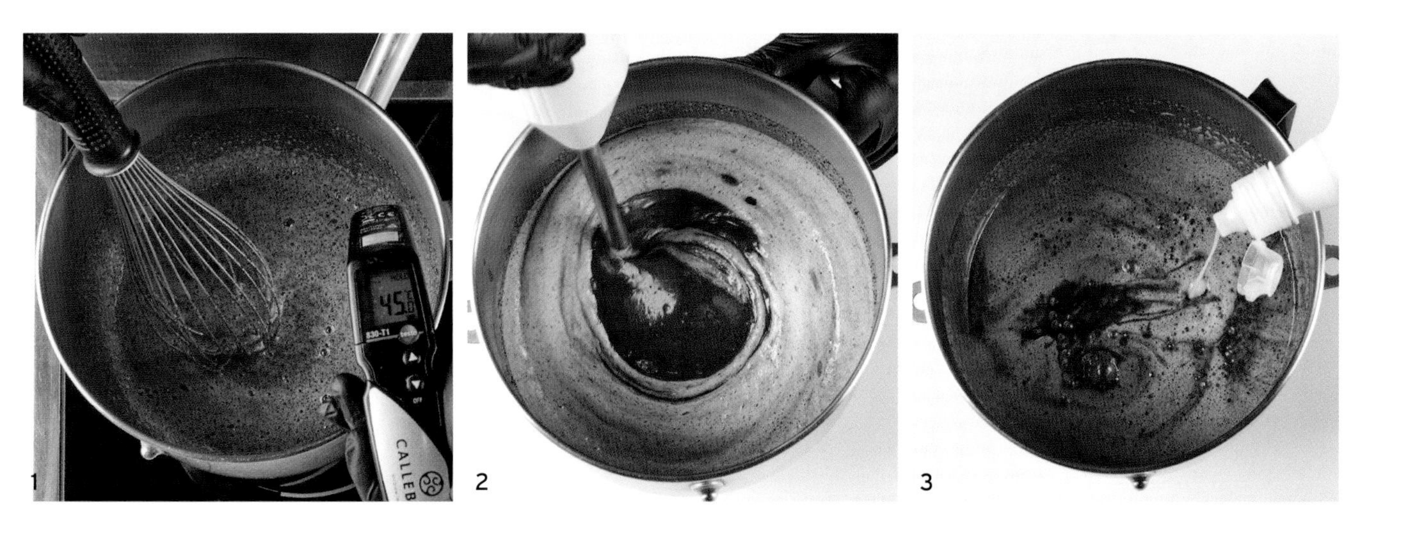
1
2
3

4
5
6

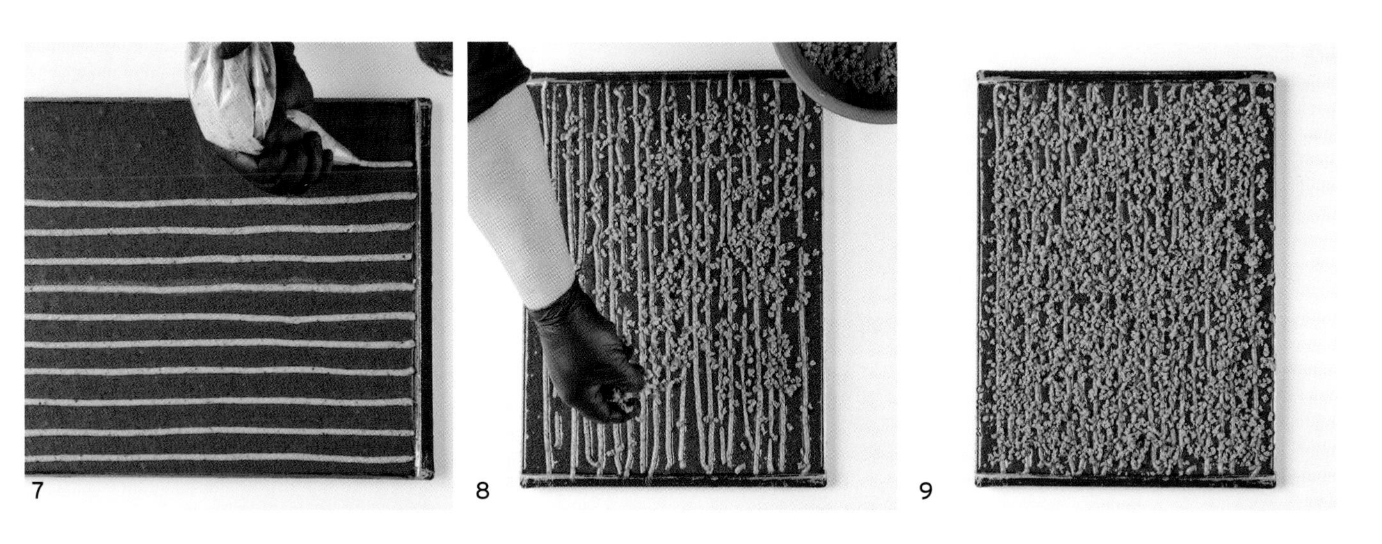
7
8
9

✤ ✤

ALUNGA EARL GREY MONTÉE CREAM

알룬가 얼그레이 몽테 크림

1 냄비에 얼그레이 티백 잎과 물을 넣고 가열한 후 체에 거른다. 다시 냄비에 걸러둔 얼그레이 티백 잎, 휘핑크림A, 우유, 전화당, 젤라틴매스를 넣고 70℃까지 가열한다.

TIP. 잘게 잘린 상태의 얼그레이 티백의 잎을 바로 사용하면 쓰고 떫은 맛이 강하게 표현되므로, 뜨거운 물에 한 번 우려낸 후 인퓨징한다.

2 끓어오르면 불에서 내려 냄비 입구를 밀착 랩핑해 10분 동안 그대로 두고 인퓨징한다.

3 밀크초콜릿이 담긴 비커에 넣고 바믹서로 블렌딩한다.

TIP. 체에 내려 얼그레이를 거른 후 블렌딩한다.

4 소금, 휘핑크림B를 넣어가며 블렌딩한다.

5 바트에 담고 밀착 랩핑한 후 냉장고에 보관한다.

6 사용하기 직전 비터로 부드럽게 풀어 사용한다.

1 Boil the tea leaves and water briefly and strain. Add the strained tea leaves, whipping cream A, milk, inverted sugar, and gelatin mass back into the pot and cook to 70°C.

TIP. Using the chopped tea leaves from the tea bags right away will release a bitter and astringent taste. So, soak in hot water to rinse, then use it to infuse.

2 When it comes to a boil, remove from the heat, cover the pot, and leave to infuse for 10 minutes.

3 Blend in a beaker with milk chocolate using an immersion blender.

TIP. Blend after straining out the tea leaves.

4 Continue to blend while adding salt and whipping cream B.

5 Pour onto a stainless-steel tray, adhere the plastic wrap to the preparation to cover, and refrigerate.

6 Gently whip using the paddle attachment before use.

70℃
1

2

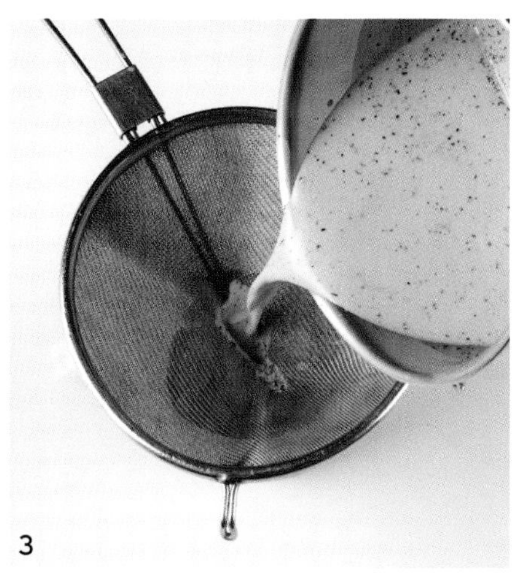
3

4

5

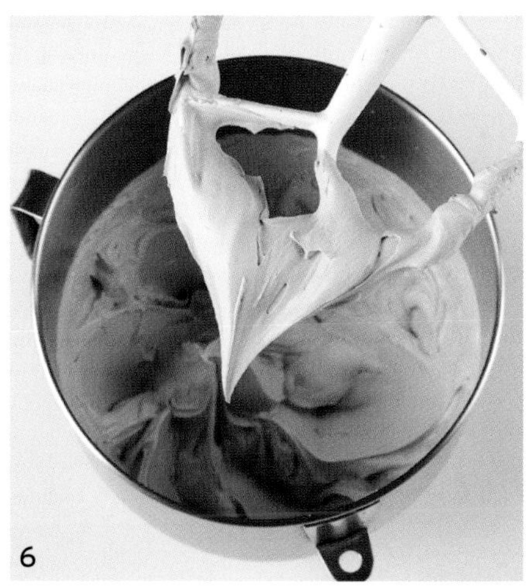
6

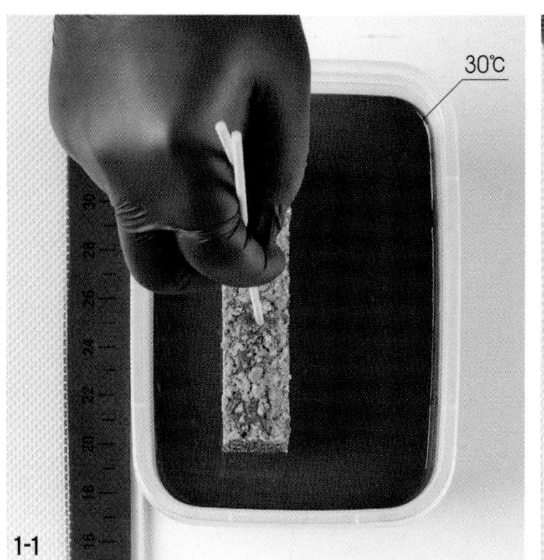

1-1

1-2

1-3

2-1

2-2

✤ ✤ ✤

FINISH

마무리

1. 12 × 2.5cm 크기로 자른 블론디 케이크(냉동 상태)에 룸 템퍼라처 글레이즈(30℃)를 코팅한다.

   TIP. 윗면을 제외한 옆면에 코팅한 후 밑면을 깔끔하게 정리해 굳힌다.

2. 블론디 케이크를 나란히 놓고 지름 1.8cm 깍지를 끼운 짤주머니에 알룬가 얼그레이 몽테 크림을 담아 일자로 파이핑한다.
3. 뜨겁게 달군 칼로 알룬가 얼그레이 몽테 크림을 깔끔하게 자른다.
4. 장식용 초콜릿을 올려 마무리한다.

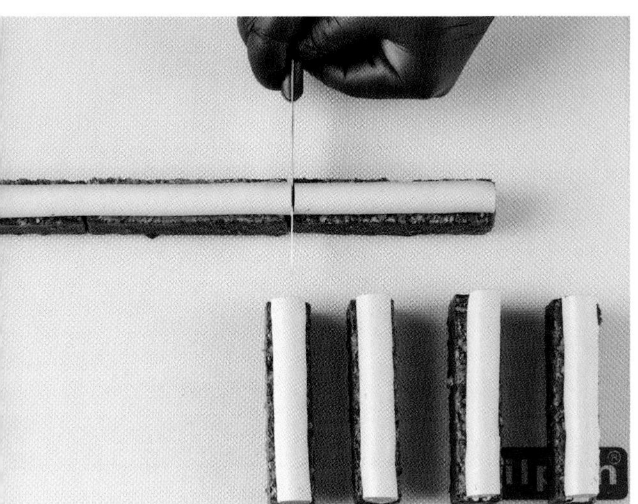

1 Prepare the frozen blondies cut to 12 x 2.5 cm. Coat them with the room temperature glaze (30°C).

TIP. Coat all the sides, excluding the top, and clean the bottom to set.

2 Arrange the blondies in a line. Put Allunga Earl Grey montée cream in a piping bag with a 1.8 cm round tip and pipe a straight line on the top.

3 Cut the cream clean using a hot knife.

4 Top with the chocolate decorations to finish.

② MATCHA BLONDIE STICK CAKE 말차 블론디 스틱 케이크

40개 분량/ 40 stick cakes

반죽 Batter (철판(42 × 32cm) 1개 분량, 1 baking tray (42 × 32 cm))

INGREDIENTS		g
달걀	Eggs	360g
설탕	Sugar	400g
전화당	Inverted sugar	60g
화이트초콜릿 (Zephyr white 34%)	White chocolate	300g
버터 (Bridel)	Butter	410g
박력분	Cake flour	225g
아몬드파우더	Almond powder	90g
베이킹파우더	Baking powder	4g
말차파우더	Matcha powder	62g
전처리한 건청포도	Pre-processed green grape	240g
전처리한 건크랜베리	Pre-processed cranberry	120g
말차 크럼블 *	Matcha crumble *	260g
TOTAL		**2531g**

- 건조 과일의 전처리 방법은 1권 p.87를 참고한다.
- Pre-treating dried fruit: Refer to Book 1, p.87, 'Pre-treating dried fruit'.

*** 만드는 법**

❶ 믹싱볼에 달걀, 설탕, 전화당을 넣고 따뜻한 물이 담긴 볼에 받쳐 휘퍼로 저어가며 45℃까지 온도를 높인다.
- 거품이 올라오지 않게 주의하면서 작업한다.

❷ 녹인 화이트초콜릿(45~50℃)과 버터(40℃)를 붓고 바믹서로 블렌딩한다.

❸ 체 친 [박력분, 아몬드파우더, 베이킹파우더, 말차파우더]를 넣고 날가루가 보이지 않을 때까지 휘퍼로 섞는다.

❹ 전처리한 건청포도와 건크랜베리를 넣고 섞는다.

❺ 테프론시트를 깐 철판(42×32cm)에 반죽을 붓고 스패출러로 평평하게 정리한다.

❻ 말차 크럼블 260g을 고르게 뿌린다.

❼ 160℃로 예열된 오븐에서 뎀퍼를 100% 열고 50분간 굽는다.
- 유산지를 올린 식힘망 위에 뒤집어 빼 윗면이 눌리지 않게 한다.

*** Procedure**

❶ Add eggs, sugar, and inverted sugar in a mixing bowl. Place it over hot water and whisk until it reaches 45°C.
- Be careful not to make bubbles.

❷ Blend with melted white chocolate (45~50°C) and butter (40°C) using an immersion blender.

❸ Mix sifted [cake flour, almond powder, baking powder, and matcha powder] with a whisk until the powder ingredients are no longer visible.

❹ Mix with the pre-processed green grapes and cranberries.

❺ Spread onto a baking tray (42 × 32 cm) lined with a Teflon sheet using an offset spatula.

❻ Evenly sprinkle 260 grams of matcha crumble.

❼ Bake in an oven preheated to 160°C and damper open at 100% for 50 minutes.
- Place it upside down on a cooling rack lined with parchment paper and remove the tray so the top doesn't get pressed.

말차 크럼블 *
Matcha Crumble

INGREDIENTS		g
버터 (Bridel)	Butter	90g
코코넛슈거	Coconut sugar	50g
설탕	Sugar	25g
소금	Salt	2g
박력분	Cake flour	150g
말차파우더	Matcha powder	30g
옥수수전분	Cornstarch	15g
물	Water	20g
TOTAL		**382g**

*** 만드는 법**

❶ 믹싱볼에 차가운 상태의 버터, 코코넛슈거, 설탕, 소금을 넣고 비터로 부드럽게 풀어준다.

❷ 체 친 [박력분, 말차파우더, 옥수수전분]을 넣고 물을 흘려가며 믹싱해 한 덩어리로 만든다.

❸ 테프론시트를 깐 철판 위에서 체(간격 약 0.5cm)에 반죽을 내려 크럼블 상태로 만들어 냉동한다.

❹ 냉동시킨 크럼블은 밀폐용기에 담아 냉동 보관하면서 사용하고, 사용하기 전 손으로 가볍게 풀어준다.

*** Procedure**

❶ Soften cold butter, coconut sugar, sugar, and salt using a paddle attachment.

❷ Mix with sifted [cake flour, matcha powder, and cornstarch] while drizzling in the water into a dough.

❸ Pass the dough through a coarse sieve (about 0.5cm mesh) onto a baking sheet lined with a Teflon sheet to make crumbles and freeze.

❹ Store the frozen crumbles in an airtight container and use as needed. Gently break them apart with your hands before using them.

룸 템퍼라처 글레이즈
Room Temperature Glaze

INGREDIENTS		g
화이트초콜릿 (Zephyr white 34%)	White chocolate	700g
식용유	Vegetable oil	200g
말차파우더	Matcha powder	30g
다진 피스타치오	Pistachios, chopped	60g
TOTAL		**990g**

*** 만드는 법**

❶ 비커에 녹인 화이트초콜릿(45~50℃), 식용유를 담고 바믹서로 블렌딩한다.

❷ 말차파우더 - 다진 피스타치오 순서로 넣고 블렌딩한다.

❸ 30℃로 맞춰 사용한다.

*** Procedure**

❶ In a beaker, blend melted white chocolate (45~50°C) and vegetable oil with an immersion blender.

❷ Blend with matcha powder and chopped pistachios in order.

❸ Use at 30°C.

말차 몽테 크림

Matcha Montée Cream

INGREDIENTS		g
휘핑크림A	Whipping cream A	500g
젤라틴매스 (×5)	Gelatin mass (×5)	55g
바닐라빈	Vanilla bean	1.5개 분량/ 1.5 pcs
화이트초콜릿 (Zephyr white 34%)	White chocolate	400g
휘핑크림B	Whipping cream B	1000g
말차파우더	Matcha powder	25g
TOTAL		**1980g**

*** 만드는 법**

❶ 냄비에 휘핑크림A, 젤라틴매스, 바닐라빈 껍질을 넣고 가열한다.
❷ 화이트초콜릿이 든 비커에 ❶을 넣고 바믹서로 블렌딩한다.
❸ 휘핑크림B, 바닐라빈 씨를 넣고 블렌딩한다.
❹ 말차파우더를 넣고 블렌딩한다.
❺ 바트에 담아 밀착 랩핑한 후 냉장고에서 약 12시간 숙성시킨다.
❻ 사용할 때는 비터로 부드럽게 풀어 사용한다.

*** Procedure**

❶ Boil whipping cream A, gelatin mass, and vanilla bean pod (without the seeds) in the pot.
❷ Blend in a beaker with white chocolate using an immersion blender.
❸ Blend with whipping cream B and vanilla bean seeds.
❹ Continue to blend with matcha powder.
❺ Pour onto a stainless-steel tray, adhere the plastic wrap to the preparation to cover, and refrigerate for 12 hours.
❻ Gently whip using the paddle attachment before use.

*** 마무리**

❶ 12 × 2.5cm 크기로 자른 블론디 케이크(냉동 상태)에 룸 템퍼라처 글레이즈(30℃)를 코팅한다.
❷ 블론디 케이크를 나란히 놓고 지름 1.8cm 깍지를 끼운 짤주머니에 말차 몽테 크림을 담아 일자로 파이핑한다.
❸ 뜨겁게 달군 칼로 말차 몽테 크림을 깔끔하게 자른다.
❹ 장식용 초콜릿을 올려 마무리한다.

*** Finish**

❶ Prepare the frozen blondies cut to 12 × 2.5 cm. Coat them with the room temperature glaze (30°C).
❷ Arrange the blondies in a line. Put matcha montée cream in a piping bag with a 1.8 cm round tip and pipe a straight line on the top.
❸ Cut the cream clean using a hot knife.
❹ Top with the chocolate decorations to finish.

NOTE.

③ STRAWBERRY BLONDIE STICK CAKE
스트로베리 블론디 스틱 케이크

40개 분량/ 40 stick cakes

반죽 Batter (철판(42 × 32cm) 1개 분량, 1 baking tray (42 × 32 cm))

INGREDIENTS		g
달걀	Eggs	370g
설탕	Sugar	370g
전화당	Inverted sugar	60g
딸기초콜릿 (Callebaut)	Strawberry chocolate	260g
버터 (Bridel)	Butter	410g
박력분	Cake flour	230g
아몬드파우더	Almond powder	90g
베이킹파우더	Baking powder	5g
전처리한 건크랜베리	Pre-processed cranberry	175g
동결건조 딸기파우더	Freeze-dried strawberry powder	19g
동결건조 라즈베리파우더	Freeze-dried raspberry powder	19g
플레인 크럼블 *	Plain crumble *	240g
TOTAL		2248g

＊만드는 법

❶ 믹싱볼에 달걀, 설탕, 전화당을 넣고 따뜻한 물이 담긴 볼에 받쳐 휘퍼로 저어가며 45℃까지 온도를 높인다.
- 거품이 올라오지 않게 주의하면서 작업한다.

❷ 녹인 딸기초콜릿(45~50℃)과 버터(40℃)를 붓고 바믹서로 블렌딩한다.

❸ 체 친 [박력분, 아몬드파우더, 베이킹파우더]를 넣고 날가루가 보이지 않을 때까지 휘퍼로 섞는다.

❹ 전처리한 건크랜베리, 동결건조 딸기파우더와 라즈베리파우더를 넣고 섞는다.

❺ 테프론시트를 깐 철판(42×32cm)에 반죽을 붓고 스패출러로 평평하게 정리한다.

❻ 플레인 크럼블 240g을 고르게 뿌린다.

❼ 160℃로 예열된 오븐에서 뎀퍼를 100% 열고 50분간 굽는다.
- 유산지를 올린 식힘망 위에 뒤집어 빼 윗면이 눌리지 않게 한다.

＊Procedure

❶ Add eggs, sugar, and inverted sugar in a mixing bowl. Place it over hot water and whisk until it reaches 45°C.
- Be careful not to make bubbles.

❷ Blend with melted strawberry chocolate (45~50°C) and butter (40°C) using an immersion blender.

❸ Mix sifted [cake flour, almond powder, and baking powder] with a whisk until the powder ingredients are no longer visible.

❹ Mix with the pre-processed cranberries, freeze-dried strawberry powder, and freeze-dried raspberry powder.

❺ Spread onto a baking tray (42 × 32 cm) lined with a Teflon sheet using an offset spatula.

❻ Evenly sprinkle 240 grams of plain crumble.

❼ Bake in an oven preheated to 160°C and damper open at 100% for 50 minutes.
- Place it upside down on a cooling rack lined with parchment paper and remove the tray so the top doesn't get pressed.

플레인 크럼블 *

Plain Crumble

INGREDIENTS		g
버터 (Bridel)	Butter	90g
코코넛슈거	Coconut sugar	50g
설탕	Sugar	25g
소금	Salt	2g
박력분	Cake flour	150g
옥수수전분	Cornstarch	15g
물	Water	9g
TOTAL		**341g**

∗ 만드는 법

❶ 믹싱볼에 차가운 상태의 버터, 코코넛슈거, 설탕, 소금을 넣고 비터로 부드럽게 풀어준다.

❷ 체 친 [박력분, 옥수수전분]을 넣고 물을 흘려가며 믹싱해 한 덩어리로 만든다.

❸ 테프론시트를 깐 철판 위에서 체(간격 약 0.5cm)에 반죽을 내려 크럼블 상태로 만들어 냉동한다.

❹ 냉동시킨 크럼블은 밀폐용기에 담아 냉동 보관하면서 사용하고, 사용하기 전 손으로 가볍게 풀어준다.

∗ Procedure

❶ Soften cold butter, coconut sugar, sugar, and salt using a paddle attachment.

❷ Mix with sifted [cake flour and cornstarch] while drizzling in the water into a dough.

❸ Pass the dough through a coarse sieve (about 0.5cm mesh) onto a baking sheet lined with a Teflon sheet to make crumbles and freeze.

❹ Store the frozen crumbles in an airtight container and use as needed. Gently break them apart with your hands before using them.

딸기잼

Strawberry Jam

INGREDIENTS		g
딸기 퓌레	Strawberry purée	420g
냉동 라즈베리	Raspberry IQF	300g
설탕	Sugar	250g
NH펙틴	NH pectin	15g
로거스트빈검	Locust bean gum	1g
레몬 퓌레	Lemon purée	50g
바닐라에센스	Vanilla essence	15g
TOTAL		**1051g**

∗ 만드는 법

❶ 볼에 딸기 퓌레, 냉동 라즈베리를 넣고 입구를 랩핑한 후 전자레인지에서 끊어 돌려가며 충분히 녹여준다.

❷ 비커로 옮겨 바믹서로 블렌딩한 후 냄비로 옮겨 가열한다.

❸ 40℃가 되면 불에서 내려 미리 섞어둔 [설탕, NH펙틴, 로거스트빈검]을 넣고 휘퍼로 저어가며 섞어준다.

❹ 다시 불에 올려 레몬 퓌레를 넣고 주걱으로 저어가며 가열한다.

❺ 펙틴 반응을 확인한 후 불에서 내려 바닐라에센스를 넣고 섞어준 후 쿨링한다.

❻ 바트에 담고 밀착 랩핑한 후 냉장 보관한다.

∗ Procedure

❶ Put strawberry purée and frozen raspberries in a bowl and wrap to cover. Melt them sufficiently in the microwave in short intervals.

❷ Blend in a beaker using an immersion blender, pour back into the pot, and heat.

❸ When it reaches 40°C, remove from the heat and whisk in the previously mixed [sugar, pectin NH, and locust bean gum].

❹ Heat them again with lemon purée while stirring with a spatula.

❺ Remove from the heat after pectin activates, mix with vanilla essence, and cool.

❻ Pour onto a stainless-steel tray, adhere the plastic wrap to the preparation to cover, and refrigerate.

룸 템퍼라처 글레이즈

Room Temperature Glaze

INGREDIENTS		g
딸기초콜릿 (Callebaut)	Strawberry chocolate	350g
루비초콜릿 (Callebaut RB2)	Ruby chocolate	350g
식용유	Vegetable oil	200g
다진 코팅 라즈베리 크리스피 (Sosa)	Wet-proof raspberry crispy, chopped	45g
TOTAL		**945g**

* 만드는 법

❶ 비커에 녹인 딸기초콜릿과 루비초콜릿(45~50℃), 식용유를 담고 바믹서로 블렌딩한다.

❷ 다진 코팅 라즈베리 크리스피를 넣고 섞는다.

❸ 30℃로 맞춰 사용한다.

* Procedure

❶ In a beaker, blend melted strawberry chocolate, ruby chocolate (45~50°C), and vegetable oil with an immersion blender.

❷ Blend with chopped raspberry crispies.

❸ Use at 30°C.

딸기 몽테 크림

Strawberry Montée Cream

INGREDIENTS		g
휘핑크림A	Whipping cream A	290g
젤라틴매스 (×5)	Gelatin mass (×5)	35g
연유	Condenced milk	50g
딸기초콜릿 (Callebaut)	Strawberry chocolate	125g
루비초콜릿 (Callebaut RB2)	Ruby chocolate	125g
휘핑크림B	Whipping cream B	530g
딸기 리큐르 (Dijon Strawberry)	Strawberry liqueur	20g
동결건조 딸기파우더	Freeze-dried strawberry powder	6g
동결건조 라즈베리파우더	Freeze-dried raspberry powder	6g
TOTAL		**1187g**

* 만드는 법

❶ 냄비에 휘핑크림A, 젤라틴매스를 넣고 젤라틴매스가 녹을 때까지 가열한 후(약 60℃) 연유를 넣는다.

❷ 딸기초콜릿, 루비초콜릿이 담긴 비커에 ❶을 붓고 바믹서로 블렌딩한다.

❸ 차가운 상태의 휘핑크림B, 딸기 리큐르, 동결건조 딸기파우더와 라즈베리파우더를 넣고 블렌딩한다.

❹ 바트에 담아 밀착 랩핑한 후 냉장고에서 12시간 동안 숙성시킨다.

❺ 사용하기 전 파이핑하기 좋은 상태로 휘핑해 사용한다.

* Procedure

❶ Heat whipping cream A and gelatin mass until the gelatin melts (about 60°C) and add condensed milk.

❷ Blend in a beaker with strawberry and ruby chocolate using an immersion blender.

❸ Continue to blend with cold whipping cream B, strawberry liqueur, freeze-dried strawberry, and raspberry powders.

❹ Pour onto a stainless-steel tray, adhere the plastic wrap to the preparation to cover, and refrigerate for 12 hours.

❺ Gently whip using the paddle attachment before use.

* **마무리**

❶ 12 × 2.5cm 크기로 자른 블론디 케이크(냉동 상태)에 룸 템퍼라처 글레이즈(30℃)를 코팅한다.

❷ 블론디 케이크를 나란히 놓고 지름 1.8cm 깍지를 끼운 짤주머니에 딸기 몽테 크림을 담아 일자로 파이핑한다.

❸ 뜨겁게 달군 칼로 딸기 몽테 크림을 깔끔하게 자른다.

❹ 지름 0.7cm 스테인리스 파이프에 열을 가한 후 딸기 몽테 크림에 대고 눌러 홈을 만든다.

- 비슷한 지름의 스테인리스 빨대를 사용해도 좋다.

❺ 만들어진 홈에 딸기잼을 채운다.

❻ 장식용 초콜릿을 올려 마무리한다.

---

* **Finish**

❶ Prepare the frozen blondies cut to 12 × 2.5 cm. Coat them with the room temperature glaze (30°C).

❷ Arrange the blondies in a line. Put strawberry montée cream in a piping bag with a 1.8 cm round tip and pipe a straight line on the top.

❸ Cut the cream clean using a hot knife.

❹ Heat a 0.7 cm diameter stainless-steel pipe and press it against the strawberry montée cream to make an indentation.

- You can also use a stainless steel straw of similar diameter.

❺ Fill the indentation with strawberry jam.

❻ Top with the chocolate decorations to finish.

④ CARAMEL BLONDIE STICK CAKE 캐러멜 블론디 스틱 케이크

40개 분량/ 40 stick cakes

반죽 Batter (철판(42 × 32cm) 1개 분량, 1 baking tray (42 × 32 cm))

INGREDIENTS		g
달걀	Eggs	290g
설탕	Sugar	320g
전화당	Inverted sugar	50g
캐러멜초콜릿 (Zephyr Caramel 35%)	Caramel chocolate	290g
버터 (Bridel)	Butter	330g
박력분	Cake flour	150g
아몬드파우더	Almond powder	70g
베이킹파우더	Baking powder	2g
헤이즐넛 캐러멜 *	Hazelnut caramel *	130g
피칸 페이스트	Pecan paste	70g
캐러멜라이즈 피칸	Caramelized pecan	265g
다진 토피 (p.504)	Chopped toffee (p.504)	120g
꽃소금	Fleur de sel	2.5g
TOTAL		**2089.5g**

*** 만드는 법**

❶ 믹싱볼에 달걀, 설탕, 전화당을 넣고 따뜻한 물이 담긴 볼에 받쳐 휘퍼로 저어가며 45℃까지 온도를 높인다.
- 거품이 올라오지 않게 주의하면서 작업한다.

❷ 녹인 캐러멜초콜릿(45~50℃)과 버터(40℃)를 붓고 바믹서로 블렌딩한다.

❸ 체 친 [박력분, 아몬드파우더, 베이킹파우더]를 넣고 날가루가 보이지 않을 때까지 휘퍼로 섞는다.

❹ 테프론시트를 깐 철판(42×32cm)에 반죽을 붓고 스패출러로 평평하게 정리한다.

❺ 헤이즐넛 캐러멜 130g, 피칸 페이스트 70g을 짤주머니에 각각 담아 지그재그로 파이핑한다.

❻ 캐러멜라이즈 피칸 - 다진 토피 - 꽃소금 순서로 뿌린다.

❼ 160℃로 예열된 오븐에서 뎀퍼를 100% 열고 40분간 굽는다.
- 유산지를 올린 식힘망 위에 뒤집어 빼 윗면이 눌리지 않게 한다.

*** Procedure**

❶ Add eggs, sugar, and inverted sugar in a mixing bowl. Place it over hot water and whisk until it reaches 45°C.
- Be careful not to make bubbles.

❷ Blend with melted caramel chocolate (45~50°C) and butter (40°C) using an immersion blender.

❸ Mix sifted [cake flour, almond powder, and baking powder] with a whisk until the powder ingredients are no longer visible.

❹ Spread onto a baking tray (42 × 32 cm) lined with a Teflon sheet using an offset spatula.

❺ Put 130 grams of hazelnut caramel and 70 grams of pecan paste each in piping bags. Pipe them in a zig-zag pattern.

❻ Sprinkle caramelized pecans, chopped toffee, and fleur de sel in order.

❼ Bake in an oven preheated to 160°C and damper open at 100% for 40 minutes.
- Place it upside down on a cooling rack lined with parchment paper and remove the tray so the top doesn't get pressed.

헤이즐넛 캐러멜 *

Hazelnut Caramel

INGREDIENTS		g
설탕	Sugar	160g
물엿	Corn syrup	260g
휘핑크림	Whipping cream	420g
헤이즐넛 페이스트 (Cacaobarry, Pure pâte Noisettes)	Hazelnut paste	48g
소금	Salt	4g
TOTAL		**892g**

*** 만드는 법**

❶ 냄비에 설탕, 물엿을 넣고 캐러멜화시킨다.
❷ 불에서 내린 후 뜨겁게 데운 휘핑크림을 넣어가며 저어준다.
❸ 다시 불에 올려 가열한 후 106℃가 되면 불에서 내려 헤이즐넛 페이스트, 소금을 넣고 섞는다.
❹ 쿨링한 후 냉장 보관해 사용한다.

*** Procedure**

❶ Caramelize sugar and corn syrup in a pot.
❷ Remove from the heat and gradually stir in the hot whipping cream.
❸ Cook to 106°C, remove from the heat, and mix with hazelnut paste and salt.
❹ Let cool and refrigerate to use.

룸 템퍼라처 글레이즈

Room Temperature Glaze

INGREDIENTS		g
캐러멜초콜릿 (Zephyr Caramel 35%)	Caramel chocolate	1000g
식용유	Vegetable oil	200g
다진 캐러멜라이즈 헤이즐넛 (p.494)	Caramelized hazelnuts, chopped (p.494)	75g
TOTAL		**1275g**

*** 만드는 법**

❶ 비커에 녹인 캐러멜초콜릿(45~50℃), 식용유를 담고 바믹서로 블렌딩한다.
❷ 다진 캐러멜라이즈 헤이즐넛을 넣고 블렌딩한다.
❸ 30℃로 맞춰 사용한다.

*** Procedure**

❶ In a beaker, blend melted caramel chocolate (45~50°C) and vegetable oil with an immersion blender.
❷ Blend with chopped caramelized hazelnuts.
❸ Use at 30°C.

캐러멜 몽테 크림

Caramel Montée Cream

INGREDIENTS		g
휘핑크림A	Whipping cream A	440g
바닐라빈 껍질	Vanilla bean pod	1.5개 분량/ 1.5 pcs
설탕	Sugar	65g
젤라틴매스 (×5)	Gelatin mass (×5)	105g
캐러멜초콜릿 (Zephyr Caramel 35%)	Caramel chocolate	380g
피칸 페이스트	Pecan paste	80g
소금	Salt	5g
휘핑크림B	Whipping cream B	1000g
TOTAL		**2075g**

＊만드는 법

❶ 냄비에 휘핑크림A, 바닐라빈 껍질을 넣고 가열한다.
❷ 다른 냄비에 설탕을 넣고 가열해 캐러멜화시킨 후 ❶을 넣어가며 휘퍼로 저어준다.
❸ 젤라틴매스를 넣고 녹을 때까지 섞어준다.
❹ 캐러멜초콜릿, 피칸 페이스트, 소금이 담긴 비커에 ❸을 붓고 바믹서로 블렌딩한다.
❺ 차가운 상태의 휘핑크림B를 넣고 블렌딩한다.
❻ 바트에 담아 밀착 랩핑한 후 냉장고에서 12시간 동안 숙성시킨다.
❼ 사용하기 전 파이핑하기 좋은 상태로 휘핑해 사용한다.

＊Procedure

❶ Boil whipping cream A and vanilla bean pod in the pot.
❷ Caramelize sugar in a separate pot and stir in ❶ while whisking.
❸ Add gelatin mass and stir until it melts.
❹ Blend in a beaker with caramel chocolate, pecan paste, and salt using an immersion blender.
❺ Continue to blend with cold whipping cream B.
❻ Pour onto a stainless-steel tray, adhere the plastic wrap to the preparation to cover, and refrigerate for 12 hours.
❼ Whip to a consistency suitable for piping.

＊마무리

❶ 12 × 2.5cm 크기로 자른 블론디 케이크(냉동 상태)에 룸 템퍼라처 글레이즈(30℃)를 코팅한다.
❷ 블론디 케이크를 나란히 놓고 지름 1.8cm 깍지를 끼운 짤주머니에 캐러멜 몽테 크림을 담아 일자로 파이핑한다.
❸ 뜨겁게 달군 칼로 캐러멜 몽테 크림을 깔끔하게 자른다.
❹ 장식용 초콜릿을 올려 마무리한다.

＊Finish

❶ Prepare the frozen blondies cut to 12 × 2.5 cm. Coat them with the room temperature glaze (30°C).
❷ Arrange the blondies in a line. Put caramel montée cream in a piping bag with a 1.8 cm round tip and pipe a straight line on the top.
❸ Cut the cream clean using a hot knife.
❹ Top with the chocolate decorations to finish.

⑤ DARK BLONDIE STICK CAKE 다크 블론디 스틱 케이크

40개 분량/ 40 stick cakes

반죽 Batter (철판(42 × 32cm) 1개 분량, 1 baking tray (42 × 32 cm))

INGREDIENTS		g
달걀	Eggs	300g
설탕	Sugar	340g
전화당	Inverted sugar	50g
다크초콜릿 (Mi-Amère 58%)	Dark chocolate	200g
버터 (Bridel)	Butter	340g
카카오파우더 (Fleur de Cao)	Cacao powder	100g
박력분	Cake flour	150g
아몬드파우더	Almond powder	80g
베이킹파우더	Baking powder	3g
헤이즐넛 캐러멜	Hazelnut caramel	130g
헤이즐넛 페이스트	Hazelnut paste	90g
청크초콜릿 (다크)	Dark chocolate chunks	180g
카카오 크럼블 *	Cacao crumble *	300g
TOTAL		**2263g**

* 만드는 법

❶ 믹싱볼에 달걀, 설탕, 전화당을 넣고 따뜻한 물이 담긴 볼에 받쳐 휘퍼로 저어가며 45℃까지 온도를 높인다.
- 거품이 올라오지 않게 주의하면서 작업한다.

❷ 녹인 다크초콜릿(45~50℃)과 버터(40℃)를 붓고 바믹서로 블렌딩한다.

❸ 체 친 [카카오파우더, 박력분, 아몬드파우더, 베이킹파우더]를 넣고 날가루가 보이지 않을 때까지 휘퍼로 섞는다.

❹ 테프론시트를 깐 철판(42×32cm)에 반죽을 붓고 스패출러로 평평하게 정리한다.

❺ 헤이즐넛 캐러멜 130g, 헤이즐넛 페이스트 90g을 각각 짤주머니에 담아 지그재그로 파이핑한다.

❻ 청크초콜릿 180g과 카카오 크럼블 300g을 고르게 뿌린다.

❼ 160℃로 예열된 오븐에서 댐퍼를 100% 열고 40분간 굽는다.
- 유산지를 올린 식힘망 위에 뒤집어 빼 윗면이 눌리지 않게 한다.

* Procedure

❶ Add eggs, sugar, and inverted sugar in a mixing bowl. Place it over hot water and whisk until it reaches 45°C.
- Be careful not to make bubbles.

❷ Blend with melted dark chocolate (45~50°C) and butter (40°C) using an immersion blender.

❸ Mix sifted [cacao powder, cake flour, almond powder, and baking powder] with a whisk until the powder ingredients are no longer visible.

❹ Spread onto a baking tray (42 × 32 cm) lined with a Teflon sheet using an offset spatula.

❺ Put 130 grams of hazelnut caramel and 90 grams of hazelnut paste each in piping bags. Pipe them in a zig-zag pattern.

❻ Sprinkle 180 grams of chocolate chunks and 300 grams of cacao crumble.

❼ Bake in an oven preheated to 160°C and damper open at 100% for 40 minutes.
- Place it upside down on a cooling rack lined with parchment paper and remove the tray so the top doesn't get pressed.

카카오 크럼블 *

Cacao Crumble

INGREDIENTS		g
버터 (Bridel)	Butter	90g
코코넛슈거	Coconut sugar	50g
설탕	Sugar	25g
소금	Salt	2g
박력분	Cake flour	150g
카카오파우더 (Extra Brute)	Cacao powder	30g
옥수수전분	Cornstarch	15g
물	Water	25g
TOTAL		**387g**

*** 만드는 법**

❶ 믹싱볼에 차가운 상태의 버터, 코코넛슈거, 설탕, 소금을 넣고 비터로 부드럽게 풀어준다.
❷ 체 친 [박력분, 카카오파우더, 옥수수전분]을 넣고 물을 흘려가며 믹싱해 한 덩어리로 만든다.
❸ 테프론시트를 깐 철판 위에서 체(간격 약 0.5cm)에 반죽을 내려 크럼블 상태로 만들어 냉동한다.
❹ 냉동시킨 크럼블은 밀폐용기에 담아 냉동 보관하면서 사용하고, 사용하기 전 손으로 가볍게 풀어준다.

*** Procedure**

❶ Soften cold butter, coconut sugar, sugar, and salt using a paddle attachment.
❷ Mix with sifted [cake flour, cacao powder, and cornstarch] while drizzling in the water into a dough.
❸ Pass the dough through a coarse sieve (about 0.5cm mesh) onto a baking sheet lined with a Teflon sheet to make crumbles and freeze.
❹ Store the frozen crumbles in an airtight container and use as needed. Gently break them apart with your hands before using them.

룸 템퍼라처 글레이즈

Room Temperature Glaze

INGREDIENTS		g
다크초콜릿 (Mi-Amère 58%)	Dark chocolate	1000g
카카오버터	Cacao butter	100g
식용유	Vegetable oil	100g
다진 아몬드 슬라이스	Almond slices, chopped	40g
TOTAL		**1240g**

*** 만드는 법**

❶ 비커에 녹인 다크초콜릿(45~50℃)과 50℃ 이하로 녹인 카카오버터, 식용유를 담고 바믹서로 블렌딩한다.
❷ 다진 아몬드 슬라이스를 넣고 블렌딩한다.
❸ 30℃로 맞춰 사용한다.

*** Procedure**

❶ In a beaker, blend melted dark chocolate (45~50°C), cacao butter melted to below 50°C, and vegetable oil with an immersion blender.
❷ Blend with chopped almond slices.
❸ Use at 30°C.

미아메르 몽테 크림

Mi-Amère Montée Cream

INGREDIENTS		g
휘핑크림A	Whipping cream A	400g
우유	Milk	120g
전화당	Inverted sugar	84g
젤라틴매스 (×5)	Gelatin mass (×5)	68g
다크초콜릿 (Mi-Amère 58%)	Dark chocolate	400g
아몬드 페이스트	Almond paste	80g
소금	Salt	1g
바닐라에센스 (Aroma Piu)	Vanilla essence	10g
휘핑크림B	Whipping cream B	840g
TOTAL		**2003g**

＊만드는 법

❶ 냄비에 휘핑크림A, 우유, 전화당, 젤라틴매스를 넣고 젤라틴매스가 녹을 때까지 가열한다.
❷ 다크초콜릿이 담긴 비커에 ❶을 붓고 바믹서로 블렌딩한다.
❸ 아몬드 페이스트 - 소금 - 바닐라에센스 - 차가운 상태의 휘핑크림B 순서대로 넣어가며 블렌딩한다.
❹ 바트에 담아 밀착 랩핑한 후 냉장고에서 12시간 동안 숙성시킨다.
❺ 사용하기 전 파이핑하기 좋은 상태로 휘핑해 사용한다.

＊Procedure

❶ Boil whipping cream A, milk, inverted sugar, and gelatin mass in the pot until the gelatin melts.
❷ Blend in a beaker with dark chocolate using an immersion blender.
❸ Add almond paste, salt, vanilla essence, and cold whipping cream B in order, to blend.
❹ Pour onto a stainless-steel tray, adhere the plastic wrap to the preparation to cover, and refrigerate for 12 hours.
❺ Whip to a consistency suitable for piping.

＊마무리

❶ 12 × 2.5cm 크기로 자른 블론디 케이크(냉동 상태)에 룸 템퍼라처 글레이즈(30℃)를 코팅한다.
❷ 블론디 케이크를 나란히 놓고 지름 1.8cm 깍지를 끼운 짤주머니에 미아메르 몽테 크림을 담아 일자로 파이핑한다.
❸ 뜨겁게 달군 칼로 미아메르 몽테 크림을 깔끔하게 자른다.
❹ 장식용 초콜릿을 올려 마무리한다.

＊Finish

❶ Prepare the frozen blondies cut to 12 × 2.5 cm. Coat them with the room temperature glaze (30°C).
❷ Arrange the blondies in a line. Put Mi-Amère montée cream in a piping bag with a 1.8 cm round tip and pipe a straight line on the top.
❸ Cut the cream clean using a hot knife.
❹ Top with the chocolate decorations to finish.

PISTACHIO POUND & CAKE

피스타치오 파운드 & 케이크

PAIRING & TEXTURE
페어링 & 텍스처

촉촉한 타입의 케이크에 단단하면서도 바스러지는 질감의 머랭 스틱을 더해 상반되는 텍스처를 동시에 느낄 수 있다.

Adding meringue sticks with a firm, but crumbly texture to a moist cake gives you a pleasant contrast of textures.

TECHNIQUE
테크닉

올인원법

- 모든 재료를 한 번에 섞어 만드는 올인원 방식으로 만들었다.
- 버터를 크림화하지 않고 녹여 사용하는 레시피이므로, 버터 대신 식물성 오일을 사용해 비건 레시피로 변형할 수 있다.

ALL-IN-ONE METHOD

- It's an all-in-one method that mixes all the ingredients in one go.
- Since the recipe uses melted butter instead of creaming it, you can make it vegan using vegetable oil instead of butter.

DESIGN
디자인

- 파운드 케이크 모양을 하고 있는 케이크(가토).
- 평범해보일 수 있는 직사각형 케이크에 머랭을 촘촘하게 올려 독특한 디자인으로 연출했다.
- 피스타치오파우더를 사용해 인위적이지 않고 자연스러운 피스타치오 본연의 색상을 표현했다.
- 매끈하게 코팅한 초콜릿 글레이즈와 거친 느낌의 머랭을 대비시켜 세련된 디자인으로 연출했다.

- It is a cake (gateau) shaped like a pound cake.
- The meringue sticks are tightly packed into the rectangular pound cake to create a unique design.
- Pistachio powder is used to create pistachio's natural, non-artificial color.
- The contrast between the smooth chocolate glaze and the rough meringue sticks creates a sophisticated design.

✦ HOW TO COMPOSE THIS RECIPE ✦

STEP 1. | **메뉴 정하기** | **파운드 케이크 모양의 케이크 (가토, Gateau)**

Decide on the menu | Pound cake shaped cake (Gateau)

STEP 2. | **메인 맛 정하기** | **피스타치오**

Choose the primary flavor | Pistachio

STEP 3. | **메인 맛(피스타치오)과의 페어링 선택하기**

Select a pairing flavor (Pistachio)

- ☑ 라즈베리 Raspberry
- ☑ 화이트 초콜릿 White Chocolate
- ☑ 건크랜베리 Dried Cranberries
- ☑ 건청포도 Green Raisins

STEP 4. | **구성하기**

Assemble

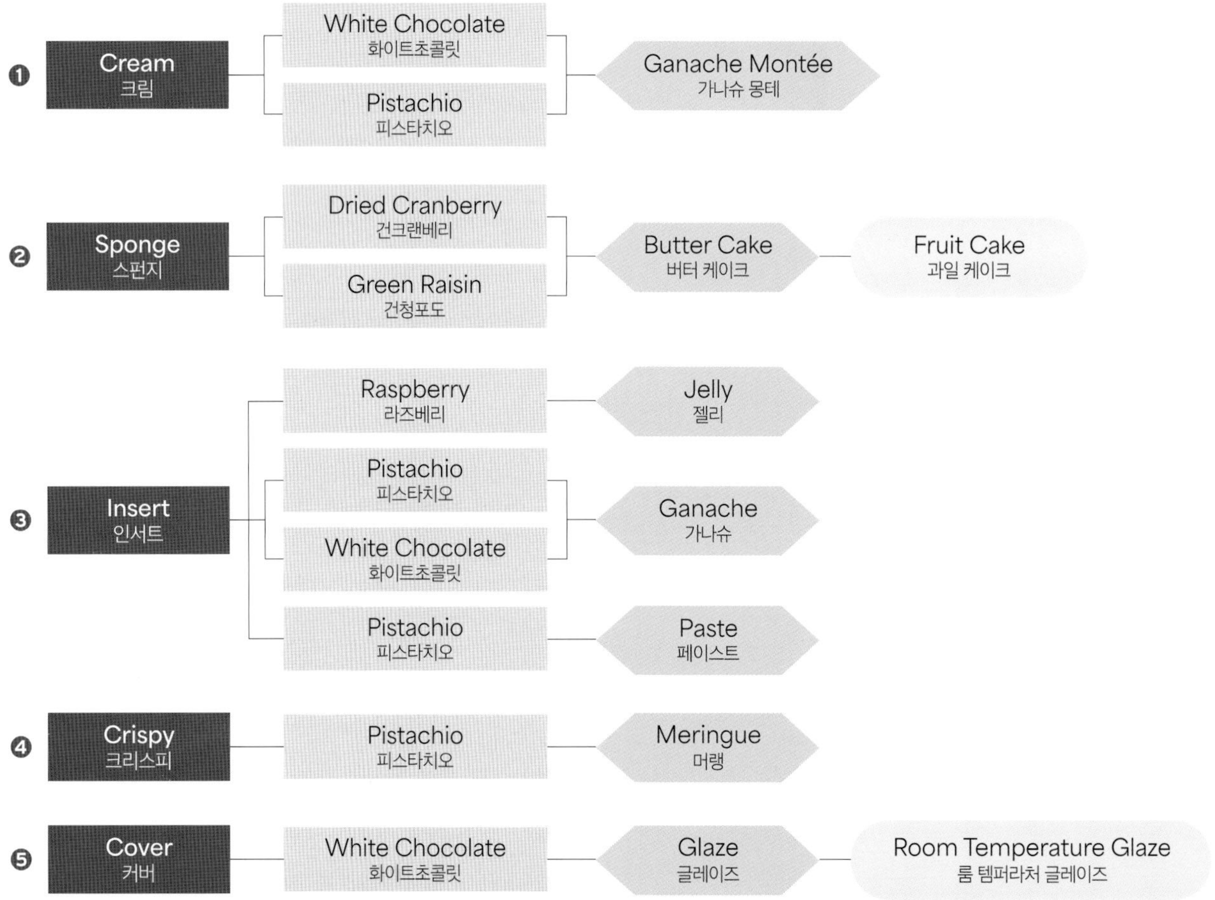

❶ Cream 크림
- White Chocolate 화이트초콜릿 + Pistachio 피스타치오 → Ganache Montée 가나슈 몽테

❷ Sponge 스펀지
- Dried Cranberry 건크랜베리 + Green Raisin 건청포도 → Butter Cake 버터 케이크 — Fruit Cake 과일 케이크

❸ Insert 인서트
- Raspberry 라즈베리 → Jelly 젤리
- Pistachio 피스타치오 + White Chocolate 화이트초콜릿 → Ganache 가나슈
- Pistachio 피스타치오 → Paste 페이스트

❹ Crispy 크리스피
- Pistachio 피스타치오 → Meringue 머랭

❺ Cover 커버
- White Chocolate 화이트초콜릿 → Glaze 글레이즈 — Room Temperature Glaze 룸 템퍼라처 글레이즈

피스타치오 머랭
Pistachio Meringue
라즈베리 젤란검 젤리
Raspberry
Gellan Gum Jelly
스타치오 몽테
Pistachio
Montée
피스타치오
가나슈
Pistachio
Ganache
피스타치오 페이스트
Pistachio Paste
피스타치오 과일 케이크
Pistachio Fruit Cake
피스타치오 룸
템퍼라처 글레이즈
Pistachio Room
Temperature Glaze

INGREDIENTS 5개 분량/ 5 cakes

피스타치오 과일 케이크
Pistachio Fruit Cake

(철판(42 × 32cm) 1개 분량)
(1 baking tray (42 × 32 cm))

INGREDIENTS		g
달걀	Eggs	280g
설탕	Sugar	300g
전화당	Inverted sugar	50g
화이트초콜릿 (Zephyr white 34%)	White chocolate	190g
버터 (Bridel)	Butter	330g
바닐라에센스 (Aroma Piu)	Vanilla essence	5g
피스타치오 페이스트	Pistachio paste	200g
박력분	Cake flour	150g
아몬드파우더	Almond powder	80g
베이킹파우더	Baking powder	4g
전처리한 건청포도	Green raisins, presoaked	200g
전처리한 건크랜베리	Dried cranberries, presoaked	100g
TOTAL		1889g

머랭 스틱 *
Meringue Sticks

INGREDIENTS		g
흰자	Egg whites	180g
이소말트파우더	Isomalt powder	100g
설탕	Sugar	80g
알부민파우더	Albumin powder	6g
슈거파우더	Sugar powder	120g
탈지분유	Skim milk powder	18g
옥수수전분	Cornstarch	17g
피스타치오파우더	Pistachio powder	40g
TOTAL		561g

피스타치오 머랭
Pistachio Meringue

INGREDIENTS		g
머랭 스틱 *	Meringue sticks *	250g
미크리오 버터 (Cacaobarry)	Mycryo butter	20g
다진 피스타치오	Pistachios, chopped	적당량/ QS
피스타치오 파우더	Pistachio powder	적당량/ QS
TOTAL		270g

라즈베리 젤란검 젤리

Raspberry Gellan Gum Jelly

INGREDIENTS		g
딸기 퓌레	Strawberry purée	180g
라즈베리 퓌레	Raspberry purée	390g
물	Water	100g
설탕	Sugar	280g
레몬즙	Lemon juice	20g
젤란검	Gellan gum	15g
TOTAL		**985g**

피스타치오 가나슈

Pistachio Ganache

INGREDIENTS		g
휘핑크림	Whipping cream	200g
전화당	Inverted sugar	16g
소르비톨	Sorbitol	90g
화이트초콜릿 (Zephyr white 34%)	White chocolate	500g
피스타치오 페이스트	Pistachio paste	60g
소금	Salt	2g
TOTAL		**868g**

피스타치오 몽테

Pistachio Montée

INGREDIENTS		g
생크림	Heavy cream	210g
젤라틴매스 (×5)	Gelatin mass (×5)	45g
전화당	Inverted sugar	45g
화이트초콜릿 (Zephyr white 34%)	White chocolate	165g
연유	Condensed milk	45g
휘핑크림	Whipping cream	480g
바닐라에센스 (Aroma Piu)	Vanilla essence	9g
골드럼 (PAN RUM)	Gold rum	9g
피스타치오 페이스트	Pistachio paste	90g
TOTAL		**1098g**

피스타치오 룸 템퍼라처 글레이즈

Pistachio Room Temperature Glaze

INGREDIENTS		g
화이트초콜릿 (Zephyr white 34%)	White chocolate	1000g
아보카도오일	Avocado oil	200g
피스타치오 페이스트	Pistachio paste	150g
다진 피스타치오	Pistachios, chopped	150g
TOTAL		**1500g**

✤

PISTACHIO FRUIT CAKE

피스타치오 과일 케이크

1 믹싱볼에 달걀, 설탕, 전화당을 넣고 중탕볼에 받쳐 저어가며 45℃로 온도를 올린다.

TIP. 거품이 나지 않게 주의하며 작업한다.

2 중탕으로 녹인 화이트초콜릿(40~45℃)과 녹인 버터(45℃)를 넣고 바믹서로 블렌딩한다.

3 바닐라에센스 - 피스타치오 페이스트 순서로 넣고 바믹서로 블렌딩한다.

4 체 친 [박력분, 아몬드파우더, 베이킹파우더]를 넣고 날가루가 보이지 않을 때까지 휘퍼로 골고루 섞는다.

5 전처리한 건청포도와 건크랜베리를 넣고 가볍게 섞는다. (최종 온도 35~40℃)

6 테프론시트를 깐 철판(42 × 32cm)에 고르게 팬닝한 후 165℃로 예열된 오븐에서 30분간 굽는다.

1 Warm eggs, sugar, and inverted sugar in a mixing bowl over a double boiler (bain-marie) to 45°C.

TIP. Heat while stirring with a whisk, careful not to let it foam.

2 Blend white chocolate melted over a double boiler (40~45°C) and melted butter(45°C) with an immersion blender.

3 Add vanilla essence, followed by pistachio paste, and continue to mix with the blender.

4 Add sifted [cake flour, almond powder, and baking powder]; mix with a whisk until no streaks of powder are visible.

5 Add the presoaked green raisins and cranberries and mix lightly. (Final temperature: 35~40°C)

6 Spread evenly on a baking tray (42 × 32cm) lined with Teflon sheet and bake in an oven preheated to 165°C for 30 minutes.

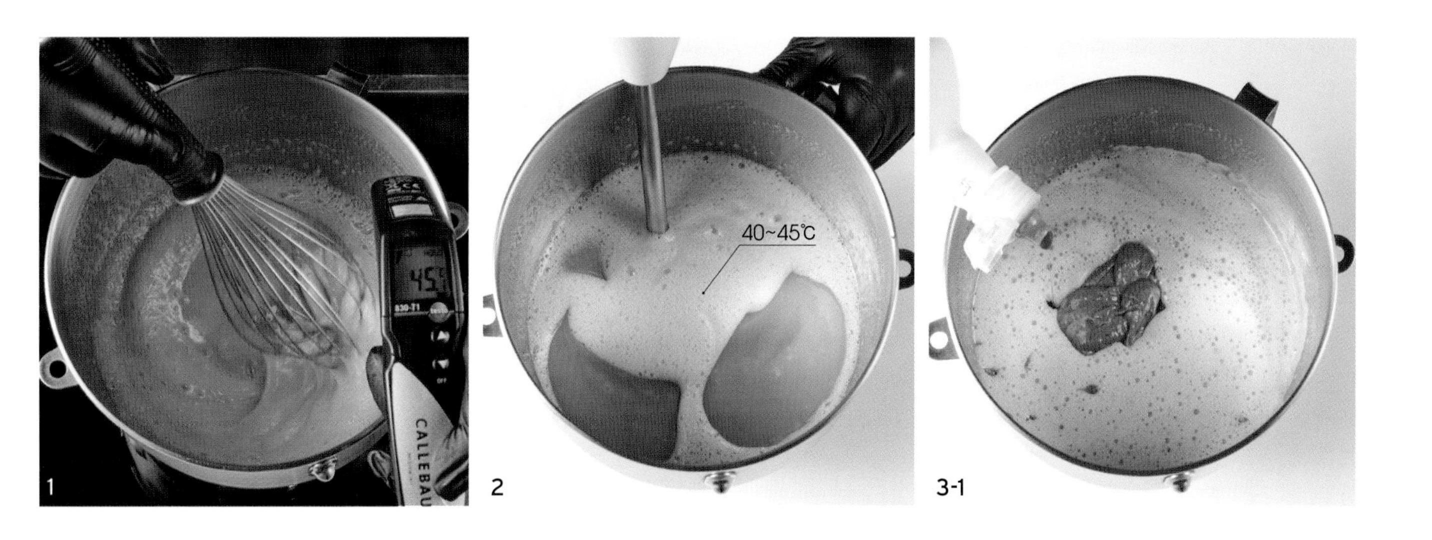
1
40~45℃
2
3-1

3-2
4
35~40℃
5

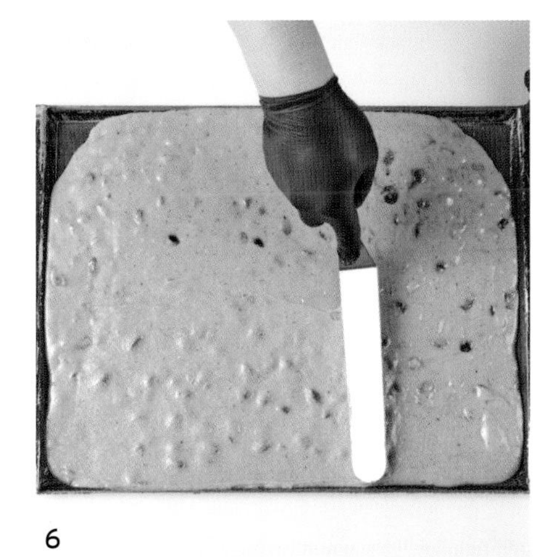
6

✤ ✤

PISTACHIO MERINGUE

피스타치오 머랭

머랭 스틱

1 믹싱볼에 흰자, 이소말트 파우더를 넣고 중탕볼에 받쳐 저어가며 50℃로 온도를 올린 후 휘핑한다.

2 거품이 안정적으로 올라오면 설탕, 알부민파우더를 넣고 단단하고 조밀한 상태가 될 때까지 휘핑한다.

3 체 친 [슈거파우더, 탈지분유, 옥수수전분, 피스타치오파우더]를 넣고 가볍게 섞는다.

4 지름 1cm 원형 깍지를 끼운 짤주머니에 담아 테프론시트 위에 길게 파이핑한다.

5 피스타치오파우더(분량 외)를 뿌린다.

6 110℃로 예열된 오븐에서 1시간 굽고 100℃로 낮춰 1시간 더 굽는다. 완전히 식으면 1.5cm 길이로 잘라 준비한다.

피스타치오 머랭

7 길이 1.5cm로 자른 머랭 스틱을 전자레인지에 살짝 돌려 50~55℃로 맞춘다.

8 미크리오 버터를 뿌리고 골고루 섞는다.

TIP. 미크리오 버터는 두세 번 나눠 넣어가며 버무리듯 골고루 섞는다.
미크리오 버터로 코팅하는 이유는 쇼케이스 안에서 최대한 습기를 먹지 않게 하기 위함이다. (길게 만들어 잘라 사용하는 머랭이므로 잘려진 단면이 빠르게 습기를 머금을 수 있다.)

9 다진 피스타치오, 피스타치오파우더를 뿌리고 골고루 섞는다.

Meringue Sticks

1 Warm egg whites and isomalt powder in a mixing bowl over a double boiler (bain-marie) to 50°C and whip.

2 When the foam becomes stable, add sugar and albumin powder; whip until stiff and dense.

3 Add sifted [sugar powder, skim milk powder, cornstarch, and pistachio powder] and mix as if to fold.

4 Fill a piping bag fitted with a 1 cm round tip and pipe it long on a Teflon sheet.

5 Sprinkle with pistachio powder (other than requested).

6 Bake in an oven preheated to 110°C for one hour, then reduce the temperature to 100°C and bake for another hour. Once completely cooled, cut into 1.5 cm pieces.

Pistachio Meringue

7 Briefly microwave the meringue sticks to 50~55°C.

8 Sprinkle with Mycryo butter and mix thoroughly.

TIP. Add the Mycryo butter half or a third at a time, tossing to coat evenly.
The reason for coating is to keep as much moisture out as possible while storing in the showcase. (Since the meringues are made long and cut to use, the cut sides can attract moisture quickly.)

9 Sprinkle chopped pistachios and pistachio powder; mix thoroughly.

1

2

3

4

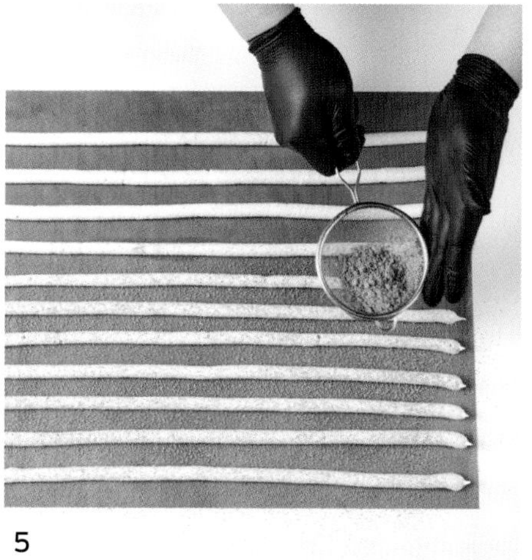
5

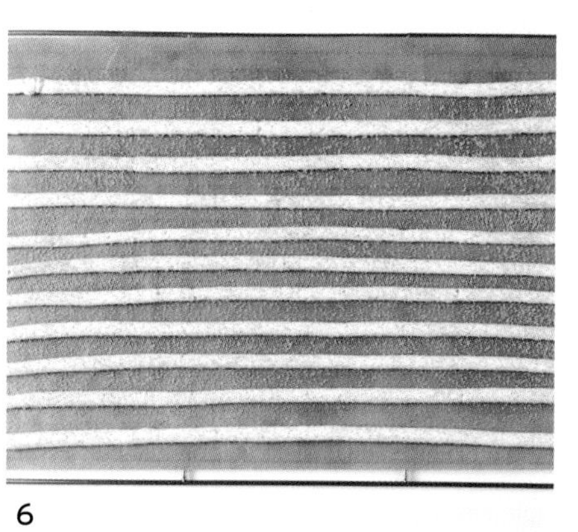
6

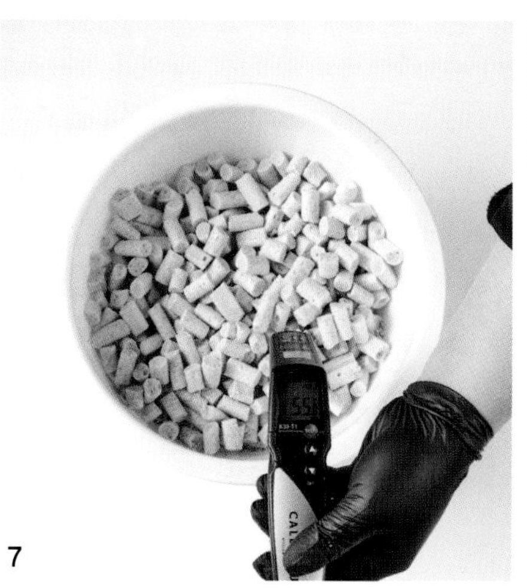
7

8

9

✤ ✤ ✤

RASPBERRY GELLAN GUM JELLY

라즈베리 젤란검 젤리

1. 비커에 모든 재료를 넣고 바믹서로 블렌딩한다.
2. 냄비로 옮겨 가열한다.

   TIP. 주걱으로 저어주며 가열하다가 끓기 시작하면 휘퍼로 바꿔 저어준다.
   사용하는 젤란검의 브랜드에 따라 입자 크기의 차이가 생긴다. 소사(Sosa) 제품의 경우 가열한 후 끓어오르면 완성되는 반면, 중국 제품의 경우 가열할 때 점도를 확인해가며 상대적으로 더 오래 끓여야 한다. 완성한 후에는 바믹서로 한 번 더 갈아준다. (입자의 크기가 소사 제품보다 크므로 바믹서로 갈아 완성한다.)
3. 비커에 옮겨 바믹서로 블렌딩한다.
4. 바트에 담아 냉장고에 보관한다.

1. Blend all the ingredients in a beaker with an immersion blender.
2. Pour it into a saucepan and cook.

   TIP. Stir with a spatula, then switch to a whisk when it starts to boil.
   Depending on the brand of gellan gum, the particle size will vary. Sosa Ingredients' gum's cooking process finishes when it boils after heating. Yet, Chinese products must be boiled relatively longer, checked during heating, and mixed once more with an immersion blender after cooking. (Because the particle size is larger than that of Sosa Ingredients, it must be ground with a blender.)
3. Transfer to a beaker and combine with an immersion blender.
4. Pour it into a tray and refrigerate.

✤ ✤ ✤ ✤

PISTACHIO GANACHE

피스타치오 가나슈

1. 냄비에 휘핑크림, 전화당, 소르비톨을 넣고 80℃까지 가열한다.
2. 화이트초콜릿, 피스타치오 페이스트, 소금이 담긴 비커에 담는다.
3. 바믹서로 블렌딩한다. (최종 온도 30~32℃)

1. Heat whipping cream, inverted sugar, and sorbitol in a saucepan to 80°C.
2. Pour it into a beaker with white chocolate, pistachio paste, and salt.
3. Combine with an immersion blender. (Final temperature: 30~32°C)

✤ ✤ ✤

1-1

1-2

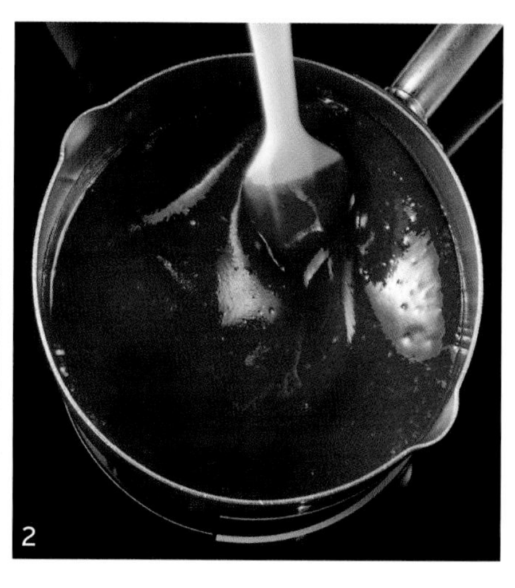

2

3

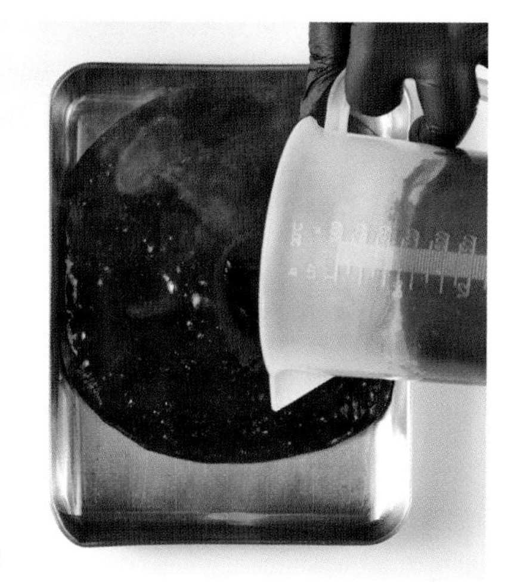

4

✤ ✤ ✤ ✤

1

2

3

✤ ✤ ✤ ✤ ✤

PISTACHIO MONTÉE

피스타치오 몽테

1 냄비에 생크림, 젤라틴매스,전화당을 넣고 70℃까지 가열한다.

2 화이트초콜릿이 담긴 볼에 넣고 바믹서로 블렌딩한다.

3 연유, 휘핑크림를 넣고 블렌딩한다.

4 바닐라에센스, 골드럼을 넣고 블렌딩하고 바트에 담아 밀착 랩핑한 후 냉장고에서 12시간 동안 숙성시킨다.

5 사용하기 직전 피스타치오 페이스트와 함께 파이핑하기 좋은 상태로 휘핑해 사용한다.

TIP. 베이스 몽테 + 몽떼 중량의 10%의 피스타치오 페이스트를 섞어 사용한다.

1 Heat heavy cream, gelatin mass, and inverted sugar in a saucepan to 70°C.

2 Pour it into a bowl with white chocolate and mix with an immersion blender.

3 Add condensed milk and whipping cream and continue to blend.

4 Blend again with vanilla essence and gold rum. Pour it into a tray and cover it with plastic wrap, make sure the wrap is in contact with the surface of the cream. Refrigerate for 12 hours.

5 Whip it with pistachio paste until it is good enough to pipe just before use.

TIP. Mix the base montée and pistachio paste of 10% of the weight of the montée to use.

✤ ✤ ✤ ✤ ✤ ✤

PISTACHIO ROOM TEMPERATURE GLAZE

피스타치오 룸 템퍼라처 글레이즈

1 녹인 화이트초콜릿(40℃)에 아보카도오일, 피스타치오 페이스트를 넣고 바믹서로 블렌딩한다.

2 다진 피스타치오를 넣고 섞은 후 30~35℃로 온도를 맞춰 사용한다.

1 Add avocado oil and pistachio paste to the melted chocolate (40°C) ; combine with an immersion blender.

2 Stir in chopped pistachios and use at 30~35°C.

✤ ✤ ✤ ✤ ✤

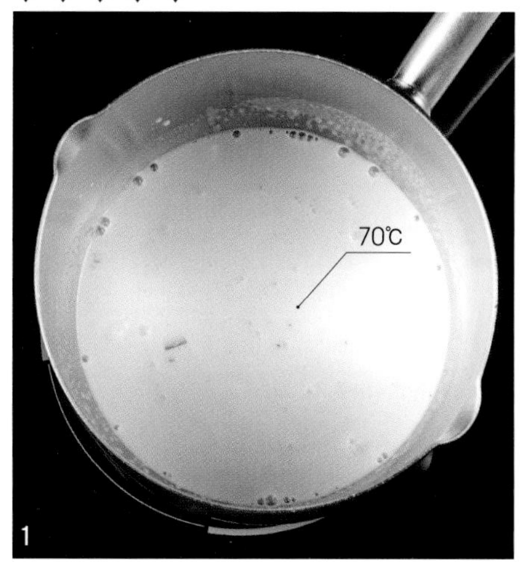

1

2

3

4

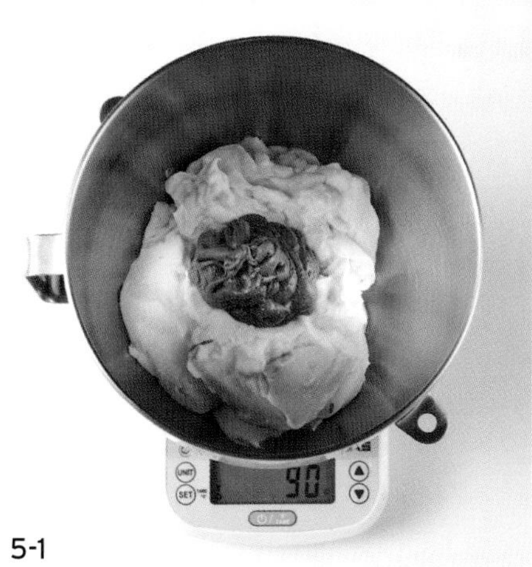

5-1

5-2

✤ ✤ ✤ ✤ ✤ ✤

1

2-1

2-2

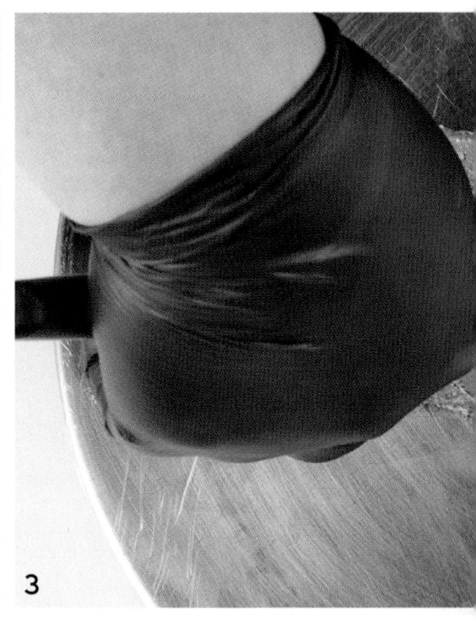

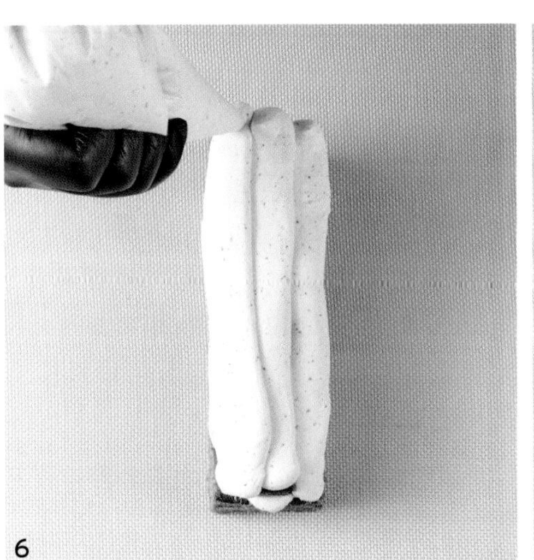

✤ ✤ ✤ ✤ ✤ ✤ ✤

FINISH 마무리

1 완전히 식힌 피스타치오 과일 케이크를 반으로 잘라 한쪽에는 피스타치오 가나슈 170g을, 한쪽에는 피스타치오 페이스트 120g을 바르고 겹친 후 냉동고에서 굳힌다.

2 6 × 19cm 크기로 자른다.

3 피스타치오 과일 케이크 윗면과 옆면에 피스타치오 페이스트 40g을 바르고 냉동실에서 얼린다.

4 피스타치오 과일 케이크 윗면에 피스타치오 몽테를 약 40g 파이핑한다.

5 2 × 19cm 크기로 자른 라즈베리 젤란검 젤리를 올린다.

6 피스타치오 몽테를 약 160g 파이핑한다.

7 얇은 필름을 이용해 윗면을 봉긋하고 깔끔하게 다듬어준 후 냉동한다.

8 냉동한 케이크의 옆면을 룸 템퍼라처 글레이즈(30~35℃)로 코팅한다.

9 피스타치오 머랭에 남은 피스타치오 몽테를 파이핑해 케이크에 붙인다.

10 다진 피스타치오와 피스타치오파우더(분량 외)를 뿌려 마무리한다.

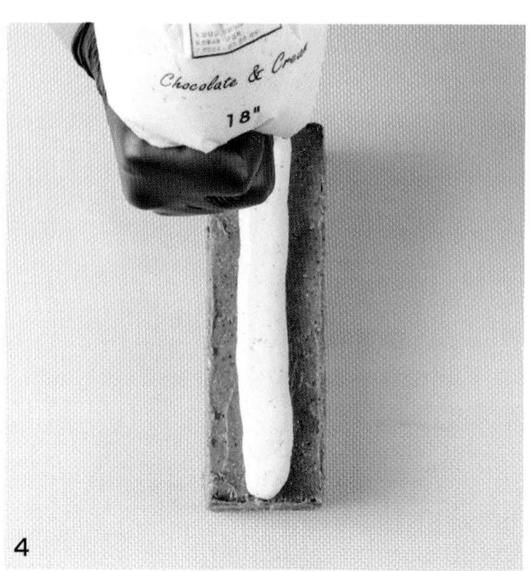

4

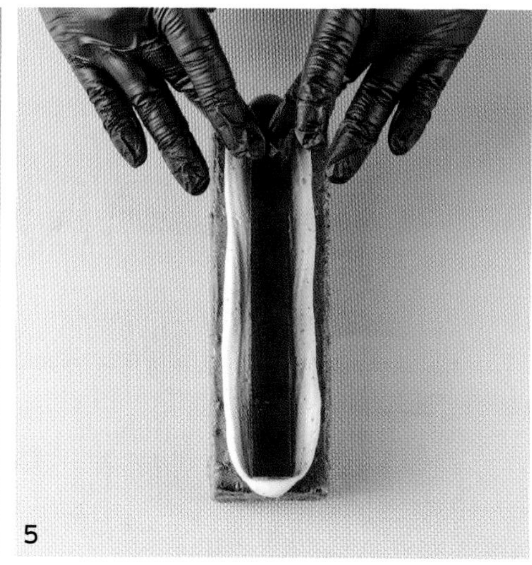
5

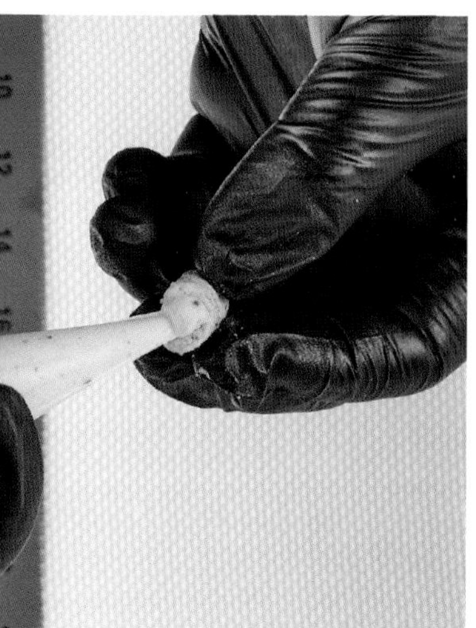

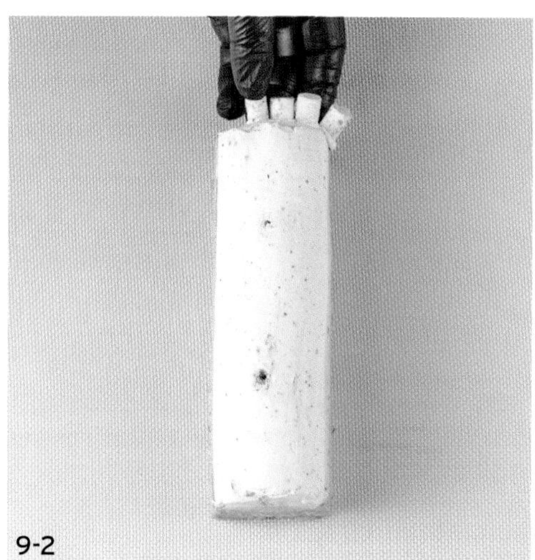
9-2

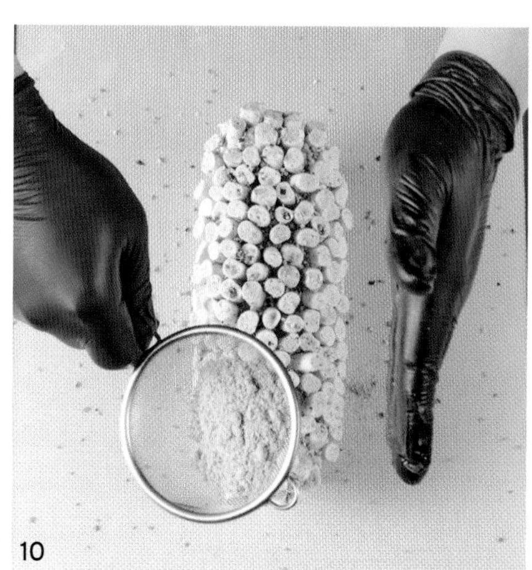
10

1 Slice the completely cooled pistachio fruit cake horizontally in half, spread one side with 170 grams of pistachio ganache and the other with 120 grams of pistachio paste. Put them back together and freeze.

2 Cut into 6 × 19 cm.

3 Spread 40 grams of pistachio paste on top and sides of the cake and freeze to set.

4 Pipe about 40 grams of pistachio montée on top of the pistachio fruit cake.

5 Top with raspberry gellan gum jelly cut into 2 × 19 cm.

6 Pipe about 160 grams of pistachio montée.

7 Trim the top with a thin acetate sheet to make a neat and smooth surface.

8 Coat the sides of the frozen cakes with the room temperature glaze (30~35°C).

9 Pipe the remaining montée on the pistachio meringue and attach it to the cake.

10 Sprinkle chopped pistachios and pistachio powder (other than requested) to finish.

9

LEMON MADELEINE TARTLET

레몬 마들렌 타르틀렛

PAIRING & TEXTURE

페어링 & 텍스처

기존 레몬 커드의 배합에서 버터는 줄이고 카카오버터를 추가했으며, 금귤 페이스트를 더해 프레시한 과일의 맛을 강조했다.

I reduced the butter in the existing lemon curd recipe, but added cacao butter, and added kumquat paste to accentuate the fresh fruit flavor.

TECHNIQUE

테크닉

- 금귤을 페이스트 형태로 만들어 사용해 감귤류가 가지고 있는 다채로운 시트러스함을 즐길 수 있다.
- 레몬 초콜릿을 그레이터로 갈아 올려 마치 치즈를 갈아 올린 요리처럼 느껴지도록 연출했다.

- You can enjoy the colorful, fresh, and tangy flavor of citrus fruits from the kumquats used in a paste form.
- I grated the lemon chocolate with a zester to make it look like a dish made with grated cheese.

DESIGN

디자인

타르트하면 흔하게 떠올리는 원형이나 사각형 셸에서 벗어난 입체적인 모양의 타르트 셸을 만들고 싶었다. 그래서 마들렌 모양 틀에 반죽을 올려 구워 입체적이면서도 자연스러운 굴곡이 있는 마들렌 모양 셸로 완성했다.

I wanted to create a three-dimensional tartlet shell that was different from the round or square shells that are more common when you imagine tartlets. So, I baked the dough on a madeleine mold, resulting in a three-dimensional, naturally curved madeleine-shaped shell.

✦ HOW TO COMPOSE THIS RECIPE ✦

STEP 1.	**메뉴 정하기** Decide on the menu	**레몬 타르틀렛** Lemon Tartlet
STEP 2.	**메인 맛 정하기** Choose the primary flavor	**레몬** Lemon

STEP 3. **메인 맛(레몬)과의 페어링 선택하기**

Select a pairing flavor (Lemon)

- ☑ 올리브 오일 Olive Oli
- ☑ 아몬드 Almond
- ☑ 금귤 Kumquat

STEP 4. **구성하기**

Assemble

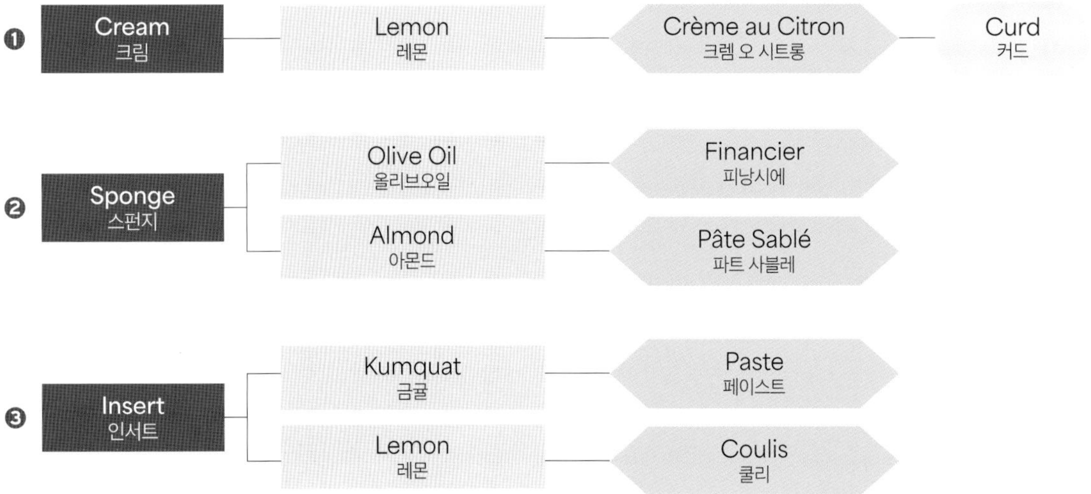

❶ Cream 크림 — Lemon 레몬 — Crème au Citron 크렘 오 시트롱 — Curd 커드

❷ Sponge 스펀지
- Olive Oil 올리브오일 — Financier 피낭시에
- Almond 아몬드 — Pâte Sablé 파트 사블레

❸ Insert 인서트
- Kumquat 금귤 — Paste 페이스트
- Lemon 레몬 — Coulis 쿨리

레몬 초콜릿
Lemon Chocolate
레몬 커드
Lemon Curd
올리브오일 피낭시에
Olive Oil Financier
레몬 쿨리
Lemon Coulis
아몬드 사블레
Almond Sablé
금귤 페이스트
Kumquat Paste

INGREDIENTS 20개 분량/ 20 lemon tartlets

아몬드 사블레
Almond Sablé

INGREDIENTS		g
버터 (Bridel)	Butter	125g
슈거파우더	Sugar powder	65g
T55밀가루	T55 flour	210g
옥수수전분	Cornstarch	15g
아몬드파우더	Almond powder	65g
소금	Salt	2g
달걀	Eggs	30g
TOTAL		**512g**

타르틀렛 코팅
Coating For Tartlets

INGREDIENTS		g
노른자	Egg yolks	45g
휘핑크림	Whipping cream	5g
설탕	Sugar	10g
물	Water	10g
TOTAL		**70g**

- 모든 재료를 섞어 사용한다.
- Mix all the ingredients to use.

올리브오일 피낭시에
Olive Oil Financier

(철판(42 × 32cm) 2/3개 분량)
(2/3 baking tray (42 × 32 cm))

INGREDIENTS		g
헤이즐넛파우더	Hazelnut powder	75g
슈거파우더	Sugar powder	165g
박력분	Cake flour	60g
베이킹파우더	Baking powder	2.5g
흰자	Egg whites	160g
꿀	Honey	20g
엑스트라 버진 올리브오일	Extra virgin olive oil	40g
누아제트 버터	Noisette butter	75g
TOTAL		**597.5g**

레몬 퓌레 *

Lemon Purée

INGREDIENTS		g
레몬 껍질	Lemon peels	40g
레몬즙	Lemon juice	170g
TOTAL		**210g**

레몬 쿨리

Lemon Coulis

INGREDIENTS		g
레몬 퓌레 *	Lemon purée *	210g
설탕	Sugar	70g
NH펙틴	NH pectin	3g
젤라틴매스 (×5)	Gelatin mass (×5)	18g
TOTAL		**301g**

레몬 커드

Lemon Curd

INGREDIENTS		g
레몬즙	Lemon juice	270g
설탕	Sugar	160g
달걀	Eggs	320g
젤라틴매스 (×5)	Gelatin mass (×5)	18g
레몬제스트	Lemon zest	5개 분량/ 5 lemons
버터 (Bridel)	Butter	100g
카카오버터	Cacao butter	100g
TOTAL		**968g**

금귤 페이스트

Kumquat Paste

INGREDIENTS		g
금귤	Kumquat	500g

ALMOND SABLÉ

아몬드 사블레

1 믹싱볼에 차가운 상태의 버터를 넣고 비터로 가볍게 풀어준다.

2 슈거파우더, 소금을 넣고 믹싱해 크림 상태로 만든 후 체 친 [T55밀가루, 옥수수전분, 아몬드파우더] 절반을 넣고 믹싱한다.

3 달걀(30℃)을 흘려 넣어가며 믹싱한다.

4 남은 **2**를 모두 넣고 날가루가 보이지 않을 때까지 믹싱한다.

5 반죽을 가볍게 치대 덩어리로 만들고 랩핑한 후 밀어 펴기 쉽게 직사각형으로 만들어 냉장고에 휴지시킨다.

6 두께 3mm로 밀어 편 후 달걀 모양 커터(6 × 8cm)로 찍어 뒤집은 마들렌 몰드(플렉시판 FP2511)에 올린다.
180℃로 예열된 오븐에서 10분간 굽고 돌려서 3분간 더 굽는다.
구워져 나온 사블레에 타르틀렛 코팅을 분사한 후 색이 날 때까지 200℃에서 2분간 더 구워 마무리한다.

TIP. 타르틀렛 코팅은 사블레에 얇게 바르거나 분사한다. 분사하는 경우 타르트 셸이 날아가지 않도록 컴프레셔 압력을 35로 낮춰 분사한다.

1 Beat to loosen cold butter in a mixing bowl.

2 Continue mixing with powdered sugar and salt until creamy. Mix with half of sifted [T55 flour, cornstarch, and almond powder].

3 Drizzle in eggs(30°C) while mixing.

4 Mix with the remaining (**2**) until no powder is visible.

5 Lightly knead the dough into a ball, wrap it in plastic wrap, make it into a rectangular shape so it's easy to roll out, and refrigerate.

6 Roll the dough into 3 mm thickness, cut it with an egg-shaped cutter (6 × 8 cm). Place on the inverted madeleine mold (Flexipan FP2511). Bake in an oven preheated to 180°C for 10 minutes, rotate the tray, and bake for another 3 minutes. Spray the coating mixture and bake for another 2 minutes at 200°C until it obtains color.

TIP. Apply a thin layer of the coating for tartlet with a brush or a spray gun. If spraying with the gun, lower the compressor pressure to 35 to prevent blowing off the tartlet shells.

OLIVE OIL FINANCIER

올리브오일 피낭시에

1 써머믹서에 체 친 [헤이즐넛파우더, 슈거파우더, 박력분, 베이킹파우더], 흰자, 꿀을 넣고 곱게 갈아준다.

2 엑스트라 버진 올리브오일, 녹인 누아제트 버터(45℃)를 넣고 모든 재료가 잘 섞이도록 갈아준다.

3 테프론시트를 깐 철판에 반죽을 붓고 평평하게 정리한다.
180℃로 예열된 오븐에서 11분간 굽고 완전히 식혀 사방 1cm 큐브 형태로 잘라 준비한다.

TIP. 사용하는 철판의 크기에 따라 가나슈 막대(각봉)로 반죽의 두께를 조절할 수 있다. 많이 부풀어오르는 반죽이 아니므로, 사용하는 두께인 1cm 정도가 되도록 팬닝한다.

1 Finely grind sifted [hazelnut powder, sugar powder, cake flour, baking powder], egg whites, and honey in the ThermoMixe

2 Thoroughly blend with extra virgin olive oil and melted noisette butter (45°C).

3 Pour the batter onto a baking tray lined with a Teflon sheet and flatten out evenly. Bake in an oven preheated to 180°C for 11 minutes. Cool completely and cut into 1 cm square cubes.

TIP. Depending on the size of the baking tray, you can adjust the thickness by using confectionery bars. This batter will not rise much during baking, so spread to about 1 cm, which is the thickness needed.

✤

1

2

3

4

5

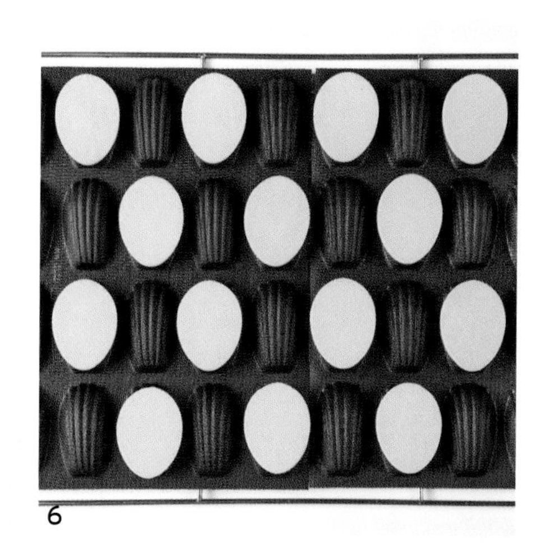

6

✤ ✤

1

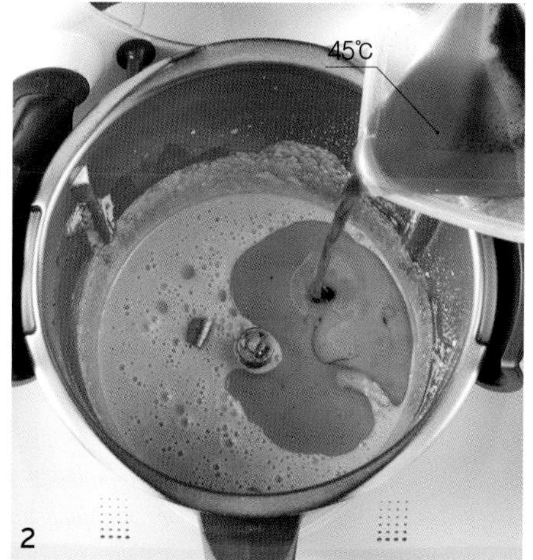

2

3

✤ ✤ ✤

LEMON COULIS

레몬 쿨리

레몬 퓌레

1 필러를 이용해 레몬 껍질을 길게 자른다.

2 끓는 물에 5번 데쳐 불순물을 제거한다.

TIP. 냄비에 레몬 껍질이 충분히 잠길 정도의 물을 넣고 레몬 껍질과 함께 가열한 후 끓어오르면 물을 버리고 찬물을 넣어 다시 가열한다. 이 작업을 5번 반복한다.

3 체에 걸러 물기를 제거한다.

4 로보쿱에 옮겨 레몬즙과 함께 갈아 레몬 퓌레를 완성한다.

레몬 쿨리

5 냄비에 완성한 레몬 퓌레, 미리 섞어둔 설탕과 NH펙틴을 넣고 휘퍼로 저어가며 가열한다.

6 펙틴의 반응을 확인하고 불에서 내린 후 젤라틴매스를 넣고 녹을 때까지 섞는다.

7 바트에 담아 밀착 랩핑한 후 냉장고에 보관한다.

Lemon Purée

1 Peel the lemon with a peeler.

2 Blanch in boiling water 5 times to remove impurities.

TIP. Add enough water in a pot to cover the lemon peels and heat it with the lemon peels. When it boils, discard the water, add cold water, and heat again. Repeat 5 times.

3 Strain to remove water.

4 Blend it with lemon juice in the Robot Coupe to finish the lemon purée.

Lemon Coulis

5 Heat the prepared lemon purée in a pot and add the previously mixed sugar and pectin while whisking.

6 After the pectin activates, remove from heat, add gelatin mass, and stir until it melts.

7 Pour into a stainless-steel tray, cover with plastic wrap, making sure the wrap is in contact with the coulis, and refrigerate.

1

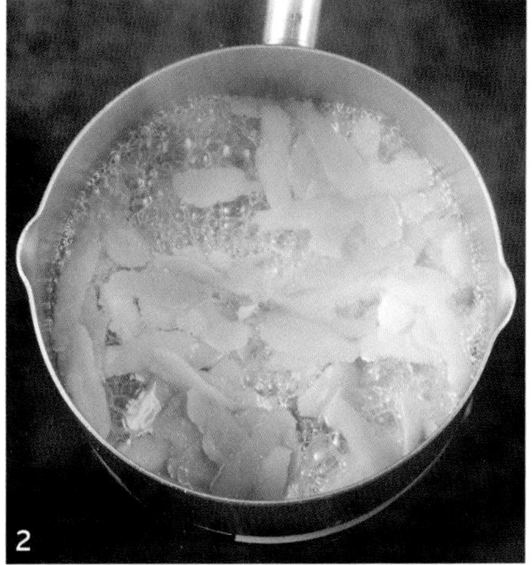
2

3

4

5

6

7

＊스팀 오븐을 사용하는 방법

: 스팀 오븐을 사용하면 좀 더 균일하고 안정적인 상태의 레몬 퓌레를 만들 수 있다.

❶ 레몬 퓌레의 3번 과정까지 작업한 레몬을 진공백에 넣어 진공 상태로 만든 후 100℃ 스팀 오븐에 1시간 동안 넣어둔다.

❷ 진공백 그대로 얼음물에 넣어 쿨링한 후 로보쿱에 옮겨 레몬즙과 함께 갈아 레몬 퓌레를 완성한다.

＊How to use a steam oven

: Using a steam oven helps make a more uniform and stable lemon purée.

❶ Put the purée from (**3**) into a vacuum-sealed bag and place it in a 100°C steam oven for 1 hour.

❷ Cool the vacuum bag in an ice bath, pour it into a Robot Coupe, and blend it with lemon juice to make the lemon purée.

LEMON CURD

레몬 커드

1. 냄비에 레몬즙, 설탕을 넣고 가열한다.
2. 달걀(30℃)이 담긴 볼에 조금씩 넣어가며 휘퍼로 저어준다.
3. 다시 냄비에 옮겨 85℃로 가열한다.
4. 불에서 내려 젤라틴매스, 레몬제스트를 넣고 섞는다.
5. 얼음물이 담긴 볼에 받쳐 40℃까지 온도를 낮춰준다.
6. 비커에 옮겨 차가운 상태의 버터와 녹인 카카오버터(45℃)를 넣고 바믹서로 블렌딩한다.
7. 체에 거른 후 밀착 랩핑해 냉장고에 보관한다.

1. Heat lemon juice and sugar in a saucepan.
2. Add into a bowl with eggs (30°C) a little bit at a time while whisking.
3. Pour back into the saucepan and cook to 85°C.
4. Remove from the heat and stir in gelatin mass and lemon zest.
5. Cool to 40°C in an ice bath.
6. Pour into a beaker and blend with cold butter and melted cacao butter (45°C) with an immersion blender.
7. Strain and pour into a stainless-steel tray, cover with plastic wrap, making sure the wrap is in contact with the curd, and refrigerate.

KUMQUAT PASTE

금귤 페이스트

1. 금귤을 끓는 물에 5번 데치고 쿨링한 후 믹서에 갈아준다.

   TIP. 금귤은 자르지 않은 홀 상태 그대로 사용한다.
   냄비에 금귤이 충분히 잠길 정도의 물과 금귤을 넣고 가열한 후 끓어오르면 물을 버리고 찬물을 넣어 다시 가열한다. 이 작업을 5번 반복한다.
   금귤 씨는 쿨링한 후 주걱으로 으깨듯 체에 내려 제거한다.

2. 진공백에 넣어 냉장 보관한다.

1. Blanch kumquats 5 times in boiling water, let cool, and grind them in a mixer.

   TIP. Do not cut kumquats, but use them as a whole.
   Add enough water to cover kumquats in a pot and heat. When it boils, discard the water, add cold water, and heat again. Repeat 5 times.
   After cooling, remove the seeds by pressing them over a sieve as if to mash them.

2. Refrigerate in a vacuum-sealed bag.

✤ ✤ ✤ ✤

1

2

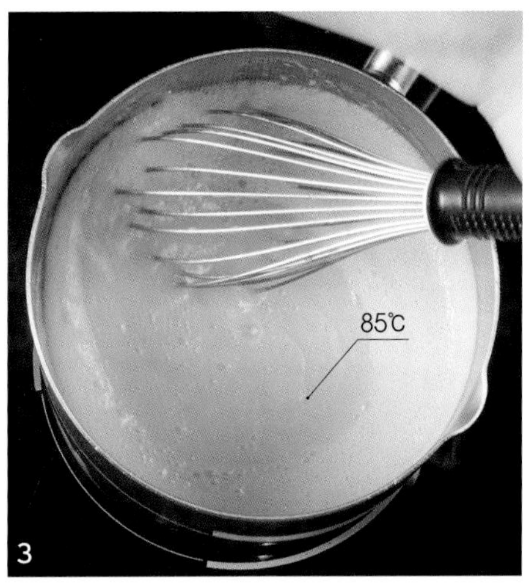

3

4

5

6

7

✤ ✤ ✤ ✤ ✤

1

2

1

2

3

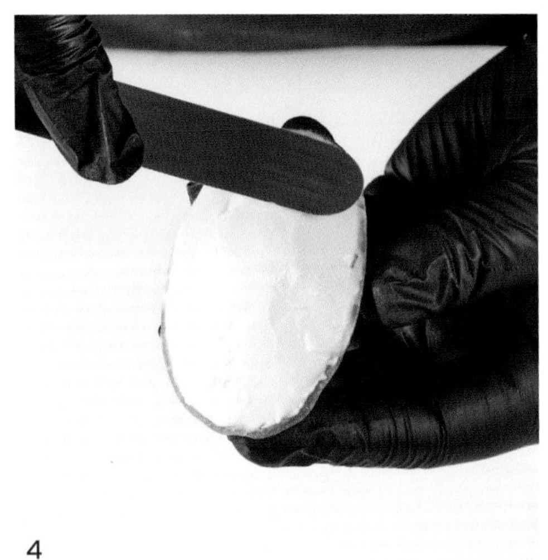

4

5

6

✤ ✤ ✤ ✤ ✤ ✤

FINISH 마무리

1 아몬드 사블레 안쪽에 녹인 화이트초콜릿(45~50℃)을 얇게 발라 코팅한다.

2 가운데에 금귤 페이스트 1g을 짠 후 바깥쪽으로 레몬 쿨리 2g을 짠다.

3 사방 1cm로 자른 올리브오일 피낭시에 5개를 올린다.

4 레몬 커드 28g을 짠 후 미니 L자 스패출러로 마들렌 배꼽 모양이 연상되도록 가운데가 볼록하게 성형한다.

5 제스터를 이용해 레몬초콜릿을 갈아준다.

TIP. 레몬초콜릿은 약 25℃ 정도에서 사용해야 적당한 길이로 갈린다. (온도가 높을수록 길게 갈리고 뭉친다.)

6 레몬 쿨리를 3군데 짠다.

7 허브를 올리고 금박으로 마무리한다.

1. Coat the inside of the almond sablé with a thin layer of melted white chocolate (45~50°C).
2. Pipe 1 gram of kumquat paste in the center and 2 grams of lemon coulis around it.
3. Put 5 pieces of olive oil financier cut into 1 cm cubes.
4. Pipe 28 grams of lemon curd and use a small offset spatula to shape it to resembles the plump madeleine.
5. Grate lemon chocolate using a zester.

   TIP. Lemon chocolate should be used at about 25°C to be grated into the right length. (It grinds in longer pieces and clumps together when used at higher temperatures.)
6. Pipe three dots with the lemon coulis.
7. Top with herbs and finish with gold leaves.

10

BUCHE DE NOEL

부시 드 노엘

PAIRING & TEXTURE
페어링 & 텍스처

마스카르포네, 초콜릿, 캐러멜, 밤을 페어링한 부시 드 노엘. 전체적으로 촉촉하고 부드러운 텍스처에 머랭 스틱의 바삭함을 더했다.

Buche de Noel paired with mascarpone, chocolate, caramel, and chestnuts. The overall texture is moist and soft, with the crunch of meringue sticks.

TECHNIQUE
테크닉

베이스가 되는 몽테 크림을 만들고, 이 몽테 크림에 다양한 재료들을 섞어 맛을 쉽게 변형할 수 있는 제품이다. 여기에서는 베이스가 되는 몽테에 캐러멜을 더하고 밤 페이스트를 섞어 사용했다.

This dessert is made with montée cream as the base, and you can easily customize the flavor by adding different ingredients. Here, I added caramel to the base cream and mixed with chestnut paste.

DESIGN
디자인

기성 몰드를 사용하지 않고도 실제 나무토막과 같은 느낌이 들 수 있도록 표현했다. 마롱 몽테 크림을 파이핑해 굳힌 후 거친 솔로 스크래치를 내어 질감을 표현했고, 초콜릿 색소로 명암을 주어 최대한 실제에 가깝게 표현했다.

I wanted it to look like a real wood log without using a commercial mold. After frosting and freezing the marron montée cream, I scratched it with a coarse brush to give its texture and used chocolate colorings to add lights and darks to make it as authentic as possible.

✦ HOW TO COMPOSE THIS RECIPE ✦

STEP 1.

메뉴 정하기	부시 드 노엘 (시즌 케이크)
Decide on the menu	Buche de Noel (Seasonal cake)

STEP 2.

메인 맛 정하기	밤
Choose the primary flavor	Chestnut

STEP 3.

메인 맛(밤)과의 페어링 선택하기

Select a pairing flavor (Chestnut)

- ☑ 머랭 Meringue
- ☑ 마스카르포네 Mascarpone
- ☑ 생크림 Heavy Cream
- ☑ 캐러멜초콜릿 Caramel Chocolate
- ☑ 헤이즐넛 Hazelnut
- ☑ 바닐라 Vanilla

STEP 4.

구성하기

Assemble

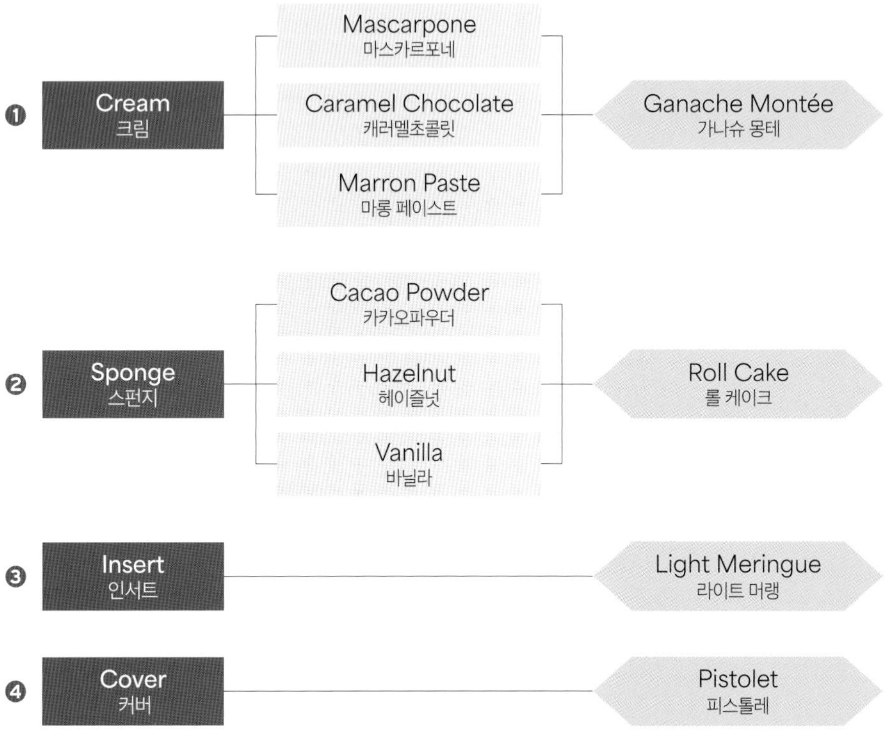

❶ Cream 크림
- Mascarpone 마스카르포네
- Caramel Chocolate 캐러멜초콜릿
- Marron Paste 마롱 페이스트
- → Ganache Montée 가나슈 몽테

❷ Sponge 스펀지
- Cacao Powder 카카오파우더
- Hazelnut 헤이즐넛
- Vanilla 바닐라
- → Roll Cake 롤 케이크

❸ Insert 인서트
- → Light Meringue 라이트 머랭

❹ Cover 커버
- → Pistolet 피스톨레

다크 & 밀크 스프레이
Dark & Milk Spray
머랭 스틱
Meringue Stick
마롱 몽테 크림
Marron Montée
Cream
마롱 페이스트
Marron Paste
롤 비스퀴
Roll Biscuit

INGREDIENTS 4개 분량/ 4 cakes

롤 비스퀴

Roll Biscuit

(철판(60 × 40cm) 2개 분량)
(2 baking trays (60 × 40 cm))

INGREDIENTS		g
노른자	Egg yolks	300g
달걀	Eggs	250g
설탕A	Sugar A	300g
물엿	Corn syrup	40g
바닐라에센스 (Aroma Piu)	Vanilla essence	10g
헤이즐넛 오일	Hazelnut oil	60g
박력분	Cake flour	200g
카카오파우더 (Extra Brute)	Cacao powder	30g
알부민파우더	Albumin powder	2g
흰자	Egg whites	270g
설탕B	Sugar B	220g
TOTAL		**1682g**

머랭 스틱

Meringue Sticks

INGREDIENTS		g
흰자	Egg whites	180g
이소말트파우더	Isomalt powder	100g
설탕	Sugar	80g
알부민파우더	Albumin powder	6g
슈거파우더	Sugar powder	120g
탈지분유	Skim milk powder	18g
옥수수전분	Cornstarch	17g
시나몬파우더	Cinnamon powder	1g
TOTAL		**522g**

*** 만드는 법**

❶ 믹싱볼에 흰자, 이소말트파우더를 넣고 따뜻한 물이 담긴 볼에 받쳐 휘퍼로 저어가며 50℃까지 중탕한다.
❷ 거품이 안정적으로 형성되면 믹싱기로 옮겨 설탕, 알부민파우더를 나눠 넣어가며 휘핑한다.
❸ 거품이 조밀하고 단단하게 형성되면 체 친 [슈거파우더, 탈지분유, 옥수수전분, 시나몬파우더]를 넣고 주걱으로 가루가 보이지 않을 때까지 섞는다.
❹ 1.5cm 원형 깍지를 이용해 테프론시트를 깐 철판에 길이 18cm, 폭 2cm로 파이핑한다.
❺ 120℃로 예열된 오븐에서 1시간 굽고, 80℃로 낮춰 1시간 더 굽는다.

*** Procedure**

❶ Warm egg whites and isomalt powder in a mixing bowl over a double boiler (bain-marie) to 50°C while stirring.
❷ Once the foam becomes stable, transfer to a mixer and whip while gradually adding sugar and albumin powder.
❸ When the meringue forms stiff peaks, add sifted [sugar powder, skim milk powder, cornstarch, and cinnamon powder]; mix with a spatula until no streaks of powder are visible.
❹ Fill a piping bag fitted with a 1.5 cm round tip and pipe 18 cm long and 2 cm wide sizes on a baking tray lined with a Teflon sheet.
❺ Bake in an oven preheated to 120°C for one hour, then reduce the temperature to 80°C and bake for another hour.

캐러멜 마스카르포네 몽테 *

Caramel Mascarpone Montée

INGREDIENTS		g
생크림	Heavy cream	500g
노른자	Egg yolks	320g
젤라틴매스 (×5)	Gelatin mass (×5)	100g
캐러멜초콜릿 (Zephyr Caramel 35%)	Caramel chocolate	240g
연유	Condensed milk	100g
마스카르포네	Mascaporne cheese	450g
휘핑크림 (Bridel)	Whipping cream	800g
아몬드 리큐르 (Dijon Almond)	Almond liqueur	20g
골드럼 (PAN RUM)	Gold rum	20g
소금	Salt	3g
TOTAL		**2553g**

* 만드는 법

❶ 냄비에 생크림을 넣고 따뜻한 정도로 가열한다.
❷ 노른자(30℃)가 담긴 볼에 ❶을 넣고 섞는다.
❸ 다시 냄비로 옮겨 75℃까지 가열한다.
❹ 불에서 내린 후 젤라틴매스를 넣고 녹을 때까지 섞는다.
❺ 캐러멜초콜릿이 담긴 비커에 ❹와 연유와 소금을 넣고 바믹서로 블렌딩한다.
❻ 마스카르포네를 넣고 블렌딩한다.
❼ 차가운 상태의 휘핑크림 - 아몬드 리큐르 - 골드럼 순서로 넣고 블렌딩한다.
❽ 바트에 담아 밀착 랩핑한 후 12시간 이상 숙성시킨다.

* Procedure

❶ Heat heavy cream in a saucepan just until warm.
❷ Add ❶ to a bowl with the egg yolks (30°C) and mix.
❸ Bring back to the saucepan and heat to 75°C.
❹ Remove from the heat, add gelatin mass, and wait until it melts.
❺ Blend ❹ and condensed milk in a beaker with caramel chocolate and salt using an immersion blender.
❻ Blend in mascarpone cheese.
❼ Blend cold whipping cream, almond liqueur, and dark rum in order.
❽ Pour it into a stainless-steel tray and cover it with plastic wrap, make sure the wrap is in contact with the cream. Refrigerate for 12 hours.

마롱 몽테 크림

Marron Montée Cream

(비스퀴 1개당 570g 사용)
(570 grams of dough per biscuit)

INGREDIENTS		g
밤 페이스트 (Imbert)	Marron paste	300g
캐러멜 마스카르포네 몽테 *	Caramel mascarpone montée *	2000g
TOTAL		**2300g**

* 만드는 법

❶ 믹싱볼에 밤 페이스트를 넣고 부드럽게 풀어준다.
❷ 캐러멜 마스카르포네 몽테를 넣고 파이핑하기 좋은 상태로 휘핑해 사용한다.

* Procedure

❶ Soften chestnut paste in a mixing bowl.
❷ Add the caramel mascarpone montée and whip until it's suitable to pipe.

다크 스프레이

Dark Spray

INGREDIENTS		g
다크초콜릿 (Mi-Amère 58%)	Dark chocolate	250g
카카오버터	Cacao butter	250g
TOTAL		**500g**

＊만드는 법

❶ 비커에 녹인 다크초콜릿(45~50℃)과 50℃ 이하로 녹인 카카오버터를 넣고 바믹서로 블렌딩한다.

❷ 50℃로 맞춰 분사한다.

＊Procedure

❶ Blend dark chocolate melted to 45~50°C and cacao butter melted to below 50°C in a beaker using an immersion blender.

❷ Spray at 50°C.

밀크 스프레이

Milk Spray

INGREDIENTS		g
밀크초콜릿 (Lactee Superieure 38%)	Milk chocolate	150g
다크초콜릿 (Guayaquil 64%)	Dark chocolate	50g
카카오버터	Cacao butter	200g
TOTAL		**400g**

＊만드는 법

❶ 비커에 녹인 밀크초콜릿과 다크초콜릿(45~50℃), 50℃ 이하로 녹인 카카오버터를 넣고 바믹서로 블렌딩한다.

❷ 50℃로 맞춰 분사한다.

＊Procedure

❶ Blend dark and milk chocolates melted to 45~50°C, and cacao butter melted to below 50°C in a beaker using an immersion blender.

❷ Spray at 50°C.

NOTE.

✤

ROLL BISCUIT

롤 비스퀴

1 볼에 노른자, 달걀, 설탕A를 넣고 뜨거운 물이 담긴 중탕볼에 받쳐 휘퍼로 저어가며 50℃까지 온도를 올린 후 믹싱기로 옮겨 고속으로 휘핑한다.

2 거품이 형성되기 시작하면 물엿을 넣고 휘핑한다.

3 바닐라에센스를 넣고 휘핑한다.

4 헤이즐넛 오일을 넣고 휘핑한다.

TIP. 머랭이 첨가될 것이므로 오일을 넣고 휘핑해도 괜찮다.

5 체 친 박력분, 카카오파우더를 넣고 섞는다.

6 다른 볼에 흰자(30℃), 설탕B, 알부민파우더를 넣고 휘핑해 머랭을 만든다.

7 **5**에 **6**을 두 번 나눠 넣어가며 섞는다.

8 테프론시트를 깐 철판(60 × 40cm)에 1cm 높이로 팬닝한 후 데크 오븐 기준 윗불 230℃, 아랫불 210℃에서 12분간 굽는다.

TIP. 여기에서는 일정함을 위해 팬닝 기계로 작업했다.

1 Warm egg yolks, eggs, and sugar A in a mixing bowl over a double boiler (bain-marie) to 50°C while stirring. Transfer to the mixer and whip at high speed.

2 When it starts to becomes foamy, add corn syrup and continue to whip.

3 Add vanilla essence and whip.

4 Add hazelnut oil and whip.

TIP. It's okay to whip with the oil because meringue will be added.

5 Fold in sifted cake flour and cacao powder.

6 In another bowl, whip egg whites (30°C), sugar B, and albumin powder to make meringue.

7 Mix (**6**) to (**5**), half at a time.

8 Spread onto a baking tray (60 × 40 cm) lined with a Teflon sheet to 1 cm thickness. Bake in a deck oven at 230°C on top and 210°C on the bottom for 12 minutes.

TIP. I used a cake depositor machine for consistency.

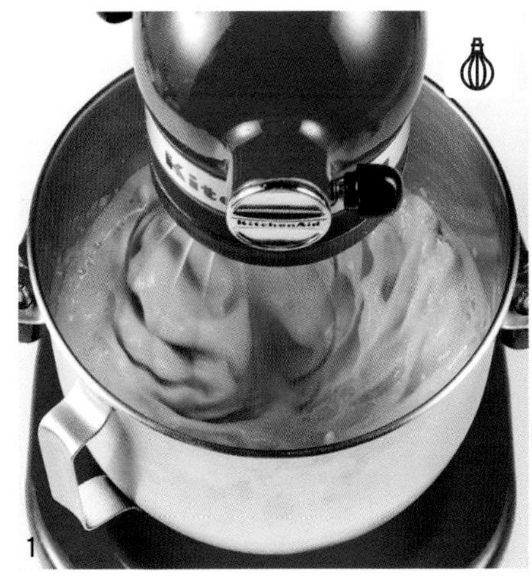
1

2

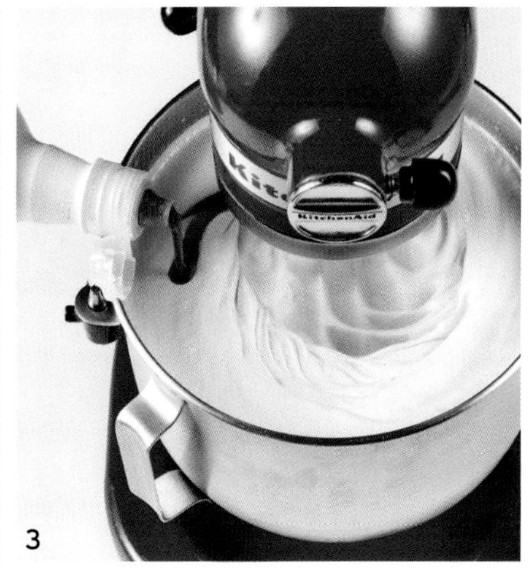
3

4-1

4-2

5

6

7

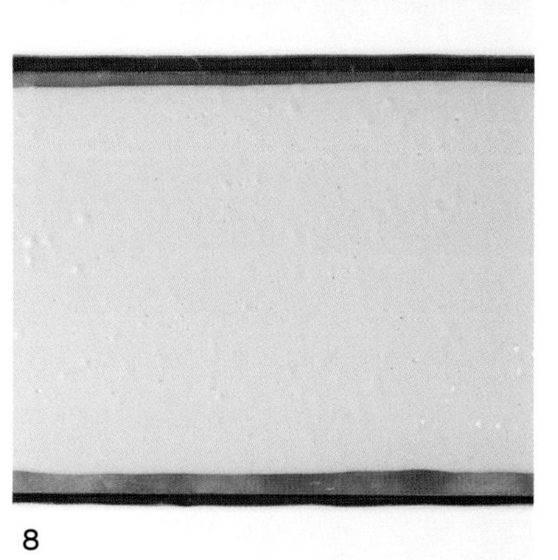
8

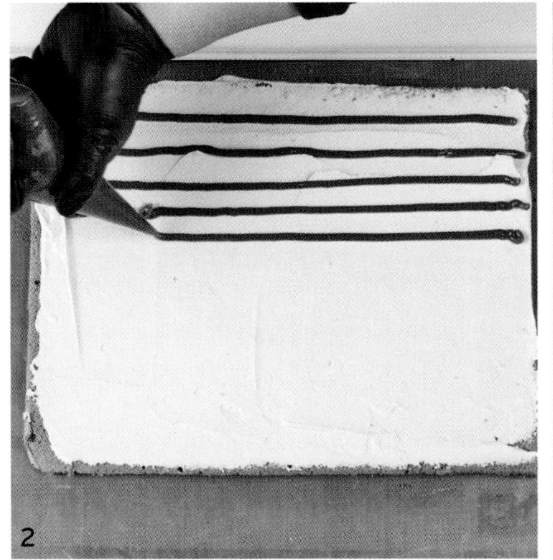

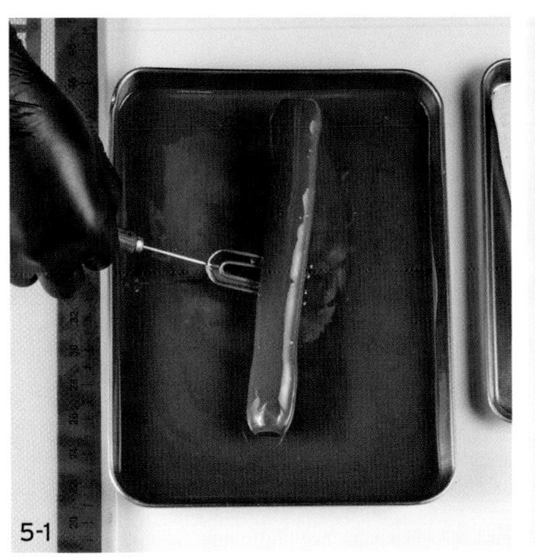

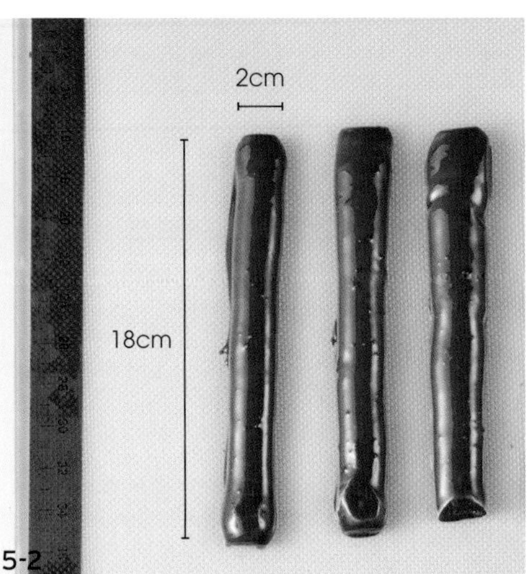

✤ ✤

FINISH 마무리

1. 29 × 40cm로 자른 롤 비스퀴에 마롱 몽테 크림 270g을 펴 바른다.
2. 밤 스프레드(분량 외) 90g을 파이핑한다.
3. 스패출러로 접는 선을 만들어가면서 돌돌 말아준다.
4. 돌돌 만 롤 비스퀴의 모양이 흐트러지지 않도록 랩핑해 냉동실에 잠시 두어 모양을 고정시킨다.
5. 머랭 스틱에 밀크 코팅초콜릿(분량 외)을 코팅한 후 굳힌다.
6. 모양을 고정시킨 롤 비스퀴를 반으로 자른다. (길이 약 19cm)
7. 마롱 몽테 크림 200g을 파이핑한다.
8. 스패출러로 윗면과 옆면으로 펴 바른다.
9. 초콜릿으로 코팅한 머랭 스틱 2개를 올린다.

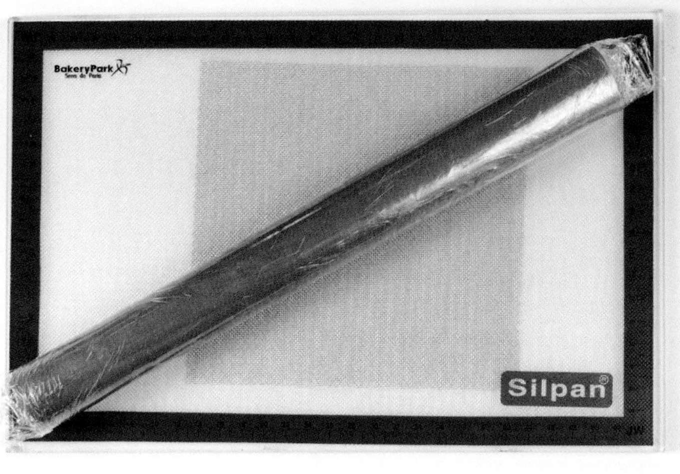

4

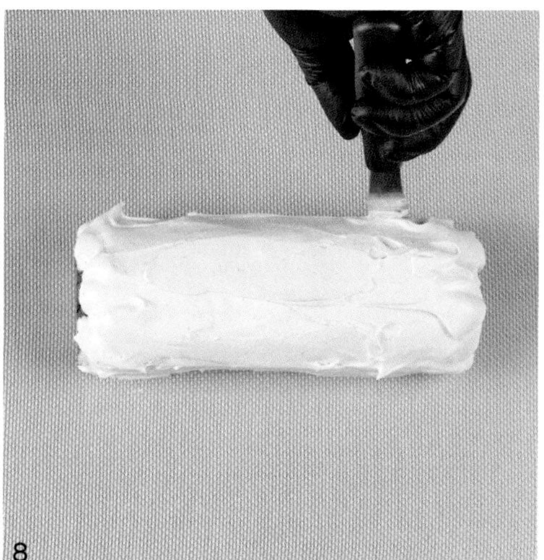

8

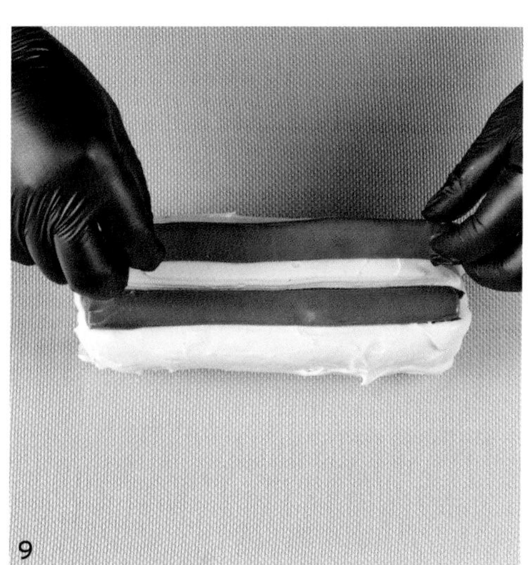

9

1 Spread 270 grams of the marron montée cream on the roll biscuit cut to 29 × 40 cm.

2 Pipe 90 grams of chestnut spread (other than requested).

3 Roll up the biscuit while making folding lines with a spatula.

4 Wrap the rolled biscuit to prevent from losing its shape and place in the freezer for a few minutes to set.

5 Coat the meringue sticks with milk coating chocolate (other than requested) and let them set.

6 Cut the wrapped rolled biscuit in half (approx. 19 cm long).

7 Pipe 200 grams of marron montée cream.

8 Spread on the top and sides with a spatula.

9 Top with 2 meringue sticks coated with chocolate.

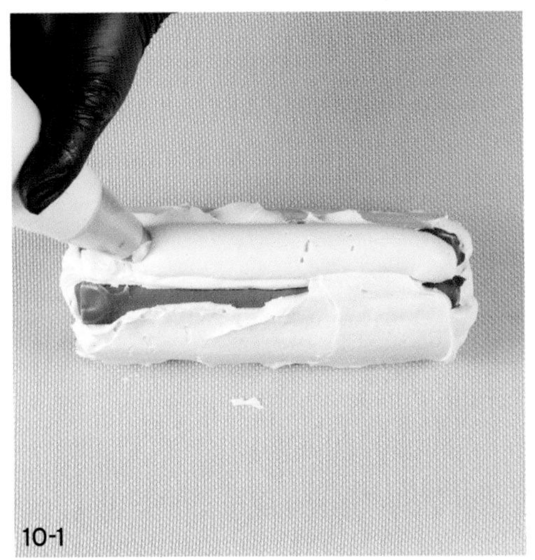
10-1

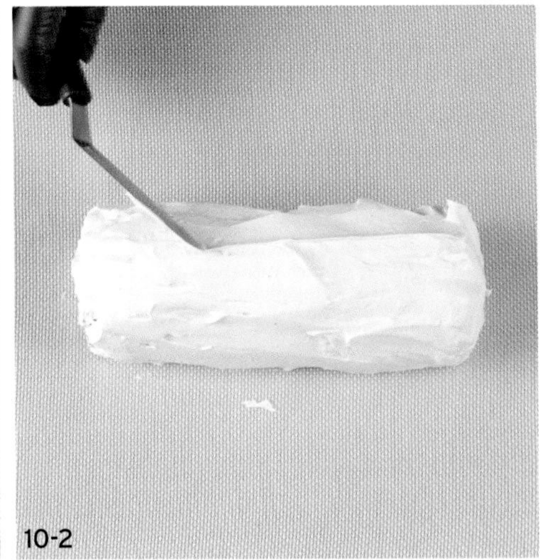
10-2

11

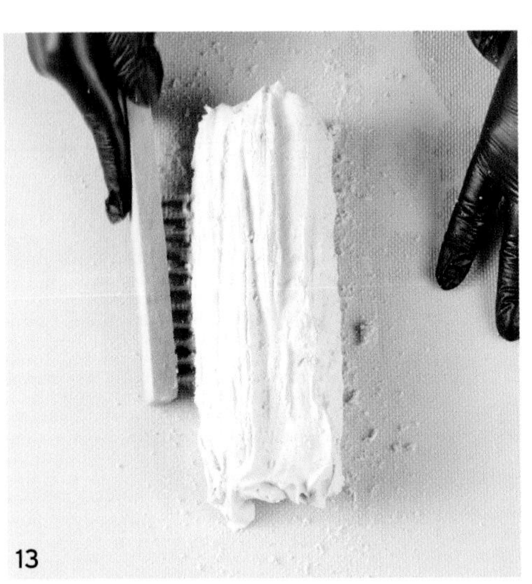
13

14

15

10 마롱 몽테 크림 100g을 파이핑하며 스패출러로 골고루 펴 바른다.

11 8 × 25cm OHP필름을 이용해 표면을 다듬는다.

12 스패출러를 마롱 몽테 크림에 닿았다 떼었다 하면서 거친 표면을 만들어준 후 냉동한다.

13 거친 솔을 이용해 통나무 껍질과 같은 질감을 연출한다.

14 손으로 문질러가며 마찰을 주어 거친 부분을 매끈하게 만든다.

15 다크 스프레이와 밀크 스프레이로 분사하고, 붓으로 카카오파우더를 골고루 묻혀준 후 열풍기로 열을 가해 명암과 음영을 표현한다.

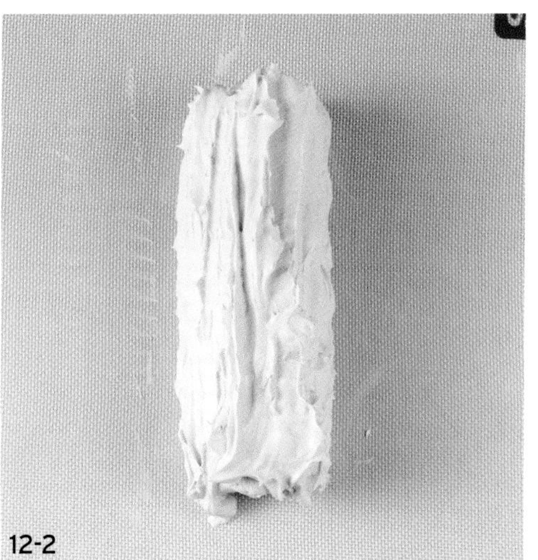

12-2

10 Pipe 100 grams of the marron montée cream and spread evenly using a spatula.

11 Use an 8 × 25 cm OHP film to trim the surface.

12 Repeat sticking and removing the spatula blade to the cream to create a rough surface. Freeze the cakes.

13 Scrape the frozen surface with a coarse brush to recreate the bark of a log.

14 Use your hands to rub against the surface to smooth out rough spots.

15 Spray with dark and milk sprays, dust with cacao powder, and apply heat with a heat gun to give a shading effect.

11

BOHEMIAN MANGO

보헤미안 망고

PAIRING & TEXTURE
페어링 & 텍스처

메인 재료인 망고와 패션푸르트, 사워크림을 페어링해 시트러스한 맛을 더욱 강조하고 코코넛과 얼그레이를 이용해 끝맛의 풍미를 더했다.

The main ingredient, mango, is paired with passion fruit and sour cream for a tanginess, and coconut and Earl Grey add flavor at the end.

TECHNIQUE
테크닉

- 얼리지 않은 인서트를 소복하게 올려 과일의 신선함을 그대로 느낄 수 있게 했다.
- 말토덱스트린이 가진 지방을 흡수하는 성질을 이용해 초콜릿 파우더를 만들고 겉면에 뿌려 피스톨레 기법과 다른 질감을 표현했다.

- I topped the dessert with a mound of unfrozen insert so you can taste the freshness of the fruits.
- The fat-absorbing properties of maltodextrin were used to make the chocolate powder, which was sprinkled on the surface to add a different texture than the spraying technique.

DESIGN
디자인

- 시판 몰드 모양으로 완성되는 제품이 아닌, 주걱과 아이스크림 스쿱으로 자연스럽게 모양을 만들어 자유분방한 디자인으로 풀어냈다.
- 사워크림 몽테를 주걱으로 떠 올린 후 아이스크림 스쿱으로 모양을 내어 인서트가 들어갈 공간을 만들어주었다.

- Instead of using commercially available molds to create the cake, I used a spatula and ice cream scoop to make a free-form design.
- After scooping the sour cream montée with a spatula, I shaped it with an ice cream scoop to make room for the insert.

✦ HOW TO COMPOSE THIS RECIPE ✦

STEP 1.

메뉴 정하기	프티 가토
Decide on the menu	Petit Gateau

STEP 2.

메인 맛 정하기	망고
Choose the primary flavor	Mango

STEP 3.

메인 맛(망고)과의 페어링 선택하기

Select a pairing flavor (Mango)

- ☑ 패션푸르트 Passion Fruit
- ☑ 레몬 Lemon
- ☑ 코코넛 Coconut
- ☑ 얼그레이 Earl Grey
- ☑ 바닐라 Vanilla
- ☑ 사워크림 Sour Cream
- ☑ 화이트초콜릿 White Chocolate

STEP 4.

구성하기

Assemble

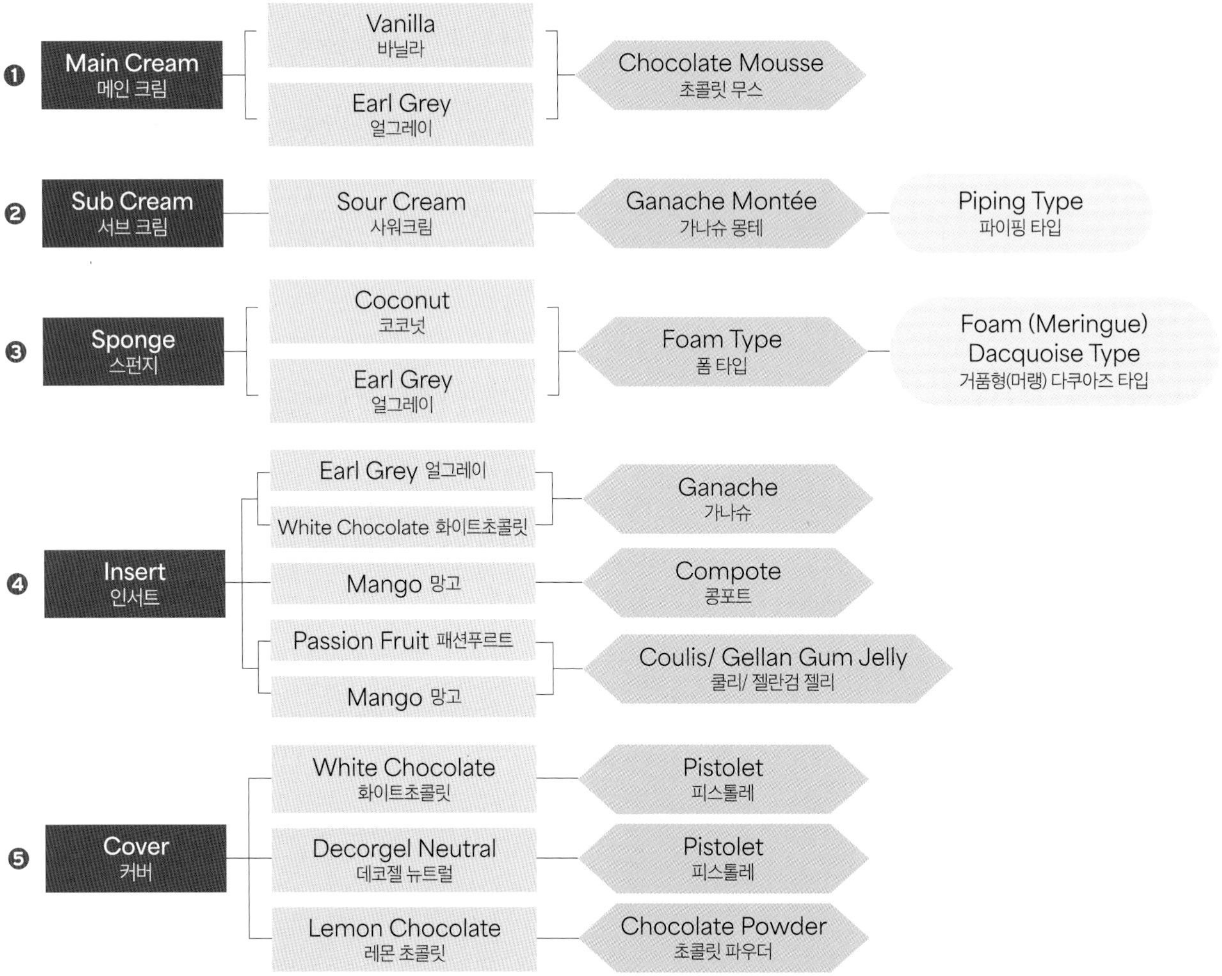

❶ Main Cream 메인 크림
- Vanilla 바닐라 / Earl Grey 얼그레이 → Chocolate Mousse 초콜릿 무스

❷ Sub Cream 서브 크림
- Sour Cream 사워크림 → Ganache Montée 가나슈 몽테 → Piping Type 파이핑 타입

❸ Sponge 스펀지
- Coconut 코코넛 / Earl Grey 얼그레이 → Foam Type 폼 타입 → Foam (Meringue) Dacquoise Type 거품형(머랭) 다쿠아즈 타입

❹ Insert 인서트
- Earl Grey 얼그레이 / White Chocolate 화이트초콜릿 → Ganache 가나슈
- Mango 망고 → Compote 콩포트
- Passion Fruit 패션푸르트 / Mango 망고 → Coulis/ Gellan Gum Jelly 쿨리/ 젤란검 젤리

❺ Cover 커버
- White Chocolate 화이트초콜릿 → Pistolet 피스톨레
- Decorgel Neutral 데코젤 뉴트럴 → Pistolet 피스톨레
- Lemon Chocolate 레몬 초콜릿 → Chocolate Powder 초콜릿 파우더

화이트 바닐라 스프레이 &
레몬 초콜릿 파우더
White Vanilla Spray &
Lemon Chocolate Powder
망고 콩포트
Mango Compote
패션푸르트 망고
젤란검 젤리
Passion Fruit Mango
Gellan Gum Jelly
더블 바닐라 얼그레이 무스
Double Vanilla
Earl Grey Mousse
얼그레이 화이트 가나슈
Earl Grey
White Ganache
코코넛 얼그레이 다쿠아즈
Coconut Earl Grey
Dacquoise
사워크림 몽테
Sour Cream Montée

INGREDIENTS

30개 분량/ 30 cakes

코코넛 얼그레이 다쿠아즈

Coconut Earl Grey Dacquoise

(철판(60 × 40cm) 1개 분량)
(1 baking tray (60 × 40 cm))

INGREDIENTS		g
코코넛롱	Shredded dried coconut	260g
박력분	Cake flour	220g
슈거파우더	Sugar powder	220g
얼그레이 티백 잎 (Twinings)	Earl Grey tea bag leaves	5g
흰자	Egg whites	490g
설탕	Sugar	260g
알부민파우더	Albumin powder	4g
구연산	Citric acid	1g
TOTAL		**1460g**

얼그레이 화이트 가나슈

Earl Grey White Ganache

INGREDIENTS		g
물	Water	45g
얼그레이 티백 잎 (Twinings)	Earl Grey tea bag leaves	7.5g
생크림	Heavy cream	100g
전화당	Inverted sugar	7.5g
화이트초콜릿 (Zephyr white 34%)	White chocolate	200g
버터 (Bridel)	Butter	20g
TOTAL		**380g**

더블 바닐라 얼그레이 무스

Double Vanilla Earl Grey Mousse

INGREDIENTS		g
생크림A	Heavy cream A	130g
전화당	Inverted sugar	20g
얼그레이 티백 잎 (Twinings)	Earl Grey tea bag leaves	3g
바닐라빈 껍질 (Madagascar)	Vanilla bean pod	1/2개 분량/ 1/2 pc
바닐라빈 껍질 (Tahiti)	Vanilla bean pod	1/2개 분량/ 1/2 pc
젤라틴매스 (×5)	Gelatin mass (×5)	16g
화이트초콜릿 (Zephyr white 34%)	White chocolate	130g
생크림B	Heavy cream B	280g
TOTAL		**579g**

사워크림 몽테

Sour Cream Montée

INGREDIENTS		g
우유	Milk	60g
전화당	Inverted sugar	20g
젤라틴매스 (×5)	Gelatin mass (×5)	50g
화이트초콜릿 (Zephyr white 34%)	White chocolate	150g
마스카르포네	Mascarpone cheese	30g
사워크림	Sour cream	150g
휘핑크림	Whipping cream	450g
레몬 리큐르 (Dijon Lemon)	Dijon lemon liqueur	15g
라임 퓌레	Lime purée	15g
TOTAL		**940g**

＊만드는 법

❶ 냄비에 우유, 전화당, 젤라틴매스를 넣고 젤라틴매스가 녹을 때까지 가열한다. (약 60℃)

❷ 화이트초콜릿이 담긴 비커에 ❶을 넣고 바믹서로 블렌딩한다.

❸ 마스카르포네, 사워크림을 넣고 블렌딩한다.

❹ 휘핑크림을 넣고 블렌딩한다.

❺ 레몬 리큐르, 라임 퓌레를 넣고 블렌딩한다.

❻ 바트에 담아 밀착 랩핑한 후 냉장고에서 12시간 동안 숙성시킨다.

❼ 사용할 때는 비터로 부드럽게 풀어 사용한다.

＊Procedure

❶ Heat milk, inverted sugar, and gelatin mass to 60°C until the gelatin mass melts.

❷ Pour into a beaker with white chocolate and mix with an immersion blender.

❸ Blend with mascarpone and sour cream.

❹ Add whipping cream and blend.

❺ Blend with lemon liqueur and lime purée.

❻ Pour into a stainless-steel tray, cover with plastic wrap, making sure the wrap is in contact with the cream, and refrigerate for 12 hours.

❼ Soften with the paddle to use.

패션푸르트 망고 젤란검 젤리

Passion Fruit Mango Gellan Gum Jelly

INGREDIENTS		g
망고 퓌레	Mango purée	60g
패션푸르트 퓌레	Passion fruit purée	120g
물	Water	40g
설탕	Sugar	100g
젤란검	Gellan gum	6.5g
레몬즙	Lemon juice	12g
TOTAL		**338.5g**

망고 쿨리 *

Mango Coulis

INGREDIENTS		g
망고 퓌레	Mango purée	220g
패션푸르트 퓌레	Passion fruit purée	130g
레몬즙	Lemon juice	70g
설탕	Sugar	180g
NH펙틴	NH pectin	11g
로거스트빈검	Locust bean gum	0.5g
주석산	Tartaric acid	0.5g
TOTAL		**612g**

망고 콩포트

Mango Compote

INGREDIENTS		g
망고 쿨리 *	Mango coulis *	300g
1cm로 깍둑썬 망고	Mango, diced into 1 cm cubes	600g
라임제스트	Lime zest	4개 분량/ 4 limes
TOTAL		**900g**

망고 내추럴 글레이즈

Mango Nature Glaze

INGREDIENTS		g
데코젤 뉴트럴	Decorgel Neutral	300g
망고 퓌레	Mango purée	3g
30° 시럽	30˚B syrup	80g
TOTAL		**383g**

*** 만드는 법**

❶ 비커에 모든 재료를 넣고 바믹서로 블렌딩한다.

❷ 50℃로 맞춰 분사한다.

*** Procedure**

❶ Blend all the ingredients in a beaker with an immersion blender.

❷ Use at 50°C.

화이트 바닐라 스프레이

White Vanilla Spray

INGREDIENTS		g
화이트초콜릿 (Zephyr white 34%)	White chocolate	350g
카카오버터	Cacao butter	350g
하얀색 색소 (이산화티타늄)	Titanium dioxide	3g
바닐라빈 씨	Vanilla bean seeds	1개 분량/ 1 pc
TOTAL		**703g**

- 국제적인 추세를 반영하여 이산화티타늄은 최소량으로 사용했다.
- Titanium dioxide was used in a minimal amount to reflect international trends.

*** 만드는 법**

❶ 비커에 녹인 화이트초콜릿(45~50℃), 50℃ 이하로 녹인 카카오버터를 넣고 바믹서로 블렌딩한다.
❷ 50℃로 온도를 맞춘 후 하얀색 색소, 바닐라빈 씨를 넣고 블렌딩한다.
❸ 50℃로 맞춰 분사한다.

*** Procedure**

❶ Blend melted white chocolate (45~50°C) and cacao butter melted to below 50°C with an immersion blender.
❷ Set to 50°C, add titanium dioxide and vanilla bean seeds. Blend with an immersion blender.
❸ Spray at 50°C.

레몬초콜릿 파우더

Lemon Chocolate Powder

INGREDIENTS		g
레몬초콜릿 (Daito Cacao)	Lemon chocolate	100g
소금	Salt	1g
바닐라빈	Vanilla bean	1개 분량/ 1 pc
말토덱스트린	Maltodextrin	100g
TOTAL		**201g**

✤

COCONUT EARL GREY DACQUOISE

코코넛 얼그레이 다쿠아즈

1 로보쿱에 코코넛롱을 넣고 곱게 갈아준다.

TIP. 코코넛롱은 오븐에서 살짝 로스팅해 사용한다.
너무 오래 갈아 기름이 새어나오지 않게 주의한다.

2 볼에 곱게 간 코코넛롱, 체 친 박력분, 슈거파우더, 얼그레이 티백 잎을 넣고 휘퍼로 섞는다.

3 믹싱볼에 흰자(30℃)를 넣고 휘핑하면서 미리 섞어둔 설탕과 알부민파우더와 구연산을 흘려 넣어가며 휘핑한다.

4 조밀하고 단단한 거품 상태가 되면 마무리한다.

5 **4**에 **2**를 두세 번 나눠 넣어가며 섞는다.

TIP. 최대한 가볍게 섞는다.

6 1cm 원형 깍지를 끼운 짤주머니에 담아 테프론시트를 깐 철판(60 × 40cm)에 대각선으로 길게 파이핑한다.

TIP. 다쿠아즈 반죽의 경우 밀어 펴 팬닝해 구우면 증기가 빠져나갈 틈이 없어 제대로 익지 않는다.

7 다쿠아즈 표면에 슈거파우더(분량 외)를 두 번 뿌린다.

TIP. 슈거파우더는 한 번 뿌린 후 스며들면 다시 한 번 뿌린다.

8 180℃로 예열된 오븐에서 20분간 구운 후 냉동한다.

1 Finely grind shredded dried coconut in the Robot Coupe.

TIP. Lightly roast the shredded coconut before use.
Be careful not to grind too long and let the oil seep out.

2 Whisk together the finely ground coconut, sifted cake flour, sugar powder, and Earl Grey tea leaves.

3 In another bowl, whip egg whites (30°C) while drizzling in the previously mixed sugar, albumin powder, and citric acid.

4 Finish when it forms dense and stiff peaks.

5 Mix in (**2**) into (**4**) in two or three batches.

TIP. Mix as lightly as possible.

6 Pipe it lengthwise diagonally with a 1 cm round tip on a baking tray (60 × 40 cm) lined with a Teflon sheet.

TIP. If you spread the dacquoise batter onto a baking sheet to bake, it will not cook properly because there is no room for steam to escape.

7 Dust sugar powder (other than requested) on dacquoise twice.

TIP. Dust the sugar powder again after the previous sugar soaks in.

8 Bake in an oven preheated to 180°C for 20 minutes and keep frozen.

1

2

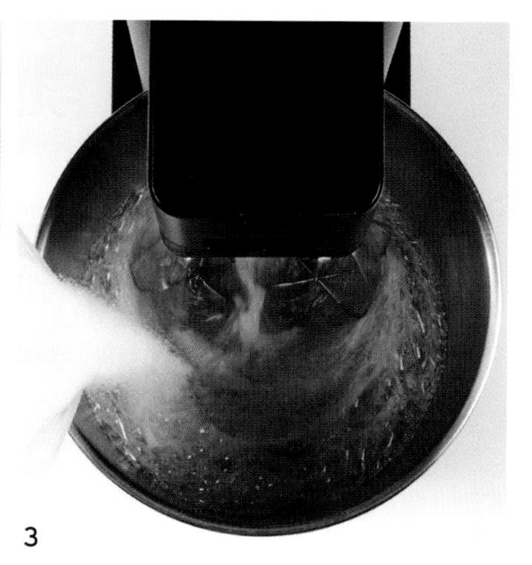
3

4

5

6

7

8

✤ ✤

EARL GREY WHITE GANACHE

얼그레이 화이트 가나슈

1 냄비에 물을 넣고 끓기 시작하면 불을 끄고 얼그레이 티백 잎을 넣은 후 향이 우러나오면 체에 걸러준다. 다시 냄비에 생크림과 전화당을 넣고 70℃가 되면 불을 끄고 한 번 거른 얼그레이 잎을 넣은 후 냄비 입구를 랩핑해 약 10분간 인퓨징한다.

TIP. 잘게 잘린 상태의 얼그레이 티백의 잎을 바로 사용하면 쓰고 떫은 맛이 강하게 표현되므로, 뜨거운 물에 한 번 우려낸 후 인퓨징한다.

2 체에 거르면서 화이트초콜릿과 버터가 담긴 비커에 붓는다.

3 바믹서로 블렌딩한다.

4 냉동한 다쿠아즈 시트에 전량을 골고루 펼쳐 바르고 굳힌 후 5.5 × 7.3cm 크기의 타원형 커터로 자른다.

1 Bring the water to a boil, remove from the heat, and add the tea leaves. When the aroma is released, strain the leaves and discard the water. Boil cream and inverted sugar to 70°C. Remove from the heat, add the strained tea leaves, cover the pot with plastic wrap, and infuse for about 10 minutes.

TIP. If you use the tea bag with finely chopped tea leaves right away, it will release a bitter and astringent taste. So, soak in hot water to rinse, and then use it to infuse.

2 Strain into a beaker with white chocolate.

3 Combine with an immersion blender.

4 Spread the entire ganache evenly on the frozen dacquoise sheet, let it set, and cut it with an oval cutter (5.5 × 7.3 cm).

✤ ✤ ✤

PASSION FRUIT MANGO GELLAN GUM JELLY

패션푸르트 망고 젤란검 젤리

1 냄비에 레몬즙을 제외한 모든 재료를 넣고 가열한다.

TIP. 바믹서로 블렌딩한 후 가열하는 것을 추천한다. 입자가 큰 젤란검을 사용하면 반응이 늦게 일어난다.

2 젤란검 반응이 일어나면(끓은 후 1~2분 사이) 불을 끄고 레몬즙을 넣고 섞는다.

3 지름 1.5cm 구형 몰드에 곧바로 채운 후 냉장고에 보관한다.

1 Heat all the ingredients except lemon juice in a pot.

TIP. I recommend cooking after mixing with an immersion blender. Because gellan gum has larger particles, it will activate more slowly.

2 When the gellan gum activates (about 1~2 minute after boiling), remove from the heat and mix with lemon juice.

3 Pour into a 1.5 cm spherical mold and refrigerate while hot.

✤ ✤

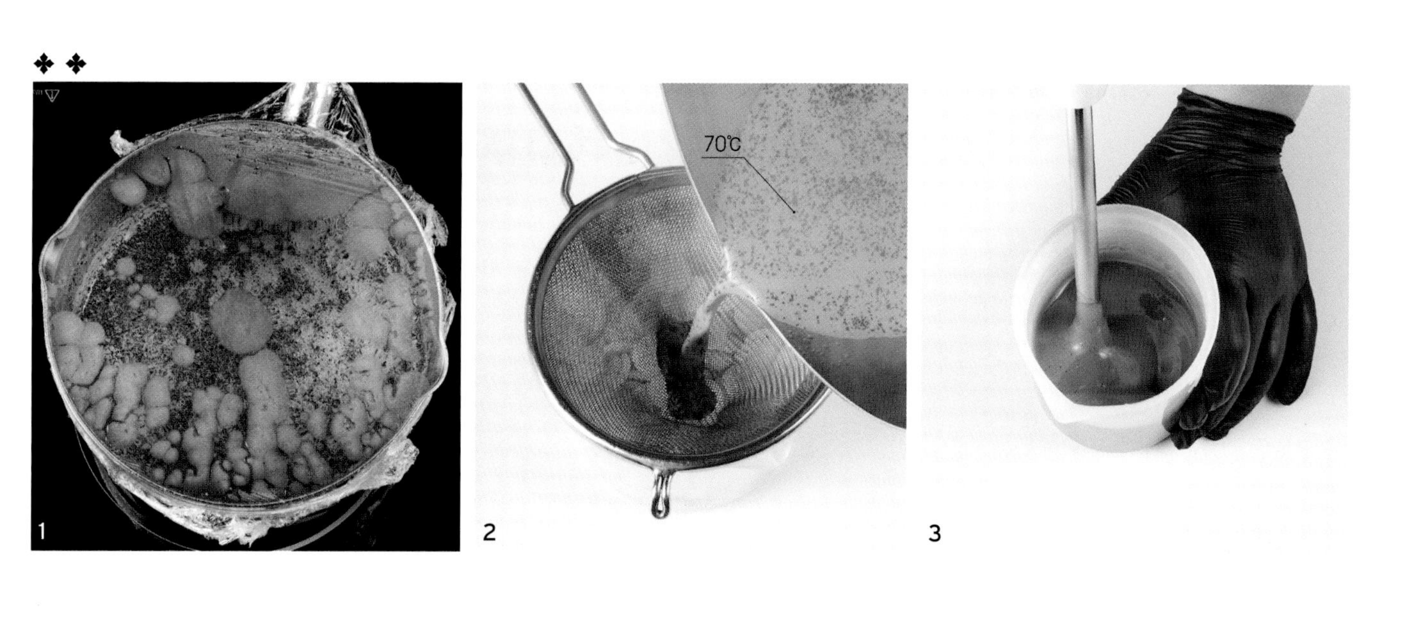

1

2

3

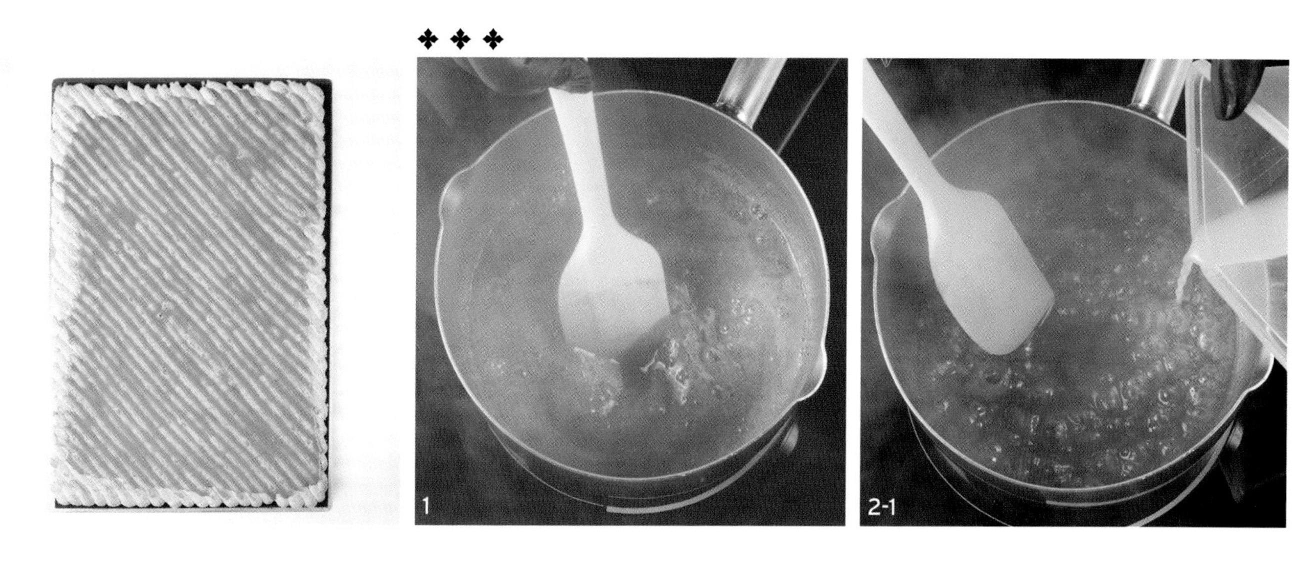

4

✤ ✤ ✤

1

2-1

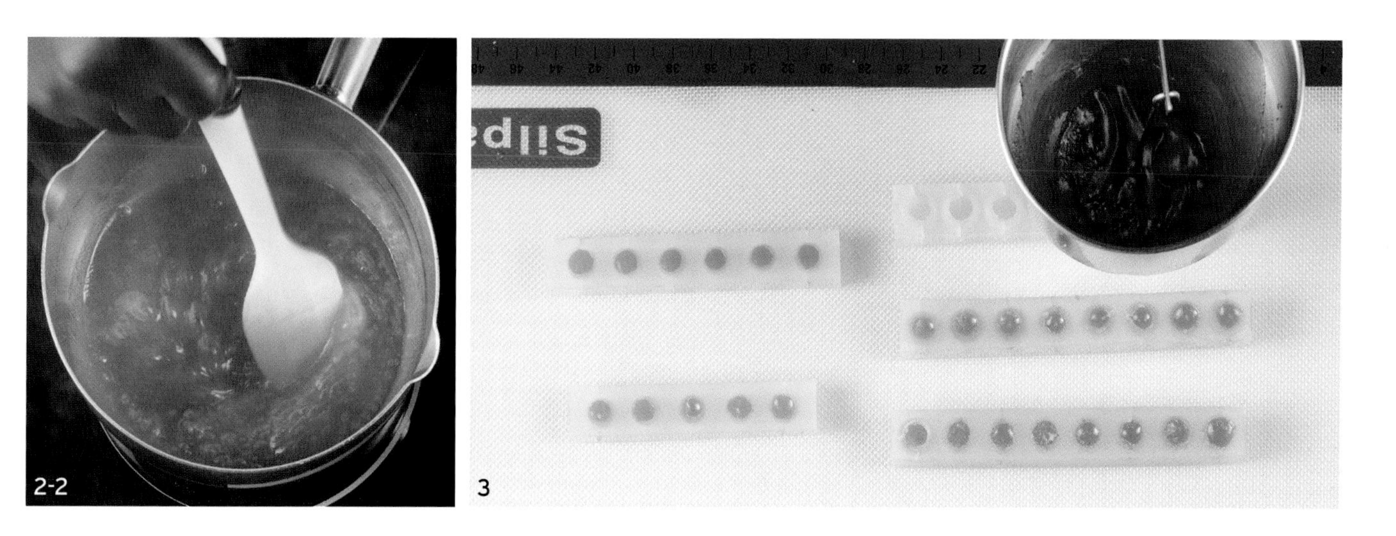

2-2

3

MANGO COMPOTE

망고 콩포트

망고 쿨리

1 냄비에 모든 재료를 넣고 가열한다.

2 펙틴 반응을 확인한 후 불에서 내린다.

3 얼음물이 담긴 볼에 받쳐 쿨링한 후 바트에 담아 밀착 랩핑해 냉장고에 보관한다.

망고 콩포트

4 볼에 만들어둔 망고 쿨리, 사방 1cm로 깍둑썰기한 망고, 라임제스트를 넣고 고르게 섞는다.

Mango Coulis

1 Heat all the ingredients in a pot.

2 Remove from the heat after checking the activation of the pectin.

3 Cool in an ice bath and pour into a stainless-steel tray, cover with plastic wrap, making sure the wrap is in contact with the coulis, and refrigerate.

Mango Compote

4 In a bowl, combine the prepared coulis, mango diced into 1 cm cubes, and lime zest.

3

✤ ✤ ✤ ✤ ✤

DOUBLE VANILLA EARL GREY MOUSSE

더블 바닐라 얼그레이 무스

1. 냄비에 생크림A, 전화당, 얼그레이 티백 잎, 바닐라빈 껍질을 넣고 가열한 후 입구를 랩핑해 30분간 인퓨징한다.
2. 불에서 내리고 바닐라빈 껍질을 제거한 후 젤라틴매스를 넣고 녹을 때까지 주걱으로 저어준다. (약 60℃)
3. 화이트초콜릿이 담긴 비커에 넣는다.
4. 바믹서로 블렌딩한 후 얼음물이 담긴 볼에 받쳐 25℃로 쿨링한다.
5. 손으로 가볍게 휘핑한(60~70%) 생크림B가 담긴 볼에 **4**를 두세 번 나눠 넣으며 섞는다.
6. 몰드에 16g씩 채운다.
7. 얼린 [다쿠아즈 + 얼그레이 화이트 가나슈]를 넣고 스패출러로 깔끔하게 정리해 급속 냉동(-35 ~ -40℃)한 후 몰드에서 빼내 냉동(-18℃) 보관한다.

1. Heat cream A, inverted sugar, tea leaves, and vanilla bean pod (without seeds). Cover and infuse for 30 minutes.
2. Remove from the heat, strain out the vanilla bean pod, and stir in the gelatin mass until it melts (about 60°C).
3. Pour into a beaker with white chocolate.
4. Combine with an immersion blender and cool to 25°C in an ice bath.
5. Lightly whip cream B (60~70%) in a bowl with a balloon whisk. Mix with (**4**) in two or three batches.
6. Fill 16 grams in the mold.
7. Insert frozen [dacquoise with Earl Grey ganache], clean the edges with an offset spatula, blast freeze (-35 ~ -40°C), remove from the mold, and keep them frozen (-18°C).

1

2-1

2-2

3

4

5-1

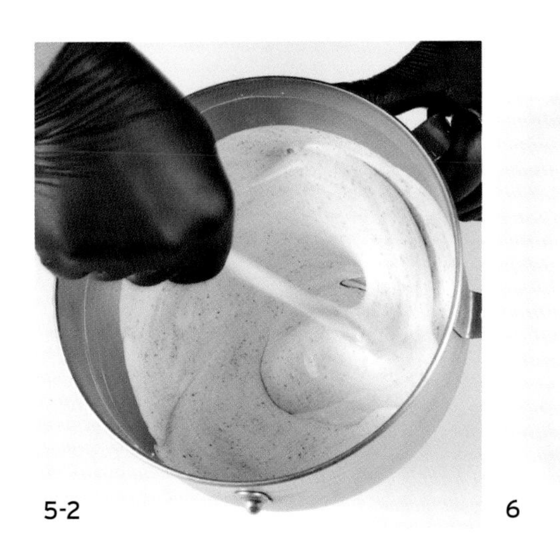

5-2

6

7

✤ ✤ ✤ ✤ ✤ ✤

LEMON CHOCOLATE POWDER

레몬초콜릿 파우더

1 중탕 볼 또는 전자레인지에서 레몬초콜릿의 온도를 50℃로 올려 녹인 후 35℃로 식히고 소금, 바닐라빈 씨를 넣고 섞는다.

2 말토덱스트린을 넣고 섞는다.

TIP. 말토덱스트린이 카카오버터의 지방을 흡수해 반죽이 단단해진다. 지방을 흡수하는 말토덱스트린의 성질을 이용해 카카오버터 외에도 다양한 맛의 초콜릿, 헤이즐넛 오일 등으로 여러 가지 파우더를 만들 수 있다.

3 반죽 상태가 되면 한 덩어리로 뭉친다.

4 넓적하게 펼친 후 랩으로 감싸 냉장고에서 굳힌다.

5 단단하게 굳으면 적당한 크기로 자르고 믹서로 곱게 갈아 냉동 보관하며 사용한다.

1 Melt the lemon chocolate in a double boiler or a microwave to 50°C.
Cool to 35°C and mix with salt and vanilla bean seeds.

2 Mix with maltodextrin.

TIP. Maltodextrin absorbs the fat from the cacao butter and makes your preparation firm. Its fat-absorbing properties can be used to make a variety of powders with different flavors of chocolate, hazelnut oil, and more, in addition to cacao butter.

3 Form the mixture into a ball.

4 Spread it wide and let it harden.

5 Once it hardens, cut into moderate sizes, finely grind, freeze, and use as needed.

1

2-1

2-2

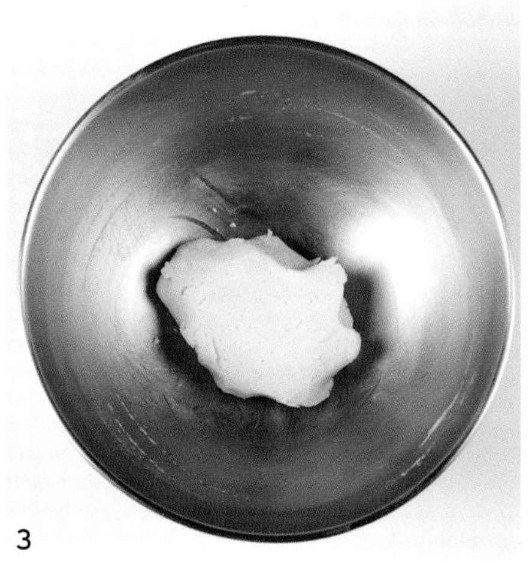

3

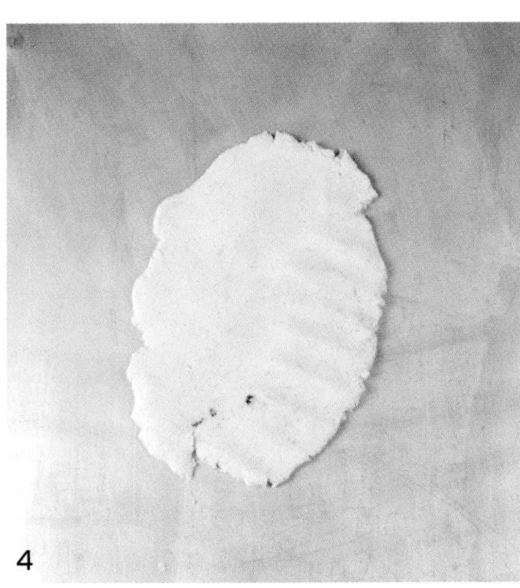

4

5-1

5-2

다양한 종류의 초콜릿을 이용해 파우더로 만들어 사용할 수 있다.

You can use different types of chocolates to make them into powders.

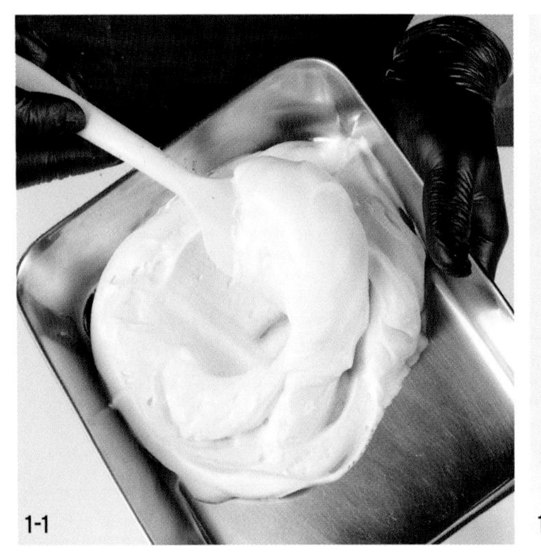
1-1

1-2

2-1

3

4

5

✤ ✤ ✤ ✤ ✤ ✤ ✤

FINISH 마무리

1 사워크림 몽테를 부드럽게 휘핑한 후 주걱으로 모양을 잡아 얼린 더블 바닐라 얼그레이 무스 위에 30g씩 올린다.

TIP. 얼린 더블 바닐라 얼그레이 무스는 화이트 바닐라 스프레이를 분사한 후 굳혀 사용한다.

2 뜨겁게 달군 아이스크림 스쿱으로 동그란 홈을 만든다.

3 망고 내추럴 글레이즈를 분사한 후 레몬초콜릿 파우더를 체 친다.

4 망고 콩포트를 소복하게 채운다.

5 라임제스트를 뿌린 후 망고 젤리 3개, 허브, 금박을 올려 마무리한다.

1 Gently whip sour cream montée, shape it with a spatula, and place 30 grams on the frozen dacquoise with mousse.

TIP. Spray the frozen double vanilla Earl Grey mousse with white vanilla spray and let harden to use.

2 Make a round indentation with a hot ice cream scoop.

3 Spray with mango natural glaze and dust with lemon chocolate powder.

4 Fill with a mound of mango compote.

5 Sprinkle with lime zest and top with 3 mango jellies, herbs, and gold leaves to finish.

12

SULHWA STRAWBERRY TART

설화 딸기 타르트

PAIRING & TEXTURE
페어링 & 텍스처

- 설화라는 품종의 딸기는 새콤하지만 시지 않고 일반 딸기에 비해 수분이 적어 과육이 단단한 편이므로 디저트의 재료로 매력적일 것이라 생각했다. 특히 단단한 과육과 새콤한 맛이 파인애플을 연상시켜 파인애플과의 페어링으로 타르트를 만들어보았다.

- One of the varieties of strawberries called Sulhwa is tart but not sour. It has firm flesh with less moisture than regular strawberries, making it attractive as an ingredient in desserts. The firm flesh and tartness reminded me of pineapple, so I decided to pair it with pineapple for this tart.

TECHNIQUE
테크닉

타르트 링에 반죽을 넣고 퐁사주하는 일반적인 방식에서 벗어나고자 고안해낸 테크닉. 실리콘 패드의 무늬를 반죽에 표현하고, 원형 에어매트 위에 올려 구워 개성 있는 무늬와 모양을 가진 타르트 셸로 만들었다.

This technique is designed to break away from the typical fonçage method of lining the dough into a tart ring. I embedded the pattern of a silicon pad on the dough and baked it in a round perforated silicon mat to create tart shells with unique patterns and shapes.

DESIGN
디자인

설화 딸기가 주재료인 만큼 제품에서도 최대한 돋보일 수 있도록 풍성하게 쌓았고, 설화 딸기 꽃을 올려 싱그러운 아름다움을 더했나.

As Sulhwa strawberries are the main ingredient, I stacked them abundantly to make the dessert stand out as much as possible. Also, Sulhwa flowers are arranged to add blossoming beauty.

✦ HOW TO COMPOSE THIS RECIPE ✦

STEP 1.

메뉴 정하기	**타르트**
Decide on the menu	Tarte

STEP 2.

메인 맛 정하기	**설화 딸기**
Choose the primary flavor	Sulhwa Strawberry

STEP 3.

메인 맛(설화 딸기)과의 페어링 선택하기

Select a pairing flavor (Sulhwa Strawberry)

- ☑ 아몬드 Almond
- ☑ 생크림 Heavy Cream
- ☑ 그릭요거트 Greek Yogurt
- ☑ 파인애플 Pineapple
- ☑ 레몬 Lemon
- ☑ 바닐라 Vanilla

STEP 4.

구성하기

Assemble

	Component	Flavor	Result
❶	Main Cream 메인 크림	Almond 아몬드	Crème d'Amande 크렘 다망드
❷	Sub Cream 서브 크림	Heavy Cream 생크림 Greek Yogurt 그릭 요거트	Cream Chantilly 크렘 샹티이
❸	Insert 인서트	Pineapple 파인애플 Strawberry (Sulhwa) 딸기(설화) Lemon 레몬 Vanilla 바닐라	Coulis 쿨리
❹	Crispy 크리스피	Almond 아몬드	Pâte Sablé 파트 사블레

크렘 샹티이
Crème Chantilly
플레인 사블레
Plain Sablé
크렘 다망드
Crème d'Amande
설화 딸기 쿨리
Sulhwa Strawberry
Coulis

INGREDIENTS

플레인 사블레

Plain Sablé

INGREDIENTS		g
버터 (Bridel)	Butter	240g
슈거파우더	Sugar powder	140g
소금	Salt	4g
달걀	Eggs	30g
노른자	Egg yolks	30g
바닐라에센스 (Aroma Piu)	Vanilla essence	10g
T55밀가루	T55 flour	400g
옥수수전분	Cornstarch	24g
아몬드파우더	Almond powder	120g
TOTAL		**998g**

*** 만드는 법**

❶ 볼에 차가운 상태의 버터(13~14℃), 슈거파우더, 소금을 넣고 비터로 믹싱해 크림화한다.
❷ 달걀과 노른자(30℃), 바닐라에센스를 나눠 넣어가며 믹싱한다.
❸ 체 친 [T55밀가루, 옥수수전분, 아몬드파우더]를 넣고 믹싱한다.
❹ 반죽을 한 덩어리로 만든다.
❺ 완성된 반죽은 랩으로 감싸 냉장고에서 휴지시킨 후 사용한다.

*** Procedure**

❶ Beat to cream the cold butter (13~14°C), sugar powder, and salt.
❷ Gradually mix in eggs, egg yolks (30°C), and vanilla essence while mixing.
❸ Mix in sifted [T55 flour, cornstarch, and almond powder].
❹ Mix until the dough comes together.
❺ Wrap the dough with plastic wrap and refrigerate before use.

크렘 다망드

Crème d'Amande

INGREDIENTS		g
통아몬드파우더	Ground whole almonds	62g
설탕	Sugar	125g
소금	Salt	1.5g
아몬드파우더 (시판)	Almond powder (Store-bought)	62g
박력분	Cake flour	12g
버터 (Bridel)	Butter	125g
달걀	Eggs	125g
골드럼 (PAN RUM)	Gold rum	12g
다진 아몬드	Chopped almonds	62g
TOTAL		**586.5g**

- 통아몬드파우더는 통아몬드를 직접 갈아 사용했다.
- Ground whole almonds are freshly ground before use.

설화 딸기 쿨리

Sulhwa Strawberry Coulis

INGREDIENTS		g
설화 딸기 (조이팜)	Sulhwa strawberries	360g
파인애플 퓌레	Pineapple purée	280g
레몬 퓌레	Lemon purée	75g
설탕	Sugar	280g
NH펙틴	NH pectin	35g
바닐라빈 껍질	Vanilla bean pod	1개 분량/ 1 pc
TOTAL		**1030g**

- 설화 딸기는 수확하는 시기에 따라 맛의 편차가 심하니 참고해 사용한다.
- Please note that the taste of Sulhwa strawberries varies greatly depending on when they're harvested.

크렘 샹티이

Crème Chantilly

INGREDIENTS		g
생크림	Heavy cream	300g
그릭요거트	Greek yogurt	120g
그래뉴당	Granulated sugar	25g
연유	Condensed milk	25g
TOTAL		**470g**

✤

PLAIN SABLÉ

플레인 사블레

1 냉장고에서 숙성한 반죽을 약 0.4cm로 밀어 편 후 무늬가 있는 실리콘 패드를 올리고 밀대로 밀어 반죽에 무늬를 내준다.

2 실리콘 패드를 제거하고 냉장고에서 30분 정도 휴지시킨다.

3 지름 13cm 원형 커터로 자른다.

4 에어매트(지름 13cm 실리코마트 Air Plus 12 - Round)의 모양에 맞춰 반죽을 정중앙에 올린다.

5 170℃로 예열된 오븐에서 15분간 굽는다.

TIP. 바닥이 될 부분이므로, 구워져 나오자마자 윗면을 평평하게 만들어준 후 식힌다.

1 Roll the refrigerated sablé into 0.4 cm, place a patterned silicon pad on top, and roll out with a rolling pin to imprint the pattern on the dough.

2 Remove the silicon pad and refrigerate for about 30 minutes.

3 Cut with a 13 cm round cutter.

4 Put the perforated silicon mold (Silikomart, Air Plus 12- Round, Ø 13 cm) upside down. Place the dough in the center on the back-side of the mold.

5 Bake for 15 minutes in an oven preheated to 170°C.

TIP. Since this part will be the bottom, flatten the top as soon as it's baked and let cool.

1

2

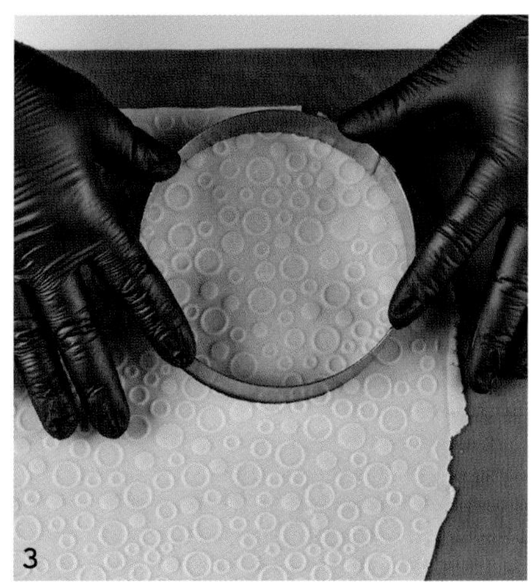
3

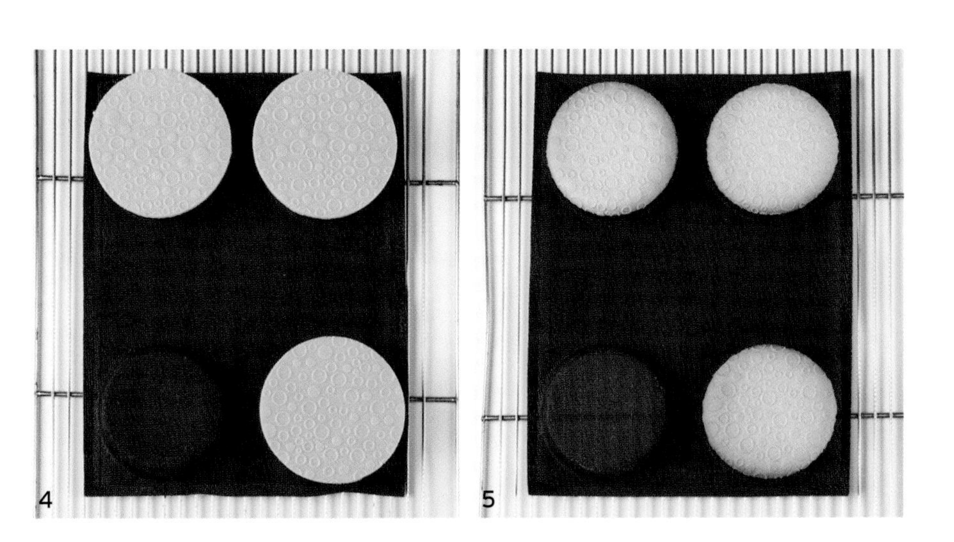
4
5

✤ ✤

CRÈME D'AMANDE

크렘 다망드

1 로보쿱에 로스팅한 통아몬드를 넣고 곱게 갈아 통아몬드파우더로 만든다.

2 설탕, 소금, 아몬드파우더, 박력분을 넣고 고르게 갈아준다.

3 포마드 상태의 버터를 넣고 갈아준다.

4 달걀(30℃)을 조금씩 나눠 넣어가며 갈아준다.

5 골드럼, 다진 아몬드를 넣고 섞는다.

6 완성된 크렘 다망드는 짤주머니에 담아 냉장 보관한다.

1 Finely grind roasted whole almond in the Robot Coupe to make the ground whole almond powder.

2 Grind with sugar, salt, almond powder, and cake flour.

3 Grind with softened butter.

4 Gradually add eggs (30°C) while grinding.

5 Continue to grind with gold rum and chopped almonds.

6 Put the finished cream in a piping bag and refrigerate.

1

2

3

4

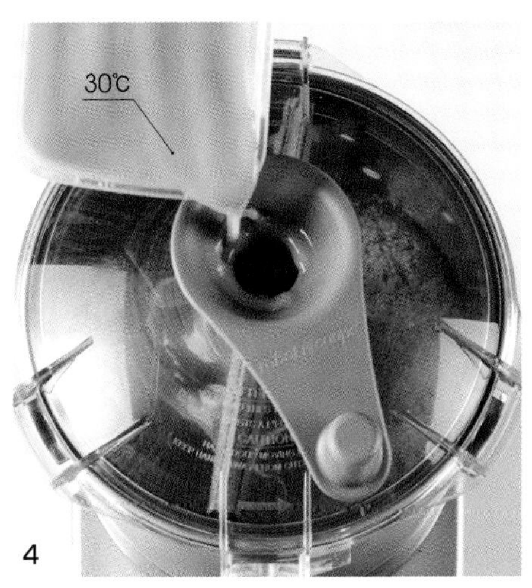

5

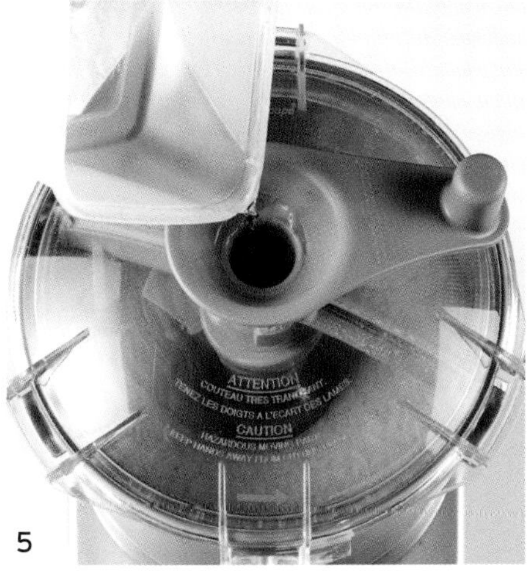

6

✤ ✤ ✤

SULHWA STRAWBERRY COULIS

설화 딸기 쿨리

1. 설화 딸기는 깨끗이 씻어 물기를 제거한 후 꼭지를 자르고 반으로 갈라 준비한다.
2. 비커에 파인애플 퓌레, 레몬 퓌레, 미리 섞어둔 설탕과 NH펙틴을 넣고 바믹서로 블렌딩한다.
3. 설화 딸기를 넣고 블렌딩한다.
4. 냄비에 옮겨 바닐라빈 껍질과 함께 가열한다.
5. 펙틴 반응을 확인하고 불에서 내려 쿨링한 후 바닐라빈 껍질을 제거하고 바트에 담아 밀착 랩핑해 냉장 보관한다.

1. Wash the strawberries thoroughly, drain, remove the stems, and cut them in half.
2. Blend pineapple purée, lemon purée, and previously mixed sugar and pectin using an immersion blender.
3. Blend with the strawberries.
4. Pour into a pot and cook with vanilla bean pod (without the seeds).
5. Remove from the heat after the pectin activates. Strain out the vanilla bean pod, pour onto a stainless-steel tray, adhere the plastic wrap to the preparation to cover, and refrigerate.

✤ ✤ ✤ ✤

CRÈME CHANTILLY

크렘 샹티이

비커에 모든 재료를 담고 바믹서로 블렌딩한 후 냉장고에서 숙성시킨다. 파이핑하기 쉬운 상태로 휘핑(80% 정도)해 냉장 온도로 차갑게 만들어 사용한다.

TIP. 비커는 미리 냉장고에 두어 차갑게 만들어두고, 생크림도 차가운 상태로 사용한다.

Blend all the ingredients in a beaker and let it rest in the refrigerator. Whip to a state that's easy to pipe (about 80%), keep it refrigerated and use cold.

TIP. Keep the beaker and the cream in the refrigerator in advance and use them cold.

✤ ✤ ✤

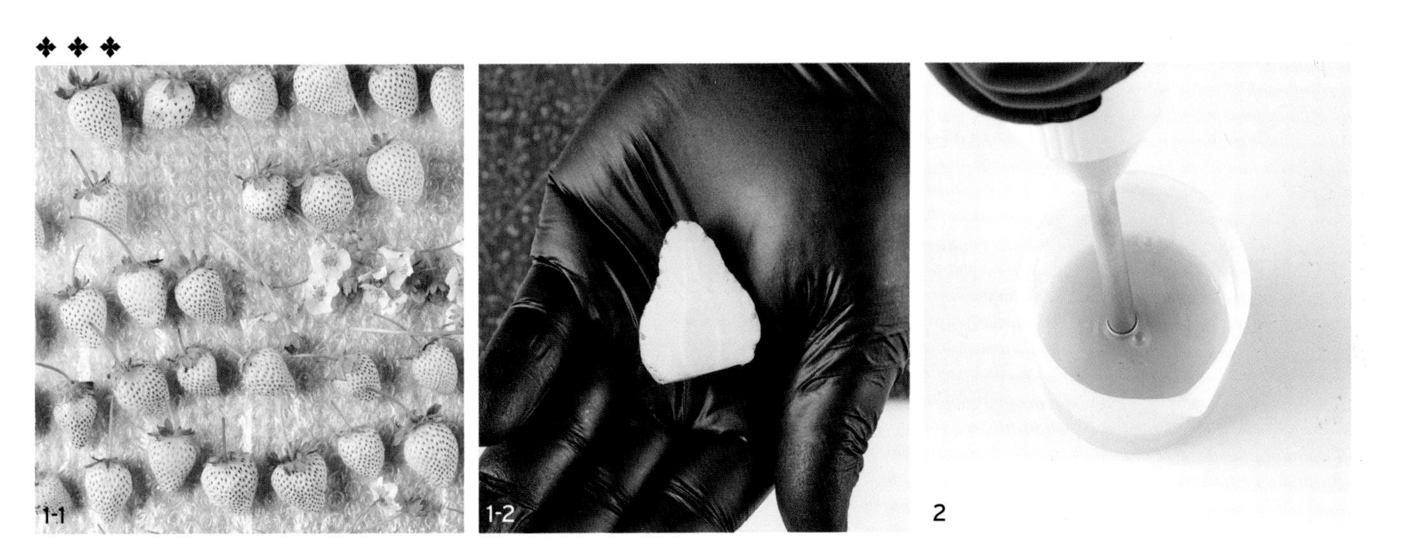
1-1 1-2 2

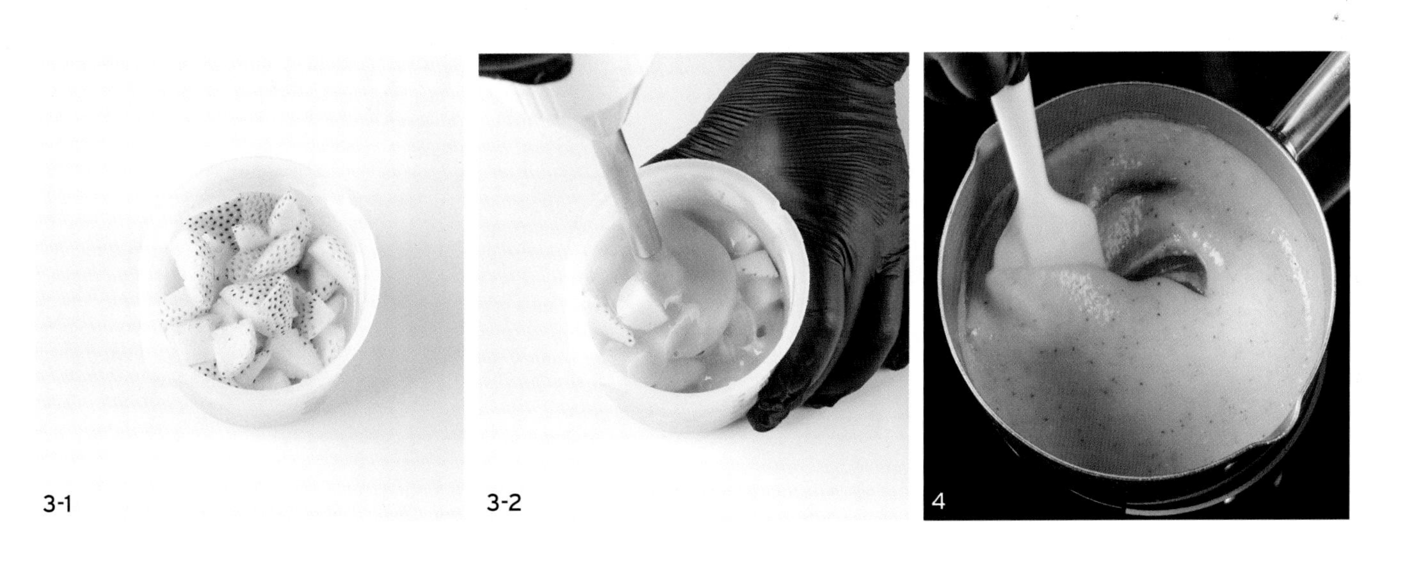
3-1 3-2 4

✤ ✤ ✤ ✤

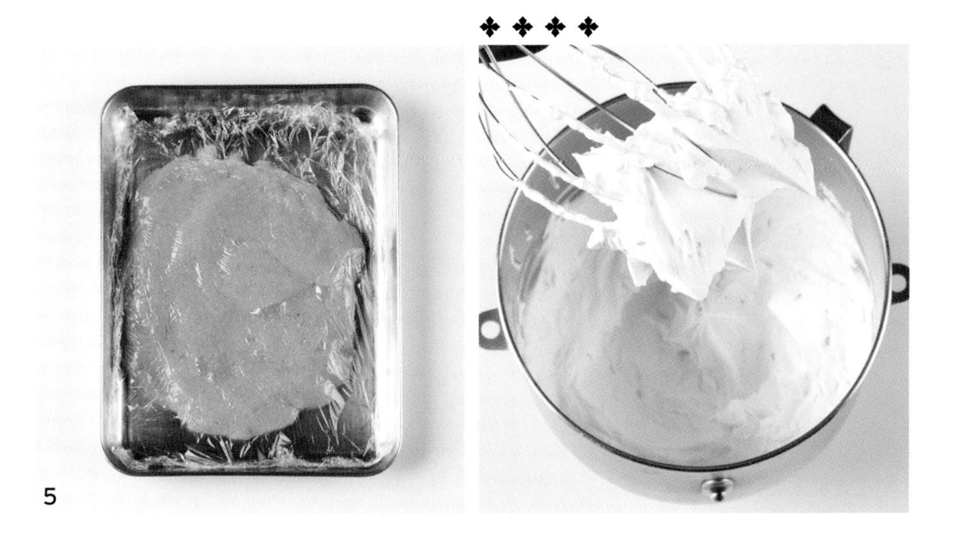
5

1

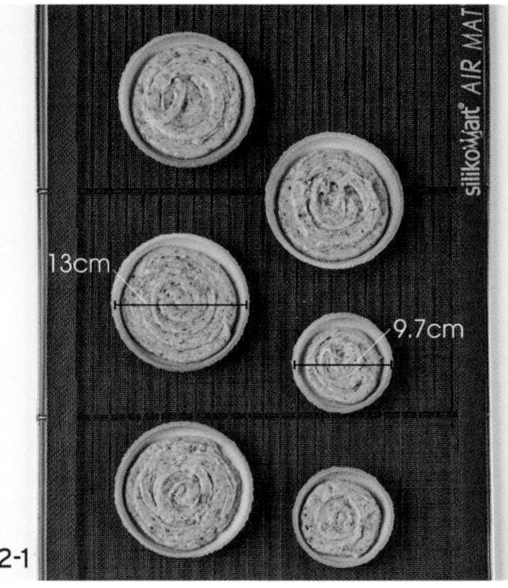

2-1

2-2

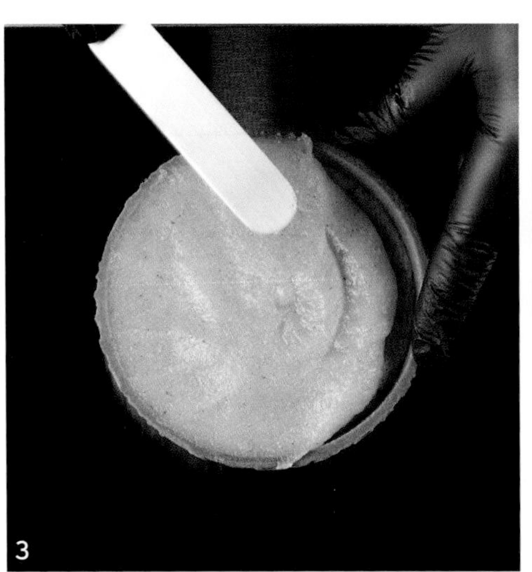
3

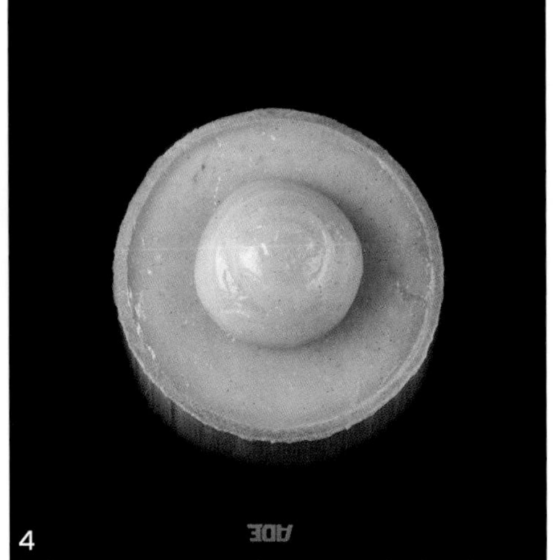
4

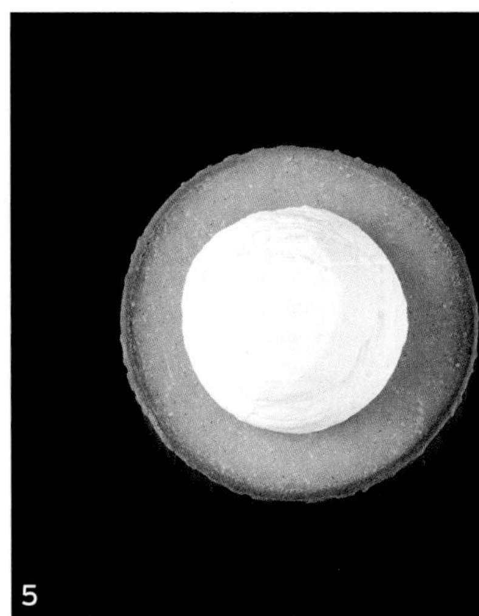
5

✤ ✤ ✤ ✤ ✤

FINISH 마무리

1. 플레인 사블레에 크렘 다망드 60g을 채우고 평평하게 정리한다.
   TIP. 플레인 사블레가 구워져 나오자마자 바로 작업한다.
2. 170℃로 예열된 오븐에서 11분간 굽고 실온에 두어 식힌다.
   TIP. 지름 9.7cm 에어매트를 사용해 미니 사이즈로 만들 수 있다.
3. 설화 딸기 쿨리 50g을 채운 후 평평하게 정리한다.
4. 정중앙에 다시 설화 딸기 쿨리 70g을 봉긋하게 파이핑한다.
5. 크렘 샹티이 50g으로 설화 딸기 쿨리를 감싸듯 봉긋하게 파이핑한다.
6. 설화 딸기를 쌓은 후 빈 공간에 설화 딸기 쿨리를 파이핑하고 데코젤 뉴트럴을 발라 윤기를 준다. 설화 딸기 꽃과 금박을 올려 마무리한다.

1 Fill with 60 grams of crème d'amande in plain sablé and spread to flatten.

TIP. Proceed as soon as the sablé comes out of the oven.

2 Bake in an oven preheated to 170°C for 11 minutes and cool at room temperature.

TIP. You can use a silicon air mat of 9.7 cm diameter to make them into mini size.

3 Fill with 50 grams of Sulhwa strawberry coulis and spread to flatten.

4 Pipe 70 grams of Sulhwa strawberry coulis in the center again in a ball.

5 Pipe 50 grams of creme chantilly around the coulis to cover.

6 Stack Sulhwa strawberries and pipe the coulis into the empty spaces. Glaze with Decorgel Neutral. Top with strawberry flowers and gold leaves to finish.

13

MY LOVE RIE

마이 러브 리에

PAIRING & TEXTURE
페어링 & 텍스처

한국 배는 수분이 많고 당도도 높아 그 자체로 충분히 맛있기 때문에 디저트의 재료로 많이 사용되지 않는 편이다. 여기에서는 한국 배를 부드러운 초콜릿 무스, 타임의 향, 유자의 향긋함과 잘 어우러지도록 페어링해보았다.

Korean pears are not often used as an ingredient in desserts because they are delicious on their own due to high moisture and sugar content. Here, I paired Korean pears with a soft chocolate mousse and the aroma of thyme and yuja.

TECHNIQUE
테크닉

얼리지 않은 신선한 인서트를 무스 위에 담아 과일 본연의 맛과 신선함을 최대한 느낄 수 있게 완성했다.

The fresh, unfrozen inserts are placed on top of the mousse to maximize the flavor and freshness of the fruit.

DESIGN
디자인

- 핫도그 번에서 영감을 얻어 볼륨감이 느껴지는 타원형 몰드를 제작해 앙증맞은 모양으로 완성했다.
- 두 가지 초콜릿 스프레이로 명암, 음영, 매트한 질감을 표현했다.
- 얇게 슬라이스하고 촘촘하게 말아 만든 배, 탱글탱글한 젤리를 올려 흔하지 않은 디자인으로 연출했다.
- 투명한 젤리, 윤기 나는 배, 반짝이는 금박이 어우러져 우아하면서도 고급스러운 느낌이 들도록 마무리했다.

- Inspired by hot dog buns, I created a voluminous oval mold and finished it in a cute shape.
- Two chocolate sprays were used to express light and dark shades and matte texture.
- I used thinly sliced and tightly rolled pears and topped with bouncy jellies for an unusual design.
- Transparent jellies, glossy pears, and glistening gold leaves combine to create an elegant and luxurious look.

✦ HOW TO COMPOSE THIS RECIPE ✦

STEP 1.

메뉴 정하기	프티 가토
Choose a menu	Petit Gateau

STEP 2.

메인 맛 정하기	배
Decide the main flavor	Pear

STEP 3.

메인 맛(배)과의 페어링 선택하기

Choose pairing with the main flavor (Pear)

- ☑ 꿀 Honey
- ☑ 유자 Yuja

- ☑ 타임 Thyme
- ☑ 레몬 Lemon

- ☑ 다크초콜릿 Dark Chocolate
- ☑ 밀크초콜릿 Milk Chocolate
- ☑ 캐러멜초콜릿 Caramel Chocolate

STEP 4.

구성하기

Configuring

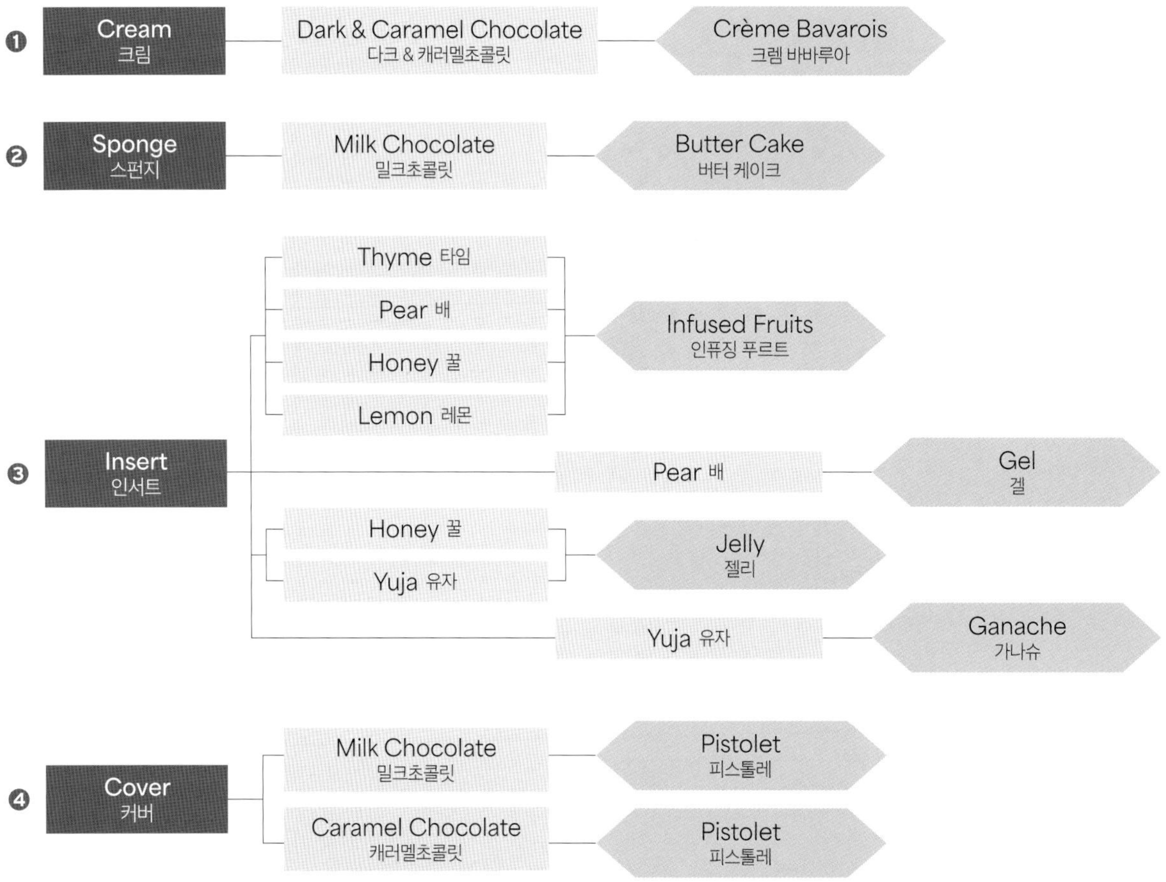

❶ Cream 크림 — Dark & Caramel Chocolate 다크 & 캐러멜초콜릿 — Crème Bavarois 크렘 바바루아

❷ Sponge 스펀지 — Milk Chocolate 밀크초콜릿 — Butter Cake 버터 케이크

❸ Insert 인서트
- Thyme 타임 / Pear 배 / Honey 꿀 / Lemon 레몬 — Infused Fruits 인퓨징 푸르트
- Pear 배 — Gel 겔
- Honey 꿀 / Yuja 유자 — Jelly 젤리
- Yuja 유자 — Ganache 가나슈

❹ Cover 커버
- Milk Chocolate 밀크초콜릿 — Pistolet 피스톨레
- Caramel Chocolate 캐러멜초콜릿 — Pistolet 피스톨레

유자 꿀 젤라틴 젤리
Yuja & Honey
Gelatin Jelly
꿀 타임 배
Honey Thyme Pear
배 겔
Pear Gel
밀크 & 제피르
캐러멜 스프레이
Milk & Zephyr
Caramel Spray
초콜릿 바바루아
Chocolate
Bavarois
유자 가나슈
Yuja Ganache
브라우니 스펀지
Brownie Sponge

INGREDIENTS 24개 분량/ 24 cakes

브라우니 스펀지
Brownie Sponge

(철판(42 × 32cm) 2개 분량)
(2 baking trays (42 × 32 cm))

INGREDIENTS		g
달걀	Eggs	160g
전화당	Inverted sugar	26g
설탕	Sugar	170g
다크초콜릿 (Fleur de cao 70%)	Dark chocolate	40g
밀크초콜릿 (Ghana 40%)	Milk chocolate	120g
버터 (Bridel)	Butter	180g
바닐라빈 씨	Vanilla bean seeds	1개 분량/ 1 pc
소금	Salt	3g
박력분	Cake flour	80g
아몬드파우더	Almond powder	40g
베이킹파우더	Baking powder	1g
TOTAL		**820g**

유자 가나슈
Yuja Ganache

INGREDIENTS		g
유자 퓌레	Yuja purée	90g
망고 퓌레	Mango purée	10g
전화당	Inverted sugar	25g
버터 (Bridel)	Butter	50g
바닐라빈 껍질	Vanilla bean pod	1/2개 분량/ 1/2 pc
화이트초콜릿 (Zephyr white 34%)	White chocolate	200g
TOTAL		**375g**

배 겔
Pear Gel

INGREDIENTS		g
배 퓌레	Pear purée	380g
레몬즙	Lemon juice	70g
트레할로스	Trehalose	50g
NH펙틴	NH pectin	10g
설탕	Sugar	100g
로거스트빈검	Locust bean gum	0.75g
젤라틴매스 (×5)	Gelatine mass (×5)	30g
TOTAL		**640.75g**

유자 꿀 젤라틴 젤리

Yuja & Honey Gelatin Jelly

INGREDIENTS		g
꿀	Honey	260g
물	Water	300g
유자 퓌레	Yuja purée	40g
젤라틴매스 (×5)	Gelatin mass (×5)	104g
TOTAL		**704g**

초콜릿 바바루아

Chocolate Bavarois

INGREDIENTS		g
크렘 앙글레이즈 (되직한 타입) (p.294)	Crème Anglaise (cooked thicker) (p.294)	250g
젤라틴매스 (×5)	Gelatin mass (×5)	45g
캐러멜초콜릿 (Zephyr Caramel 35%)	Caramel chocolate	100g
다크초콜릿 (Fleur de cao 70%)	Dark chocolate	150g
배 리큐르 (Dijon Poires William)	Pear liqueur	30g
생크림	Heavy cream	420g
TOTAL		**995g**

꿀 타임 배

Honey Thyme Pear

INGREDIENTS		g
꿀	Honey	140g
물	Water	140g
설탕	Sugar	140g
레몬즙	Lemon juice	35g
레몬제스트	Lemon zest	2개 분량/ 2 lemons
타임	Thyme	5g
배	Pear	3개/ 3 ea
TOTAL		**460g**

밀크 스프레이

Milk Spray

INGREDIENTS		g
밀크초콜릿 (Alunga 41%)	Milk chocolate	400g
카카오버터	Cacao butter	400g
TOTAL		**800g**

- 볼에 모든 재료를 넣고 녹여 바믹서로 블렌딩한 후 50℃로 맞춰 사용한다.
- Melt all the ingredients in a bowl, combine with an immersion blender, and use at 50°C.

제피르 캐러멜 스프레이

Zephyr Caramel Spray

INGREDIENTS		g
캐러멜초콜릿 (Zephyr Caramel 35%)	Caramel chocolate	400g
카카오버터	Cacao butter	400g
TOTAL		**800g**

- 볼에 모든 재료를 넣고 녹여 바믹서로 블렌딩한 후 50℃로 맞춰 사용한다.
- Melt all the ingredients in a bowl, combine with an immersion blender, and use at 50°C.

BROWNIE SPONGE

브라우니 스펀지

1. 믹싱볼에 달걀(30℃), 전화당, 설탕을 섞어 넣고 미색으로 뽀얗게 올라올 때까지 휘핑한다.
2. 녹인 다크초콜릿과 밀크초콜릿(45~50℃), 녹인 버터(45℃), 바닐라빈 씨, 소금을 넣고 섞는다.
3. 체 친 [박력분, 아몬드파우더, 베이킹파우더]를 넣고 섞는다.
4. 테프론시트를 깐 철판(42 × 32cm)에 고르게 팬닝한 후 바닥에 쳐 기포를 제거한다.
5. 170℃로 예열된 오븐에서 15분간 굽고, 식으면 냉동 보관한다.

1. Whip eggs (30°C), inverted sugar, and sugar in a mixing bowl until it aerates and turns pale.
2. Mix with melted dark and milk chocolates (45~50°C), melted butter (45°C), vanilla bean seeds, and salt.
3. Mix with sifted [cake flour, almond powder, and baking powder].
4. Spread evenly on a baking tray (42 × 32 cm) lined with a Teflon sheet and tap the tray on the table to remove air bubbles.
5. Bake in an oven preheated to 170°C for 15 minutes and freeze when cooled.

YUJA GANACHE

유자 가나슈

유자 가나슈

1. 냄비에 유자 퓌레, 망고 퓌레, 전화당, 버터, 바닐라빈 껍질을 넣고 주걱으로 저어가며 버터가 녹을 때까지 가열한다. (약 55~60℃)
2. 화이트초콜릿이 담긴 비커에 넣고 바믹서로 블렌딩해 유자 가나슈를 완성한다.

조립

3. 브라우니 스펀지에 25℃로 온도를 맞춘 유자 가나슈를 두께 약 0.2cm(약 230g)로 펴 바르고 냉장고에서 굳힌다.

   TIP. 테프론시트를 벗겨낸 면에 유자 가나슈를 바른다. 매끈한 면에 바르면 상대적으로 가나슈 흡수가 덜 되고, 스펀지 껍질 부분이 벗겨지면서 가나슈와 분리될 수 있기 때문이다.

4. 7 × 5.5cm 타원형 커터로 자른 후 냉동한다.

Yuja Ganache

1. Heat yuja purée, mango purée, inverted sugar, butter, and vanilla bean pod while stirring constantly with a spatula until the butter melts (about 55~60°C).
2. Pour the ganache into a beaker with white chocolate and combine with an immersion blender to finish.

Assemble

3. Spread yuja ganache at 25°C on the brownie sponge to about 0.2 cm (about 230 g) and set in a refrigerator.

   TIP. Spread the ganache on the side where the Teflon sheet has been peeled off. The smoother side will likely absorb the ganache, and the skin may peel off and separate.

4. Cut with a 7 × 5.5 cm oval cutter and freeze.

✤

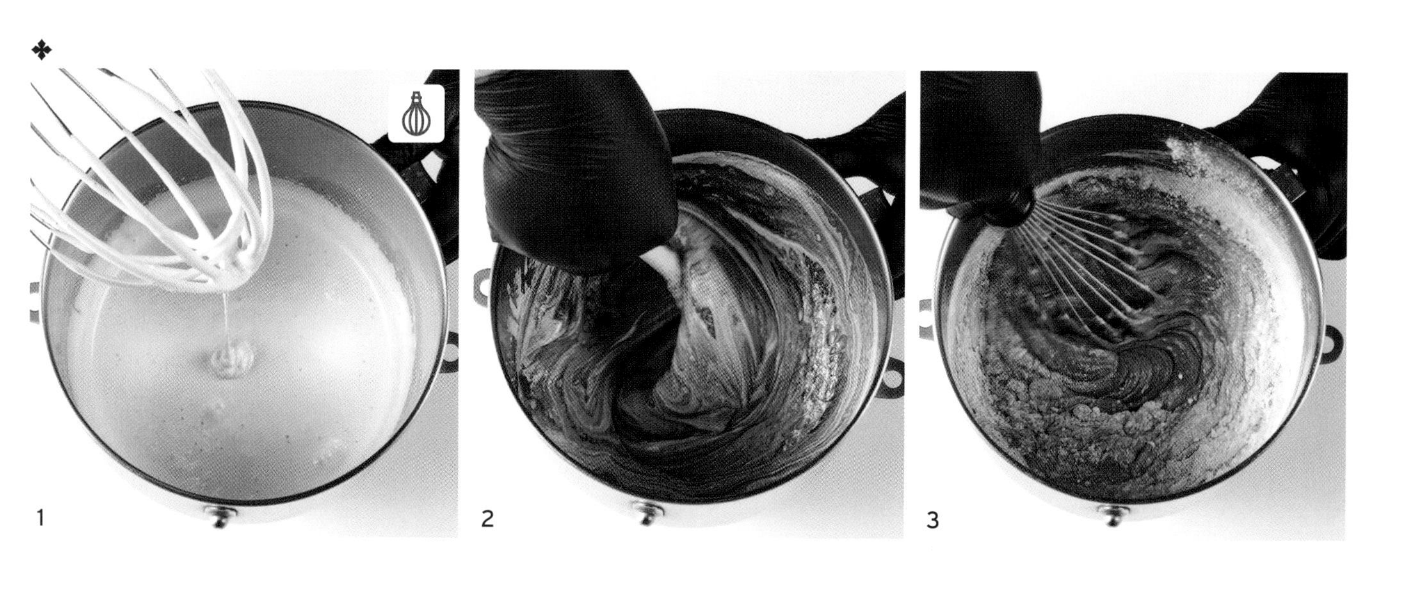
1 2 3

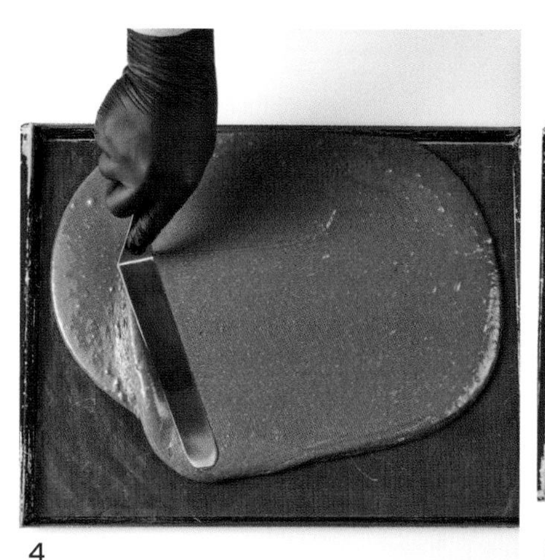
4

5

✤✤

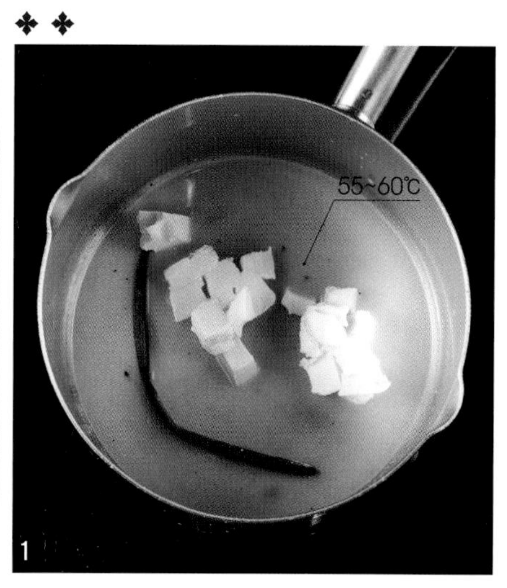

1

2

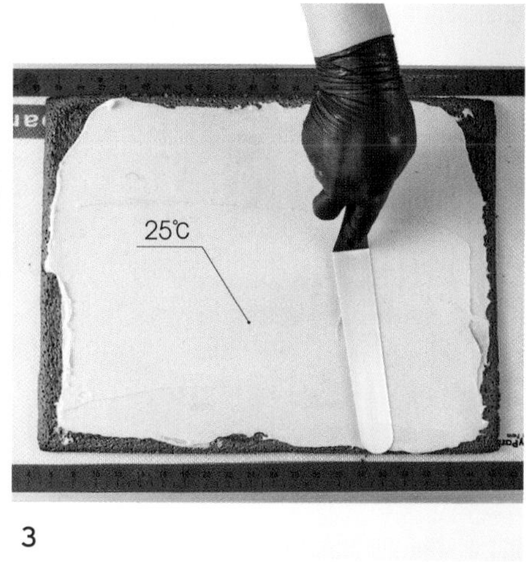

3

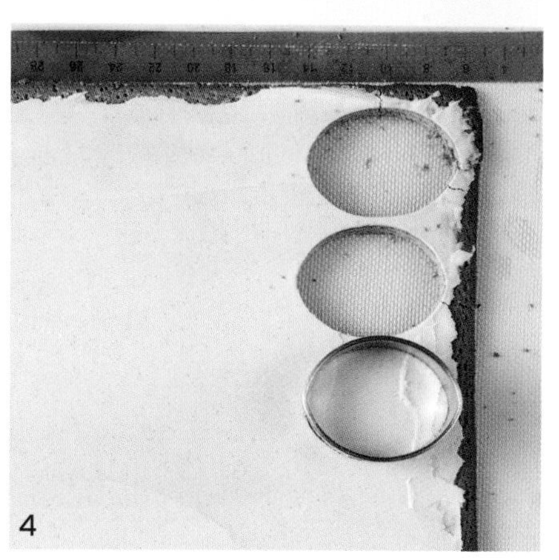
4

PEAR GEL

배 겔

1 냄비에 배 퓌레, 레몬즙을 넣고 30℃로 가열한다.

2 미리 섞어둔 [트레할로스, NH펙틴, 설탕, 로거스트빈검]을 넣고 알맞은 농도로 맞춰가며 가열한다.

3 끓어오르기 시작하고 펙틴 반응이 일어나면 젤라틴매스를 넣고 녹을 때까지 가열한다.

4 얼음물이 담긴 볼에 받쳐 10~12℃로 쿨링한 후 바트에 담고 밀착 랩핑해 냉장고에 보관한다.

1 Heat pear purée and lemon juice to 30°C.

2 Stir in previously mixed [trehalose, pectin NH, sugar, and locust bean gum].

3 Once it boils and pectin activates, add gelatin mass and continue to heat until it melts.

4 Cool in an ice bath to 10~12°C, pour onto a stainless-steel tray lined with vinyl, cover with plastic wrap, making sure the wrap is in contact with the gel, and refrigerate.

YUJA & HONEY GELATIN JELLY

유자 꿀 젤라틴 젤리

1 냄비에 모든 재료를 넣고 젤라틴매스가 녹을 때까지 가열한다. (약 60℃)

2 얼음물이 담긴 볼에 받쳐 주걱으로 저어가며 10~12℃로 쿨링한다.

3 비닐을 깐 바트에 담아 냉장 보관한다.

TIP. 기포가 보인다면 토치를 이용해 제거한다.
이 젤리는 깍둑썰어 장식하는 용도이므로 표면이 매끈하게 굳을 수 있도록 밀착 랩핑하지 않는다.

1 Heat all the ingredients until the gelatin melts. (about 60°C)

2 Cool in an ice bath to 10~12°C while stirring with a spatula.

3 Pour onto a stainless-steel tray lined with vinyl and refrigerate.

TIP. If you see air bubbles, remove them with a blow torch.
This jelly is intended to be cut into cubes and used for decoration. Don't let plastic wrap contact the jelly so the surface can be jellified smoothly.

✤ ✤ ✤

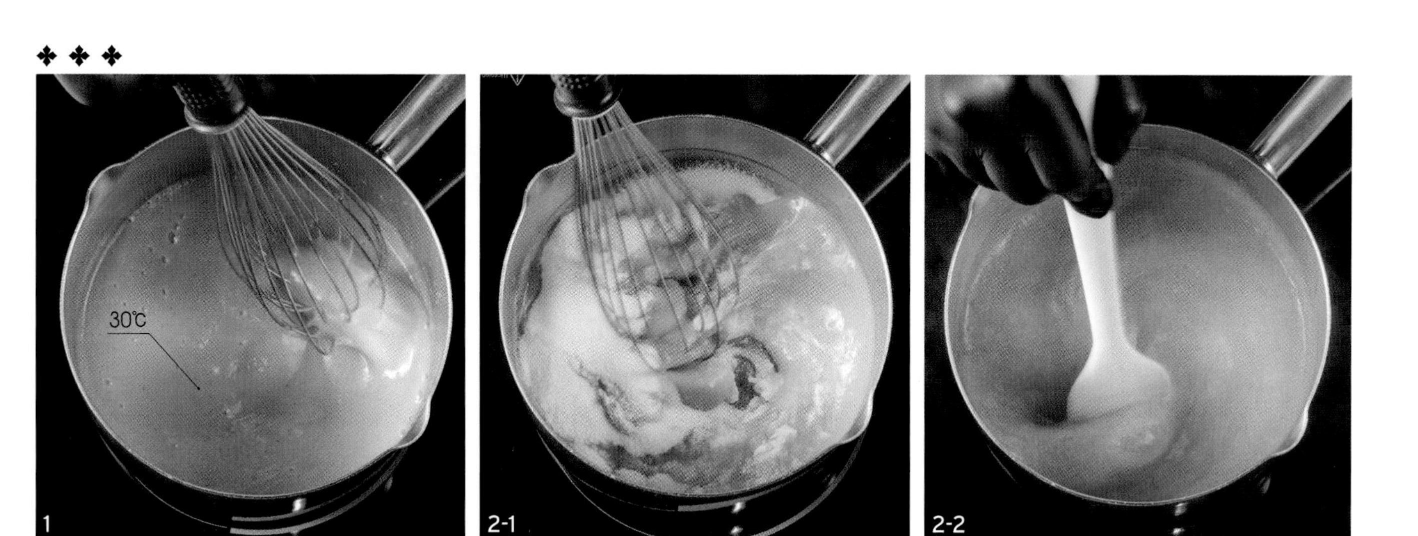

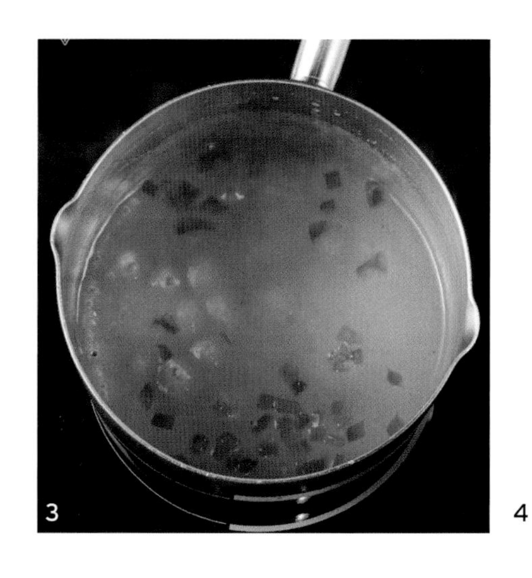

4

✤ ✤ ✤ ✤

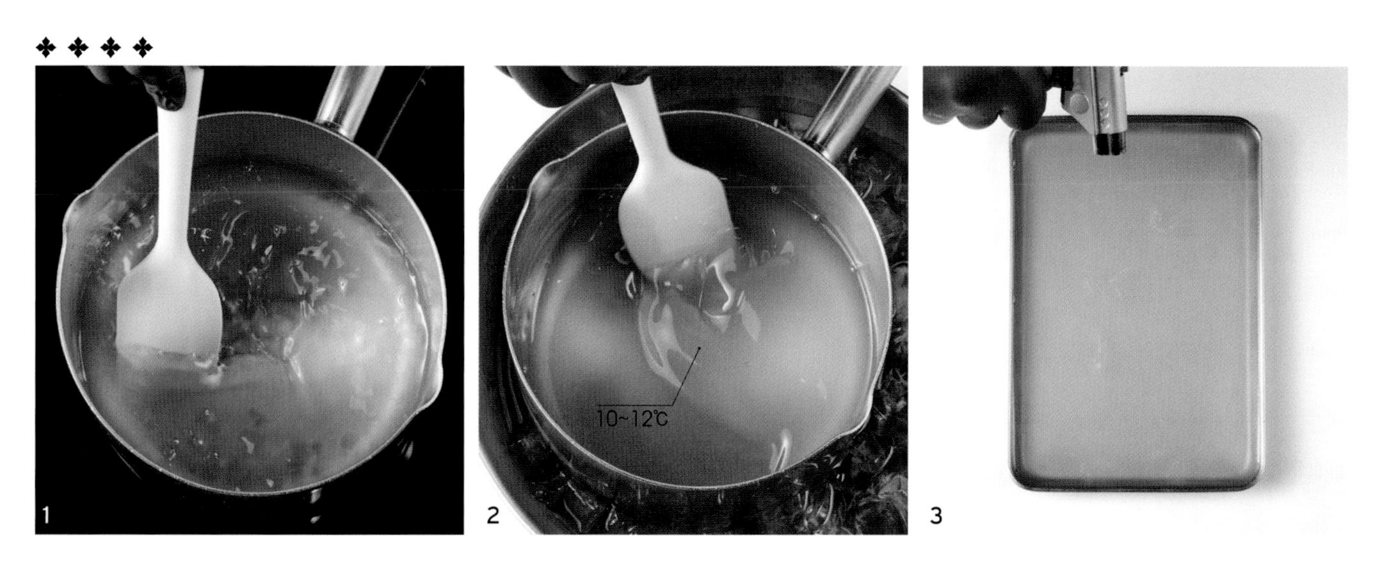

✤ ✤ ✤ ✤ ✤

CHOCOLATE BAVAROIS

초콜릿 바바루아

초콜릿 바바루아

1 냄비에 크렘 앙글레이즈, 젤라틴매스를 넣고 젤라틴매스가 녹을 때까지 주걱으로 저어가며 60~65℃까지 가열한다.

2 캐러멜초콜릿과 다크초콜릿이 담긴 비커에 담는다.

3 바믹서로 블렌딩한다.

4 배 리큐르를 넣고 블렌딩한다.

5 볼에 생크림을 넣고 60~70% 정도로 손으로 가볍게 휘핑한다.

6 **5**에 **4**를 두 번 나눠 넣어가며 고르게 섞는다.

조립

7 몰드에 32g씩 채운다.

8 얼린 [브라우니 스펀지 + 유자 가나슈]를 넣는다.

9 스패출러로 깔끔하게 정리한 후 급속냉동고(-35 ~ -40℃)에서 빠르게 얼린다.

Chocolate Mousse

1 Heat crème anglaise and gelatin mass, stirring constantly with a spatula, until the gelatin melts (about 60~65°C).

2 Pour into a beaker with dark and caramel chocolates.

3 Combine using an immersion blender.

4 Blend with pear liqueur.

5 In a separate bowl, whisk to lightly whip the cream to 60~70%.

6 Mix (**4**) evenly in two batches with (**5**).

Assemble

7 Fill 32 grams in the mold.

8 Embed the frozen insert [brownie sponge + yuja ganache].

9 Clean the edges with a spatula and freeze quickly in a blast freezer (-35 ~ -40°C).

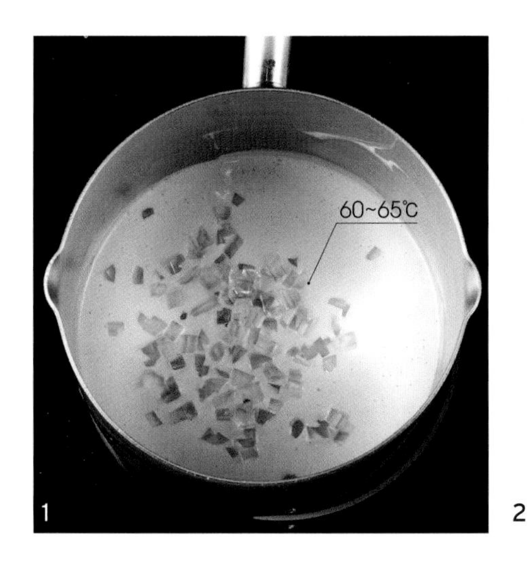

1

2

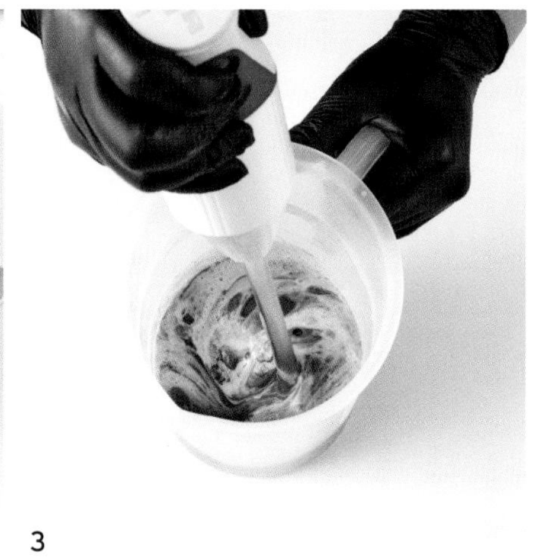

3

4

5

6

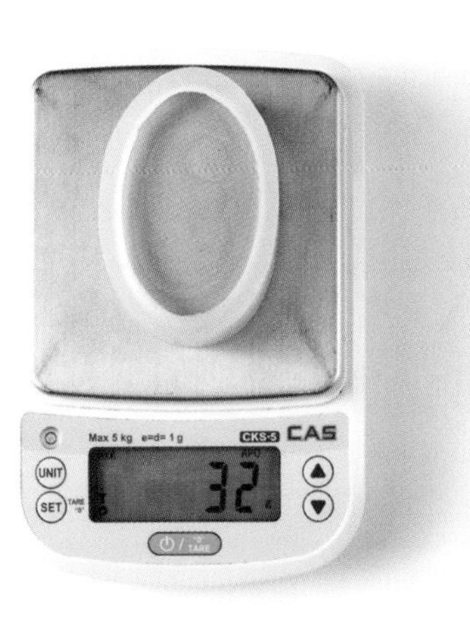

7

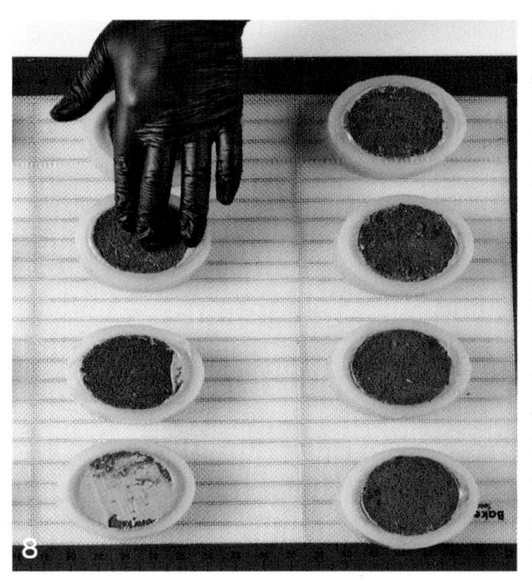

8

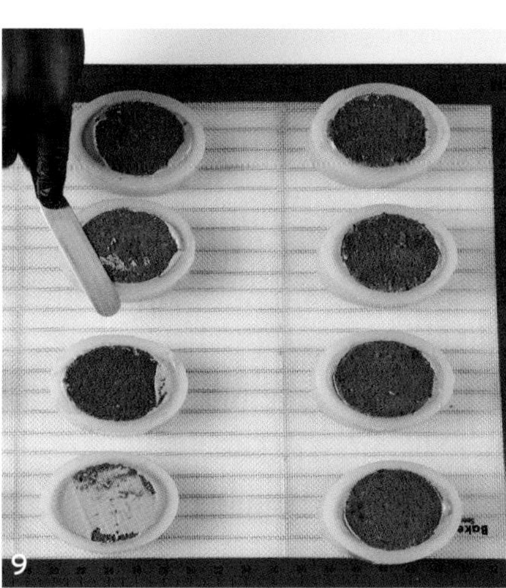

9

HONEY THYME PEAR

꿀 타임 배

1. 냄비에 물, 꿀, 설탕, 레몬즙, 레몬제스트를 넣고 가열한다.
2. 끓어오르면 타임을 넣고 불에서 내린다.
3. 얼음물이 담긴 볼에 받쳐 쿨링한다.
4. 과일 슬라이서를 이용해 배를 얇고 길게 썬다.
5. 진공백에 **3**과 **4**를 넣고 진공 상태로 만들어 냉장고에 보관한다.

   TIP. 물기를 제거한 후 돌돌 말아 랩에 감싸 냉장 보관해 모양을 고정시킨 후 1.5cm 길이로 잘라 사용한다.

1. Heat water, honey, sugar, lemon juice, and lemon zest.
2. Once it boils, add thyme and remove from heat.
3. Cool in an ice bath.
4. Slice the pears thin and long using a fruit peeler.
5. Vacuum seal (**3**) and (**4**) in a vacuum bag and refrigerate.

   TIP. Remove moisture, roll up, wrap in plastic, and refrigerate to set. Cut into 1.5 cm slices to use.

FINISH

마무리

1. 무스가 얼면 몰드에서 빼낸 후 냉동고(-18℃)에 보관한다.
2. 두 가지 스프레이를 이용해 분사하며 명암과 음영을 표현한다.
3. 패인 부분에 배 겔을 16g 채운 후 돌돌 말아 자른 꿀 타임 배 6개, 사방 1cm 크기로 자른 유자 꿀 젤라틴 젤리 7~8개를 올린다. 라임 제스트를 뿌리고 허브와 금박으로 장식해 마무리한다.

1. When the mousse is completely frozen, remove from the mold and store in a freezer at -18°C.
2. Spray with the two spray mixtures to create contrast and shading.
3. Fill the indented area with 16 grams of pear gel. Then top with 6 slices of rolled honey thyme pears and 7~8 pieces of yuja & honey gelatin jellies cut into 1 cm cubes. Decorate with herbs and gold leaves to finish.

✤ ✤ ✤ ✤ ✤ ✤

1

2

3

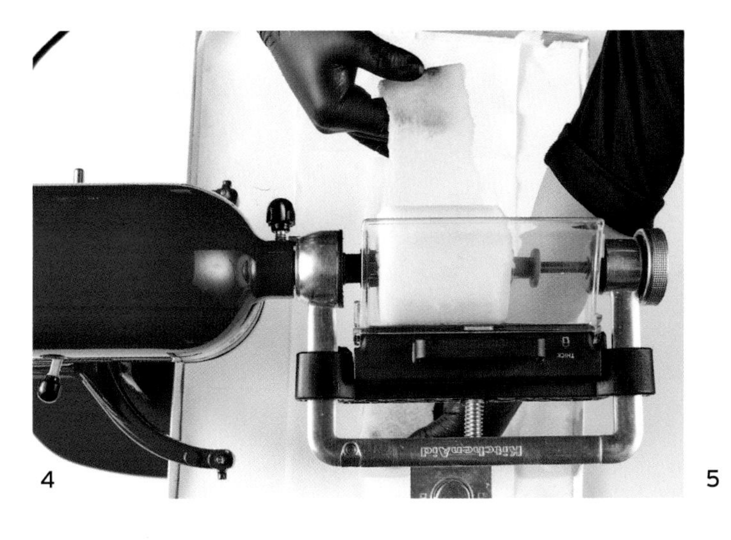
4

5

✤ ✤ ✤ ✤ ✤ ✤ ✤

1

2

3

14

CHERRY OATMEAL GATE

체리 오트밀 게이트

PAIRING & TEXTURE
페어링 & 텍스처

우유만으로는 충족시키기 어려운 곡류 특유의 풍부한 맛을 오트밀로 인퓨징해 추가했다.

Infusing oatmeal adds a rich taste of grains that is hard to achieve with milk alone.

TECHNIQUE
테크닉

영화 <인셉션>에 나오는 '클라우드 게이트' 조형물에서 영감을 받아 만든 디자인으로, 스테인리스 표면을 연출하고자 펙틴 미러 글레이즈를 사용해 글로시한 느낌을 강조했다.

Inspired by the 'Cloud Gate' sculpture from the movie Inception, I used pectin mirror glaze to create the stainless-steel-like surface, emphasizing the glossy look.

DESIGN
디자인

완성도 높은 디자인 자체가 곧 데커레이션이라고 생각해 다른 어떤 추가적인 장식 없이 아름다운 굴곡의 모양과 글레이즈만으로 마무리했다.

Thinking that the high-quality design itself is a decoration, I finished it with only beautiful curves and glaze without any additional decoration.

✦ HOW TO COMPOSE THIS RECIPE ✦

STEP 1.

메뉴 정하기	프티 가토
Decide on the menu	Petit Gateau

STEP 2.

메인 맛 정하기	체리
Choose the primary flavor	Cherry

STEP 3.

메인 맛(체리)과의 페어링 선택하기

Select a pairing flavor (Cherry)

- ☑ 화이트초콜릿 White Chocolate
- ☑ 오트밀 Oatmeal
- ☑ 아몬드 Almond

STEP 4.

구성하기

Assemble

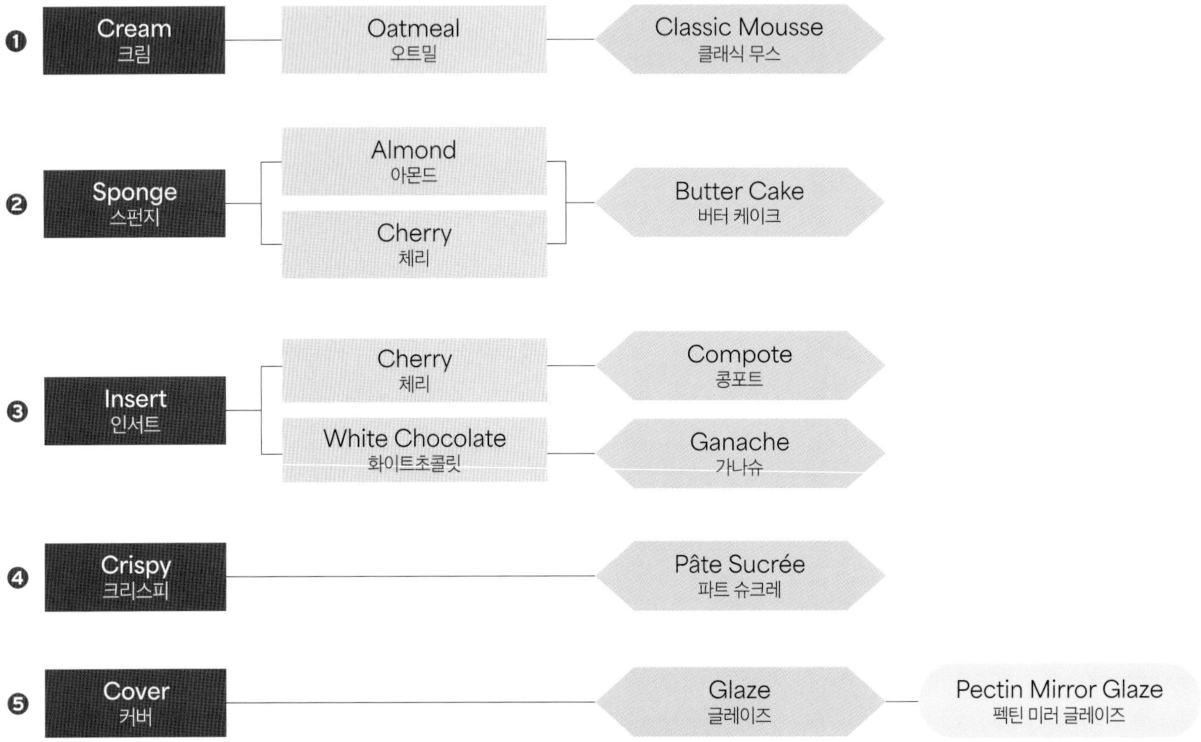

❶	Cream 크림	Oatmeal 오트밀	Classic Mousse 클래식 무스	
❷	Sponge 스펀지	Almond 아몬드 / Cherry 체리	Butter Cake 버터 케이크	
❸	Insert 인서트	Cherry 체리	Compote 콩포트	
		White Chocolate 화이트초콜릿	Ganache 가나슈	
❹	Crispy 크리스피		Pâte Sucrée 파트 슈크레	
❺	Cover 커버		Glaze 글레이즈	Pectin Mirror Glaze 펙틴 미러 글레이즈

체리 콩포트
Cherry Compote
오트밀 무스
Oatmeal Mousse
핑크 펙틴 미러 글레이즈
Pink Pectin
Mirror Glaze
체리 가나슈
Cherry Ganache
체리 아몬드 스펀지
Cherry Almond Sponge
파트 슈크레
Pâte Sucrée

INGREDIENTS 30개 분량/ 30 cakes

체리 아몬드 스펀지

Cherry Almond Sponge

(철판(42 × 32cm) 1개 분량)
(1 baking tray (42 × 32 cm))

INGREDIENTS		g
노른자	Egg yolks	80g
슈거파우더	Sugar powder	60g
소금	Salt	1g
버터 (Bridel)	Butter	80g
흰자	Egg whites	100g
설탕	Sugar	100g
알부민파우더	Albumin powder	2g
박력분	Cake flour	45g
아몬드파우더	Almond powder	25g
동결건조 체리 (dice, SOM)	Freeze-dried cherries	20g
TOTAL		**513g**

체리 가나슈

Cherry Ganache

INGREDIENTS		g
체리 퓌레 (Boiron)	Cherry purée	60g
레몬 퓌레 (Boiron)	Lemon purée	20g
버터 (Bridel)	Butter	50g
전화당	Inverted sugar	30g
설탕	Sugar	10g
바닐라빈 껍질	Vanilla bean pod	1/2개 분량/ 1/2 pc
화이트초콜릿 (Zephyr white 34%)	White chocolate	260g
TOTAL		**430g**

오트밀 베이스 *

Oatmeal Base

INGREDIENTS		g
아몬드밀크 (덴마크, 무가당)	Almond milk	250g
생크림	Heavy cream	300g
구운 오트밀	Rosted oatmeal	50g
TOTAL		**600g**

- 아몬드밀크는 브랜드마다 색이 다르므로 참고해 사용한다.
- Note that almond milk varies in color depending on the brand.

오트밀 무스

Oatmeal Mousse

INGREDIENTS		g
오트밀 베이스 *	Oatmeal base *	370g
전화당	Inverted sugar	50g
젤라틴매스 (×5)	Gelatin mass (×5)	46g
화이트초콜릿 (Zephyr white 34%)	White chocolate	120g
아몬드 리큐르 (Dijon Almond)	Almond liqueur	40g
생크림	Heavy cream	500g
TOTAL		**1126g**

체리 쿨리 *

Cherry Coulis

INGREDIENTS		g
체리 퓌레 (Boiron)	Cherry purée	350g
레몬 퓌레 (Boiron)	Lemon purée	100g
전화당	Inverted sugar	70g
설탕	Sugar	170g
NH펙틴 (SOSA)	NH pectin	13g
로거스트빈검	Locust bean gum	0.1g
레몬즙	Lemon juice	55g
레몬에센스 (Aroma Piu)	Lemon essence	0.5g
체리 리큐르 (Kirsh)	Cherry liqueur	18g
TOTAL		**776.6g**

체리 콩포트

Cherry Compote

INGREDIENTS		g
체리 쿨리 *	Cherry coulis *	600g
체리 (생과)	Fresh cherries	480g
TOTAL		**1080g**

파트 슈크레
Pâte Sucrée

INGREDIENTS		g
버터 (Bridel)	Butter	200g
설탕	Sugar	80g
슈거파우더	Sugar powder	80g
달걀	Eggs	80g
T55밀가루	T55 flour	400g
아몬드파우더	Almond powder	45g
바닐라빈 씨	Vanilla bean seeds	1/2개 분량/ 1/2 pc
TOTAL		**885g**

*** 만드는 법**

❶ 볼에 차가운 상태의 버터(13~14℃), 설탕, 슈거파우더를 넣고 비터로 믹싱해 크림화한다.
❷ 달걀(30℃)을 나눠 넣어가며 믹싱한다.
❸ 체 친 T55밀가루와 아몬드파우더, 바닐라빈 씨를 넣고 믹싱한다.
❹ 날가루가 보이지 않고 반죽이 한 덩어리가 될 때까지 치댄다.
❺ 완성된 반죽은 랩으로 감싸 냉장고에서 휴지시킨다.
❻ 휴지시킨 반죽을 2.5mm 두께로 밀어 편 후 5.5cm 원형 커터로 자른다.
❼ 180℃로 예열된 오븐에서 10분간 굽는다.

*** Procedure**

❶ Cream the cold cubed butter (13~14°C), sugar, and sugar powder with a paddle.
❷ Gradually add eggs (30°C) and mix.
❸ Mix with sifted T55 flour, almond powder, and vanilla bean seeds.
❹ Knead until the powders are not visible and the dough comes together.
❺ Cover the dough with plastic wrap and refrigerate.
❻ Roll out the refrigerated dough into 2.5 mm and cut with a 5.5 cm round cutter.
❼ Bake in an oven preheated to 180°C for 10 minutes.

체리 시럽
Cherry Syrup

INGREDIENTS		g
물	Water	500g
체리 퓌레	Cherry purée	150g
설탕	Sugar	250g
레몬	Lemon	1/4개/ 1/4 lemon
월계수잎	Bay leaf	1장/ 1 leaf
TOTAL		**900g**

*** 만드는 법**

❶ 냄비에 모든 재료를 넣고 설탕이 완전히 녹을 때까지 가열한다. (최종 당도 30 Brix)
❷ 얼음물이 담긴 볼에 받쳐 쿨링한 후 체에 걸러 사용한다.

*** Procedure**

❶ Heat all the ingredients in a pot until the sugar dissolves completely. (Final sweetness: 30 Brix)
❷ Cool in an ice bath and strain to use.

핑크 펙틴 미러 글레이즈

Pink Pectin Mirror Glaze

INGREDIENTS		g
물	Water	600g
물엿	Corn syrup	370g
설탕A	Sugar A	130g
NH펙틴	NH pectin	22g
구연산삼나트륨	Trisodium citrate	6g
설탕B	Sugar B	1100g
젤라틴매스 (×5)	Gelatin mass (×5)	250g
붉은색 색소	Red color	1g
TOTAL		**2479g**

＊만드는 법

❶ 냄비에 물, 물엿을 넣고 가열한다.
❷ 40℃가 되기 전에 미리 섞어둔 [설탕A, NH펙틴, 구연산삼나트륨]을 넣고 섞는다.
❸ 한 번 끓어오르면 설탕B를 두세 번 나눠 넣으면서 섞는다.
❹ 설탕이 모두 녹으면 다시 한번 끓어오를 때까지 가열한다.
❺ 불에서 내려 젤라틴매스를 넣고 주걱으로 저어가며 녹인다.
❻ 색소를 넣고 블렌딩한다.
❼ 바트에 붓고 밀착 랩핑해 냉장 보관한다.
❽ 사용하기 전날 실온에 꺼내두어 기포가 생기지 않게 녹인 후 블렌딩해 30~32℃로 맞춰 사용한다.

＊Procedure

❶ Heat water and corn syrup in a pot.
❷ Before the mixture reaches 40°C, with previously mixed [sugar A, pectin NH, and trisodium citrate].
❸ Once it boils, stir in sugar B in two or three batches.
❹ When the sugar is completely dissolved, heat until the mixture comes to a boil once more.
❺ Remove from heat, add gelatin mass, and stir until completely melted.
❻ Blend with food colorings.
❼ Pour onto a stainless-steel tray, cover with plastic wrap, making sure the wrap is in contact with the glaze, and refrigerate.
❽ Leave the glaze at room temperature the day before use to prevent air bubbles from forming; blend and use at 30~32°C.

✤

CHERRY ALMOND SPONGE

체리 아몬드 스펀지

1 볼에 노른자(30℃), 슈거파우더를 넣고 휘퍼로 섞는다.

2 소금, 녹인 버터(45℃)를 넣고 섞는다.

3 다른 볼에 흰자, 설탕, 알부민파우더를 넣고 휘핑해 조밀하고 단단한 상태의 머랭으로 완성한다.

TIP. 1번과 동시에 작업한다.

4 완성한 머랭 1/3과 체 친 박력분과 아몬드파우더를 넣고 섞는다.

5 남은 머랭에 **4**를 모두 섞는다.

6 동결건조 체리를 넣고 가볍게 섞는다.

7 테프론시트를 깐 철판(42 × 32cm)에 팬닝한다.

8 180℃로 예열된 오븐에서 약 8분간 굽는다. 구워져 나오자마자 체리 시럽을 바르고 식혀 냉동한다.

1 Combine egg yolks (30°C) and sugar powder in a bowl with a whisk.

2 Mix with salt and melted butter (45°C).

3 In a separate bowl, whip egg whites, sugar, and albumin powder to make a dense and stiff meringue.

TIP. Work at the same time as (1).

4 Combine 1/3 of the meringue with sifted cake flour and almond powder.

5 Mix the remaining meringue with (**4**).

6 Add the freeze-dried cherries and mix lightly.

7 Spread onto a pan (42 × 32 cm) line with a Teflon sheet.

8 Bake in an oven preheated to 180°C for about 8 minutes. As soon as it's out of the oven, brush with cherry syrup, let cool, and keep frozen.

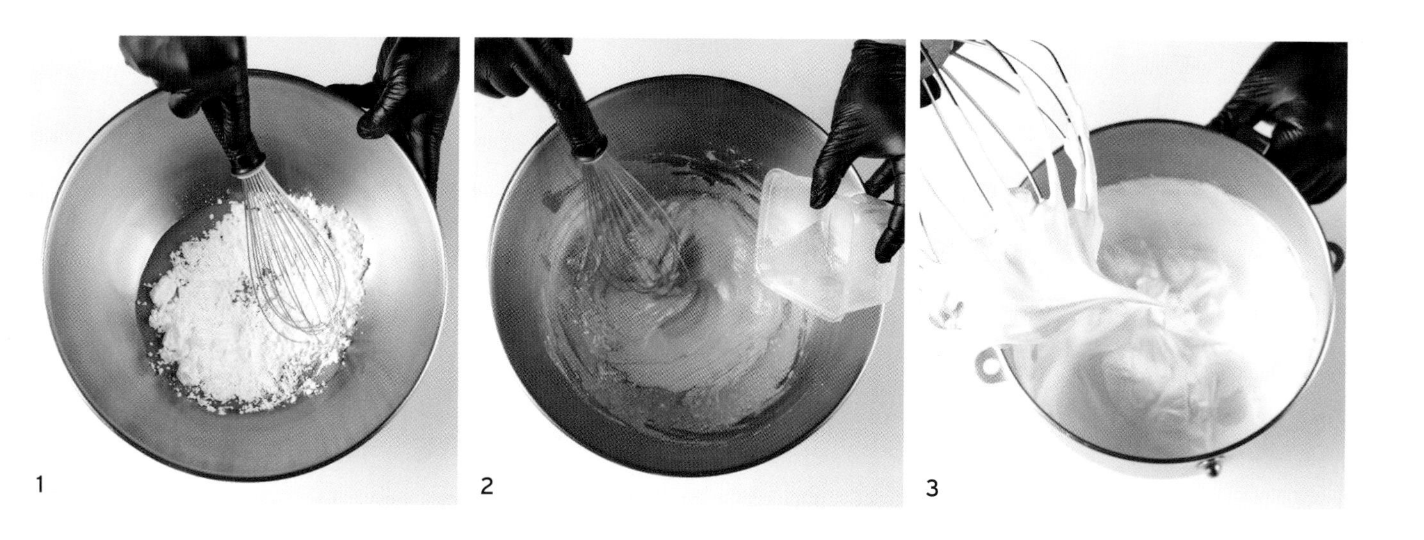

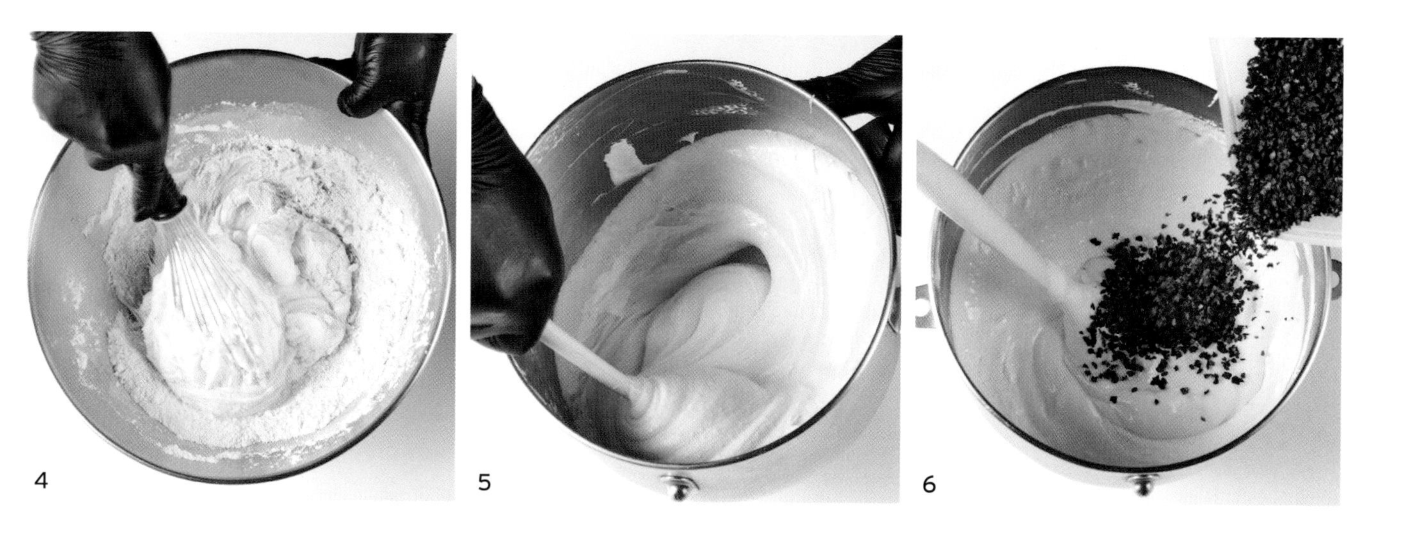

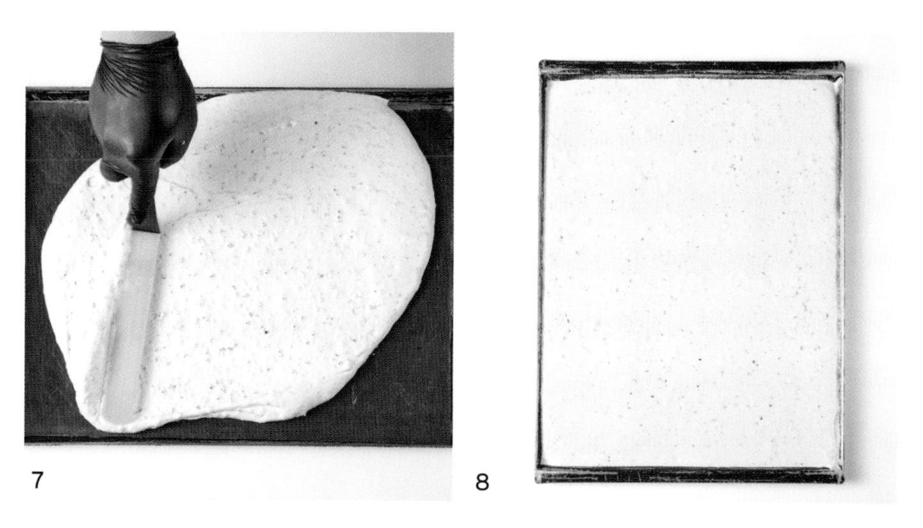

CHERRY GANACHE

체리 가나슈

체리 가나슈

1 냄비에 체리 퓌레, 레몬 퓌레, 버터, 전화당, 설탕, 바닐라빈 껍질을 넣고 65℃로 가열한다.

2 화이트초콜릿이 담긴 비커에 붓고 바믹서로 블렌딩한다.

3 얼음물이 담긴 볼에 받쳐 30~35℃로 쿨링해 사용한다.

조립

4 체리 아몬드 스펀지에 체리 가나슈 200g을 고르게 펴 바른다.

5 체리 가나슈가 완전히 굳으면 지름 5.5cm 원형 커터로 자른 후 냉동 보관한다.

Cherry Ganache

1 Heat cherry purée, lemon purée, butter, inverted sugar, sugar, and vanilla bean pod (without the seeds) in a pot to 65°C.

2 Pour into a beaker with white chocolate and combine with an immersion blender.

3 Cool in an ice bath to 30~35°C to use.

Assemble

4 Evenly spread 200 grams of cherry ganache on the cherry almond sponge.

5 When the cherry ganache is completely set, cut it with a 5.5 cm round cutter and store it frozen.

1

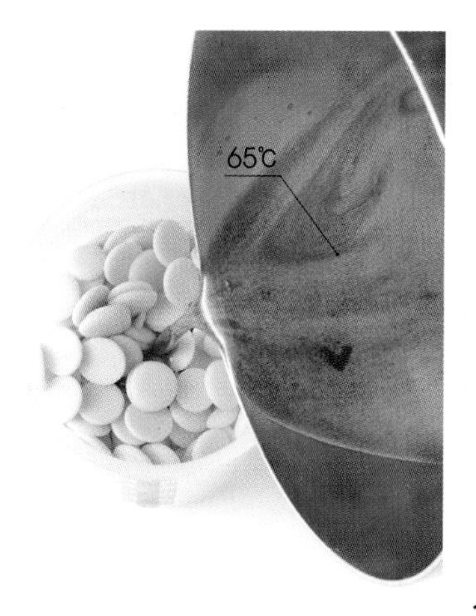

2-1

2-2

3

4

5

✤ ✤ ✤

OATMEAL MOUSSE

오트밀 무스

오트밀 베이스

1 냄비에 모든 재료를 넣고 가열한 후, 끓어오르면 불을 끄고 냄비 입구에 랩을 씌워 약 10분간 인퓨징한다.

TIP. 오트밀은 150℃에서 5분간 구워 사용한다.

2 바믹서로 블렌딩한다.

3 체에 거른다.

4 중량을 체크한 후 생크림(분량 외)을 추가해 370g으로 맞춰 준비한다.

오트밀 무스

5 냄비에 만들어둔 오트밀 베이스(185g), 전화당, 젤라틴매스를 넣고 젤라틴매스가 녹을 때까지 60~65℃로 가열한다.

6 화이트초콜릿이 담긴 비커에 붓고 바믹서로 블렌딩한다.

7 아몬드 리큐르를 넣고 블렌딩한 후 쿨링해 23~25℃로 맞춘다.

8 손으로 가볍게 휘핑(60~70% 정도)한 생크림에 넣고 섞는다.

9 얼음물이 담긴 볼에 받쳐 15℃로 쿨링한 후 몰드에 30~32g씩 채운다.

Oatmeal Base

1 Heat all the ingredients in a pot. Once it boils, remove from heat, cover the opening of the pot with plastic wrap, and infuse for about 10 minutes.

TIP. Roast the oatmeal at 150°C for 5 minutes to use.

2 Combine using an immersion blender.

3 Strain over a sieve.

4 Check the weight and add more cream (other than requested) to bring it to 370 grams.

Oatmeal Mousse

5 Heat the prepared oatmeal base (185 g), inverted sugar, and gelatin mass in a pot to 60~65°C until the gelatin melts.

6 Pour into a beaker with white chocolate and combine using an immersion blender.

7 Blend with almond liqueur and cool to 23~25°C.

8 Mix with lightly whipped cream (60~70%).

9 Cool in an ice bath to 15°C and fill 30~32 grams in the mold.

1 2 3

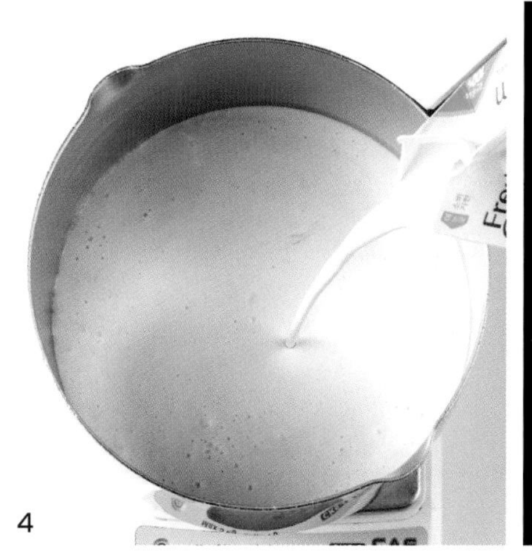

4

5

6

7

8

9

✤ ✤ ✤ ✤

CHERRY COMPOTE

체리 콩포트

체리 쿨리

1 냄비에 체리 리큐르를 제외한 모든 재료를 넣고 휘퍼로 저어가며 가열한다.

2 펙틴의 반응을 확인하고 불을 끈 후 체리 리큐르를 넣고 섞는다.

3 쿨링한 후 바트에 담아 밀착 랩핑해 냉장 보관한다.

체리 콩포트

4 만들어둔 체리 쿨리 260g을 바믹서로 부드럽게 풀어준다.

5 16등분한 체리 220g을 넣고 섞는다.

조립

6 몰드에 체리 콩포트를 35g씩 채운다.

7 얼린 [체리 아몬드 스펀지 + 체리 가나슈]를 올린 후 급속 냉동(-35 ~ -40℃)한다.

8 몰드에서 빼낸 후 오트밀 무스가 담긴 몰드에 넣는다.

9 스패출러로 깔끔하게 정리해 급속 냉동(-35 ~ -40℃)한 후 몰드에서 빼내 냉동(-18℃) 보관한다.

Cherry Coulis

1 Heat all the ingredients, except cherry brandy, in a pot while stirring with a whisk.

2 When the pectin activates, remove from heat and mix with cherry brandy.

3 Let cool, pour onto a stainless-steel tray, and cover with plastic wrap to cover, and refrigerate.

Cherry Compote

4 Gently loosen 260 grams of the prepared cherry coulis with an immersion blender.

5 Mix with 220 grams of fresh cherries cut into 16 pieces.

Assemble

6 Fill 35 grams of cherry compote in the mold.

7 Top with frozen [cherry almond sponge + cherry ganache] and blast freeze (-35 ~ -40°C).

8 Remove from the mold and insert in the mold with oatmeal mousse.

9 Clean up with a spatula, blast freeze (-35 ~ -40°C), remove from the mold and keep them frozen (-18°C).

1

2

3

4

5

6

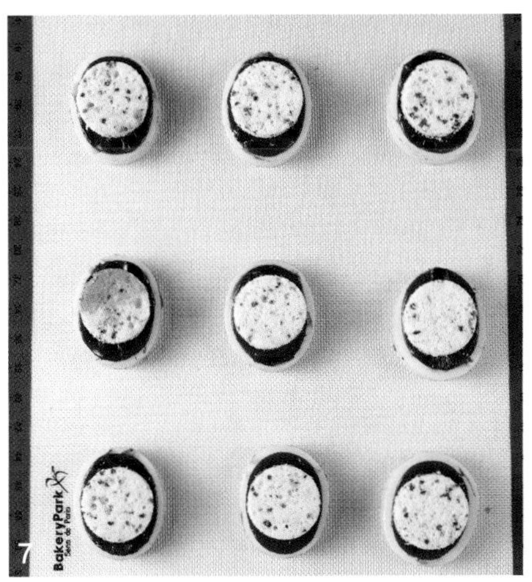

7

8

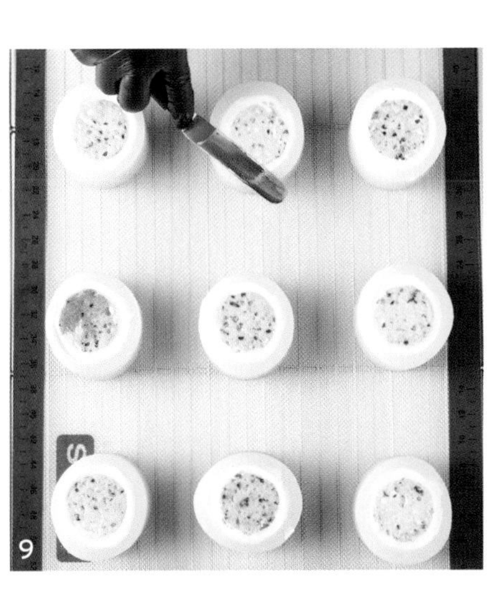

9

1-1

1-2

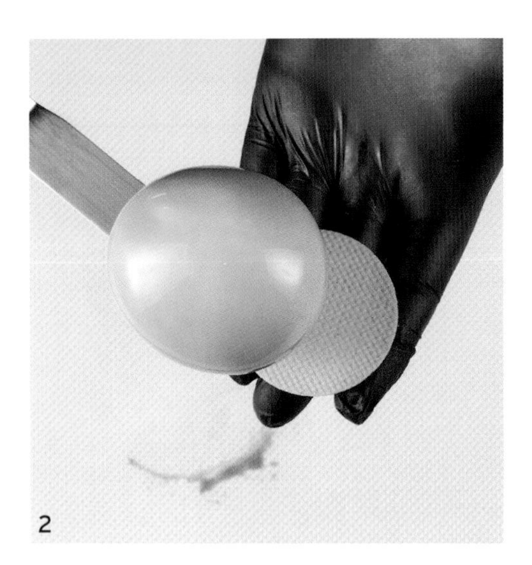
2

✤ ✤ ✤ ✤ ✤

FINISH 마무리

1 냉동한 무스를 몰드에서 빼낸 후 핑크 펙틴 미러 글레이즈로 코팅한다.

2 파트 슈크레에 올린 후 금 펄을 뿌려 마무리한다.

1 Remove the frozen mousse from the mold and coat it with pink pectin mirror glaze.

2 Place it on pâte sucrée and decorate with gold pearls to finish.

How to use the pre-made pink pectin mirror glaze

미리 만들어둔 핑크 펙틴 미러 글레이즈 사용하는 방법

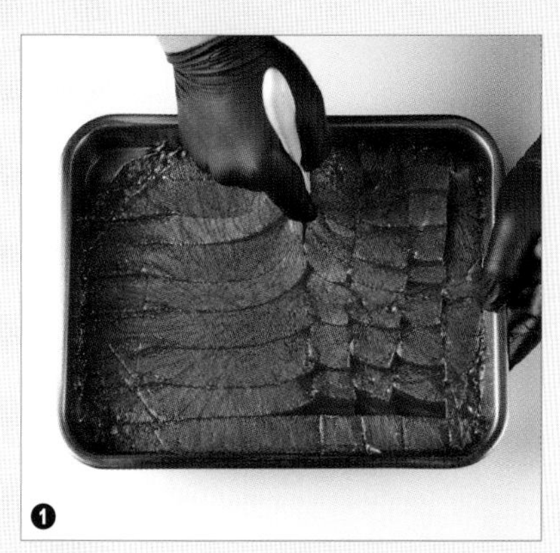

❶

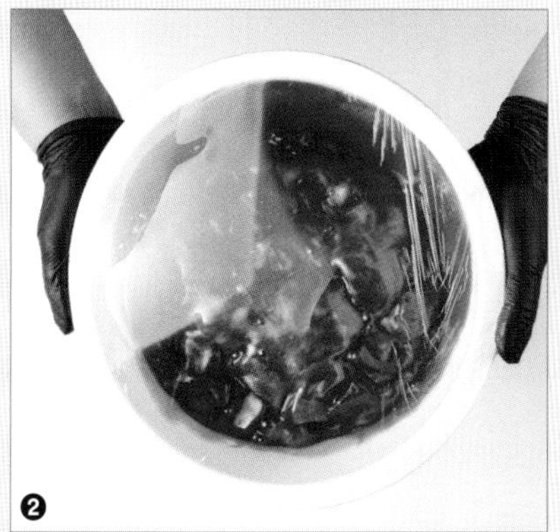

❷

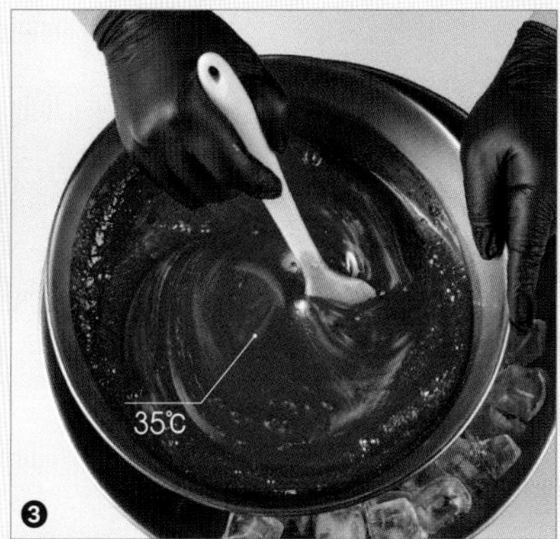

❸

❹

❶ 굳은 글레이즈를 주걱으로 갈라 조각으로 만들어준다.

❷ 볼에 담고 공기가 들어가지 않게 입구를 완전히 랩핑한 후 전자레인지에서 50℃까지 녹여준다.

❸ 얼음물이 담긴 볼에 받쳐 35℃이하로 쿨링한다.

❹ 바믹서로 블렌딩하여 30~32℃ 온도로 맞춰 사용한다.

- 표면에 기포가 보인다면 토치를 이용해 기포를 없애준다.

❶ Break the hardened glaze into pieces with a spatula.

❷ Put it in a bowl, wrap the opening completely to prevent air from entering, and then microwave to melt to 50°C.

❸ Cool in an ice bath to below 35°C.

❹ Blend with an immersion blender and use at 30~32°C.

- If you see air bubbles on the surface, use a blow torch to remove them.

15

UNIQUE CUBE PUMPKIN

유니크 큐브 펌킨

PAIRING & TEXTURE

페어링 & 텍스처

무스에서 흔히 사용하지 않는 재료인 '호박'을 주재료로 건무화과와 다크초콜릿을 이용해 페어링한 실험적인 제품이다.

This is an experimental product that uses danhobak*, an ingredient not commonly used in mousse, as the main ingredient, paired with dried figs and dark chocolate.

* Danhobak: It's a variety of winter squash, also known as 'kabocha' or 'Japanese pumpkin.'

TECHNIQUE
테크닉

- 2가지 피스톨레 기법(분사, 점사)을 사용해 질감의 차이가 느껴지도록 완성했다.
- 피스톨레 작업 후 카카오파우더를 뿌리고 부분적으로 열기를 주어 하나의 제품 안에서 다양한 질감이 보여지는 재미를 느낄 수 있게 연출했다.

- I used two different spraying techniques (fine and coarse) to achieve the difference in texture.
- After spraying, I sprinkled cacao powder and partially heated it to create a fun effect of showing various textures in one product.

DESIGN
디자인

자칫 평범해보일 수 있는 정사각형 디자인에 약간의 변화를 주어 심플하면서도 유니크한 모양으로 완성했다.

It's a simple yet unique design with a twist to the otherwise ordinary square shape.

✦ HOW TO COMPOSE THIS RECIPE ✦

STEP 1.	메뉴 정하기	프티 가토
	Decide on the menu	Petit Gateau

STEP 2.	메인 맛 정하기	호박
	Choose the primary flavor	Pumpkin

STEP 3. **메인 맛(호박)과의 페어링 선택하기**

Select a pairing flavor (Pumpkin)

- ☑ 다크초콜릿 Dark Chocolate
- ☑ 밀크초콜릿 Milk Chocolate
- ☑ 캐러멜초콜릿 Caramel Chocolate
- ☑ 피칸 Pecan
- ☑ 건크랜베리 Dried Cranberry
- ☑ 시나몬 Cinnamon
- ☑ 건무화과 Dried Figs
- ☑ 아몬드 Almond

STEP 4. **구성하기**

Assemble

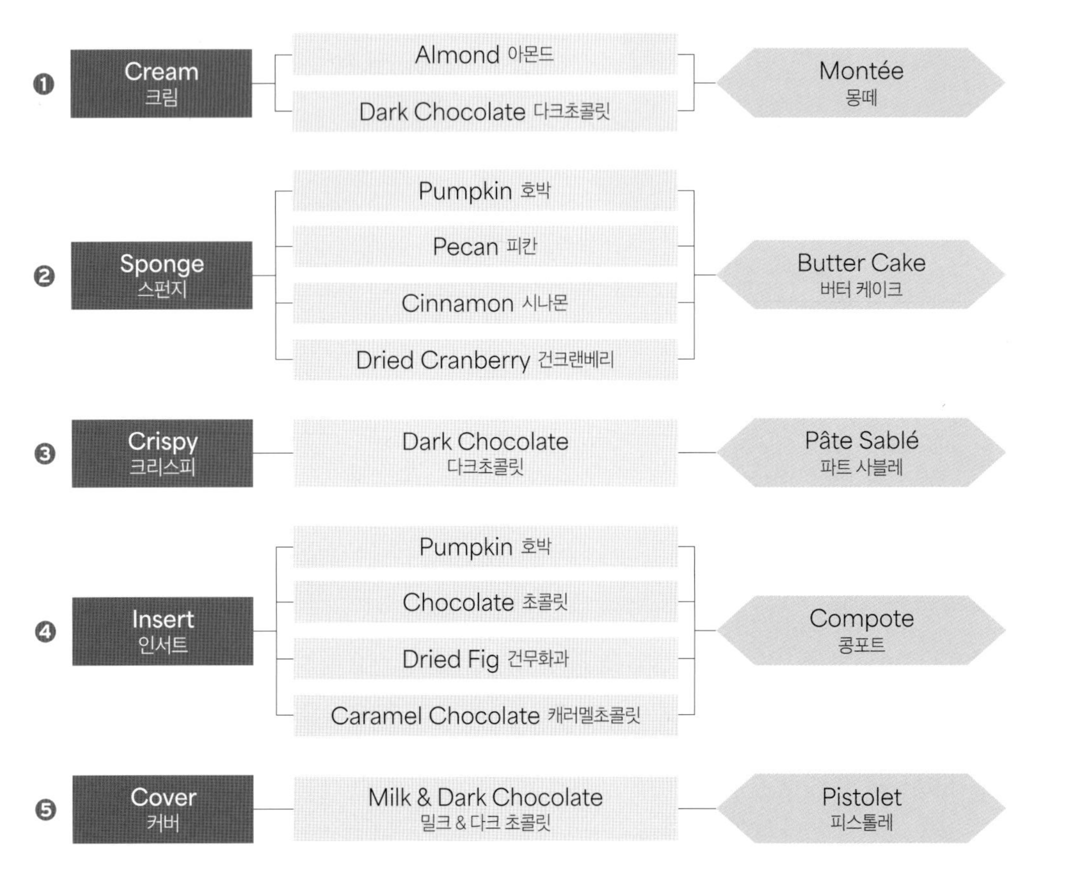

밀크 / 다크 스프레이
Milk / Dark Spray
미아메르 몽테
Mi-Amère Montée
코팅 초콜릿
Coating Chocolate
호박 케이크
Pumpkin Cake
호박 무화과 크림
Pumpkin Fig Cream
리얼 초콜릿 사블레
Real Chocolate Sablé

INGREDIENTS 20개 분량/ 20 cakes

호박 케이크
Pumpkin Cake

(철판(60 × 40cm) 1개 분량)
(1 baking tray (60 × 40 cm))

INGREDIENTS		g
호박 퓌레 (100%)	Pumpkin purée (100%)	350g
설탕	Sugar	375g
소금	Salt	3g
바닐라에센스 (Aroma Piu)	Vanilla essence	10g
달걀	Eggs	250g
식용유	Vegetable oil	220g
중력분	All purpose flour	406g
베이킹파우더	Baking powder	12g
베이킹소다	Baking soda	6g
시나몬파우더	Cinnamon powder	8g
구워 다진 피칸	Roasted pecans, chopped	150g
전처리해 다진 건크랜베리	Pre-treated dried cranberries, chopped	150g
TOTAL		**1940g**

- 건조 과일의 전처리 방법은 1권 p.87를 참고한다.
- Pre-treating dried fruit: Refer to Book 1, p.87, 'Pre-treating dried fruit'.

호박 크림 *
Pumpkin Cream

INGREDIENTS		g
호박 퓌레 (100%)	Pumpkin purée (100%)	440g
설탕	Sugar	50g
NH펙틴	NH pectin	3g
구연산삼나트륨	Trisodium citrate	0.5g
로거스트빈검	Locust bean gum	1.5g
꿀	Honey	25g
바닐라에센스 (Aroma Piu)	Vanilla essence	5g
연유	Condensed milk	20g
소금	Salt	0.5g
젤라틴매스 (×5)	Gelatin mass (×5)	8g
캐러멜초콜릿 (Zephyr Caramel 35%)	Caramel chocolate	30g
TOTAL		**583.5g**

호박 무화과 크림

Pumpkin Fig Cream

INGREDIENTS		g
호박 크림 *	Pumpkin cream *	400g
전처리한 다진 건무화과	Pre-treated dried figs, chopped	200g
TOTAL		600g

미아메르 몽테

Mi-Amère Montée

INGREDIENTS		g
휘핑크림A	Whipping cream A	200g
우유	Milk	60g
전화당	Inverted sugar	42g
젤라틴매스 (×5)	Gelatin mass (×5)	34g
다크초콜릿 (Mi-Amère 58%)	Dark chocolate	200g
아몬드 페이스트	Almond paste	40g
소금	Salt	0.5g
바닐라에센스 (Aroma Piu)	Vanilla essence	10g
휘핑크림B	Whipping cream B	420g
TOTAL		1006.5g

리얼 초콜릿 사블레

Real Chocolate Sablé

INGREDIENTS		g
T55밀가루	T55 flour	200g
카카오파우더 (Extra Brute)	Cacao powder	16g
베이킹파우더	Baking powder	3g
베이킹소다	Baking soda	1g
버터 (Bridel)	Butter	160g
황설탕	Brown sugar	130g
노른자	Egg yolks	80g
다크초콜릿 (Guayaquil 64%)	Dark chocolate	40g
소금	Salt	3g
TOTAL		**633g**

* 만드는 법

❶ 푸드프로세서에 체 친 [T55밀가루, 카카오파우더, 베이킹파우더, 베이킹소다] 절반과 차가운 상태의 버터, 황설탕을 넣고 비터로 믹싱한다.
❷ 믹싱볼에 옮겨 노른자(30℃)을 나눠 넣어가며 보슬보슬한 상태로 믹싱한다.
❸ 녹인 다크초콜릿, 소금을 넣고 믹싱한다.
❹ 남은 가루 재료를 넣고 믹싱한다.
❺ 반죽을 한 덩어리로 만든다.
❻ 랩으로 감싸 냉장고에서 휴지시킨다.
❼ 휴지시킨 반죽을 2.5mm 두께로 밀어 편 후 5.5cm 모서리가 둥근 정사각형 커터로 자른다.
❽ 철판 - 타공매트 - 사블레 - 타공매트 순서로 올린 후 160℃로 예열된 오븐에서 14분간 굽는다.

* Procedure

❶ In a food processor, mix half of sifted [T55 flour, cacao powder, baking powder, baking soda], cold cubed butter, and brown sugar.
❷ Pour into a mixing bowl; gradually add egg yolks (30°C), and mix into a crumbly state.
❸ Mix with melted dark chocolate and salt.
❹ Add the remaining powder ingredients and mix.
❺ Combine until the dough comes together.
❻ Wrap with plastic wrap and refrigerate.
❼ Roll out the refrigerated dough into 2.5 mm thickness and cut with a 5.5 cm square cutter with rounded corners.
❽ Place in order of baking sheet - perforated silicon mat - sablé - perforated silicon mat. Bake in an oven preheated to 160°C for 14 minutes.

밀크 스프레이 (코팅 겸용)

Milk Spray (also for coating)

INGREDIENTS		g
밀크초콜릿 (Alunga 41%)	Milk chocolate	500g
카카오버터	Cacao butter	500g
TOTAL		**1000g**

- 볼에 모든 재료를 넣고 녹여 바믹서로 블렌딩한 후 50℃로 맞춰 사용한다.
- Melt all the ingredients in a bowl, combine with an immersion blender, and use at 50°C.

다크 스프레이

Dark Spray

INGREDIENTS		g
다크초콜릿 (Mi-Amère 58%)	Dark chocolate	200g
카카오버터	Cacao butter	200g
TOTAL		**400g**

- 볼에 모든 재료를 넣고 녹여 바믹서로 블렌딩한 후 50℃로 맞춰 사용한다.
- Melt all the ingredients in a bowl, combine with an immersion blender, and use at 50°C.

NOTE.

✤

PUMPKIN CAKE

호박 케이크

1 로보쿱에 호박 퓌레, 설탕, 소금, 바닐라에센스를 넣고 블렌딩한다.

2 달걀(30℃)을 넣고 블렌딩한다.

3 식용유를 흘려 넣어가며 블렌딩한다.

4 볼에 옮겨 체 친 [중력분, 베이킹파우더, 베이킹소다, 시나몬파우더]와 함께 섞는다.

5 구워 다진 피칸, 전처리해 다진 건크랜베리를 넣고 섞는다.

6 테프론시트를 깐 철판(60 × 40cm)에 고르게 팬닝한 후 데크 오븐 기준 윗불 180℃, 아랫불 170℃에서 25분간 굽는다.

7 완전히 식으면 길이 3.7cm 정사각형 커터로 자른다.

1 Blend pumpkin purée, sugar, salt, and vanilla essence in the Robot Coupe.

2 Blend with eggs (30°C).

3 Drizzle in the oil while blending.

4 Pour into a mixing bowl and mix with sifted [all-purpose flour, baking powder, baking soda, and cinnamon powder].

5 Mix with chopped roasted pecan and chopped rehydrated dried cranberries.

6 Spread evenly onto a baking tray (60 × 40 cm) lined with a Teflon sheet. Bake in a deck oven with 180°C on the top and 170°C on the bottom for 25 minutes.

7 Cool completely and cut with a 3.7 cm square cutter.

1

2

3-1

3-2

4

5

6

7

✤ ✤

PUMPKIN FIG CREAM

호박 무화과 크림

호박 크림

1 비커에 호박 퓌레, 미리 섞어둔 [설탕, NH펙틴, 구연산삼나트륨, 로거스트빈검], 꿀, 바닐라에센스, 연유, 소금을 넣고 바믹서로 블렌딩한다.

2 냄비로 옮겨 휘퍼로 저어가며 가열한다.

TIP. 호박죽을 끓이듯 저어가며 오래 끓여 수분을 날려준다. 강한 불에서 끓이면 내용물이 튀어 오를 수 있으므로 중간 불에서 오래 끓인다.

3 펙틴 반응을 확인한 후 불에서 내려 젤라틴매스를 넣고 녹을 때까지 섞는다.

4 캐러멜초콜릿이 담긴 비커에 넣고 바믹서로 블렌딩한다.

5 얼음물이 담긴 볼에 받쳐 10℃로 쿨링한다.

호박 무화과 크림

6 볼에 호박 크림(800g)과 전처리한 다진 건무화과(400g)를 넣고 섞는다.

조립

7 호박 케이크 위에 호박 무화과 크림 24g을 올린다.

8 스패출러로 평평하게 정리한 후 냉동한다.

Pumpkin Cream

1 In a beaker, combine pumpkin purée, pre-mixed [sugar, pectin NH, trisodium citrate and locust bean gum], honey, vanilla essence, condensed milk, and salt with an immersion blender.

2 Pour into a pot and heat while stirring.

TIP. Cook it for a long time to reduce, as if cooking pumpkin porridge. Boiling over high heat can cause the content to splatter, so simmer over medium heat for a long time.

3 After the pectin reacts, remove from the heat, add gelatin mass, and stir until it melts.

4 Pour into a beaker with caramel chocolate and combine using an immersion blender.

5 Cool in an ice bath to 10°C.

Pumpkin Fig Cream

6 Combine pumpkin cream (800 g) and chopped pre-treated dried figs (400 g) in a bowl.

Assemble

7 Top 24 grams of pumpkin fig cream on the pumpkin cake.

8 Spread evenly with a spatula and freeze.

1

2

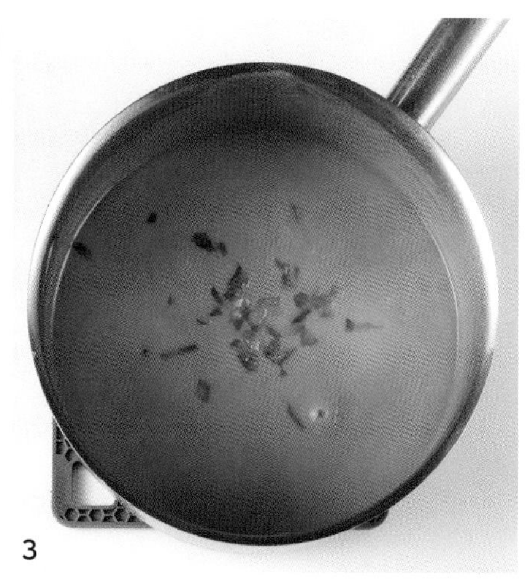

3

4

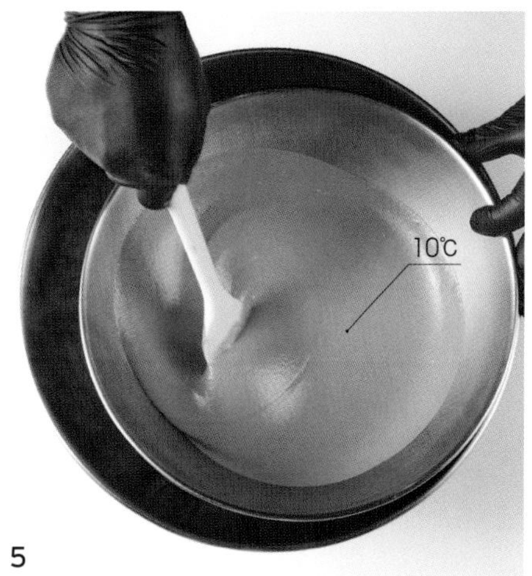

5

6

7

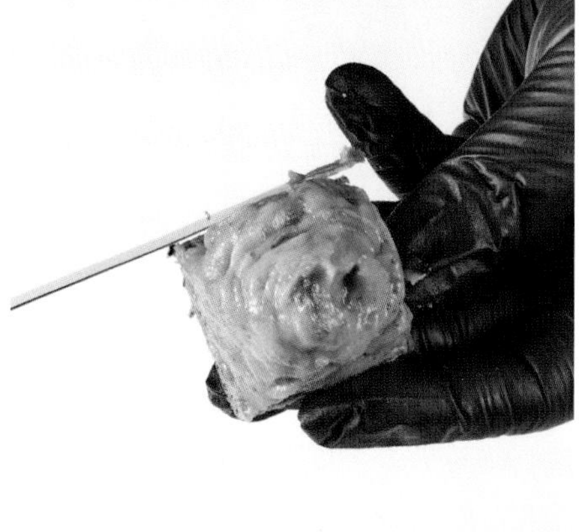

8-1

8-2

✤ ✤ ✤

MI-AMÈRE MONTÉE

미아메르 몽테

미아메르 몽테

1 냄비에 휘핑크림A, 우유, 전화당, 젤라틴매스를 넣고 70℃로 가열한다.

2 다크초콜릿이 담긴 비커에 넣고 바믹서로 블렌딩한다.

3 아몬드 페이스트 - 소금 - 바닐라에센스 - 휘핑크림B 순서로 넣고 블렌딩한다.

4 바트에 붓고 밀착 랩핑한 후 냉장고에서 숙성시킨다.

5 사용하기 직전 믹싱볼에 비터로 부드럽게 휘핑해 사용한다.

조립

6 몰드에 50g씩 채운다.

7 얼린 [호박 케이크 + 호박 무화과 크림]을 넣는다.

8 스패출러로 평평하게 정리해 급속 냉동(-35 ~ -40℃)한 후 몰드에서 빼내 냉동(-18℃) 보관한다.

Mi-Amère Montée

1 Heat whipping cream A, milk, inverted sugar, and gelatin mass to 70°C.

2 Pour into a beaker with dark chocolate and combine with an immersion blender.

3 Blend in order with almond paste, salt, vanilla essence, and whipping cream B.

4 Pour onto a stainless-steel tray, cover with plastic wrap, making sure the wrap is in contact with the cream, and refrigerate.

5 Whip gently with a paddle in a mixing bowl just before use.

Assemble

6 Fill 50 grams in the mold.

7 Insert the frozen [pumpkin cake + pumpkin fig cream].

8 Clean up with a spatula, freeze in the blast freezer (-35 ~ -40°C), remove from the mold, then store in the freezer (-18°C).

70℃
1
2
3

4

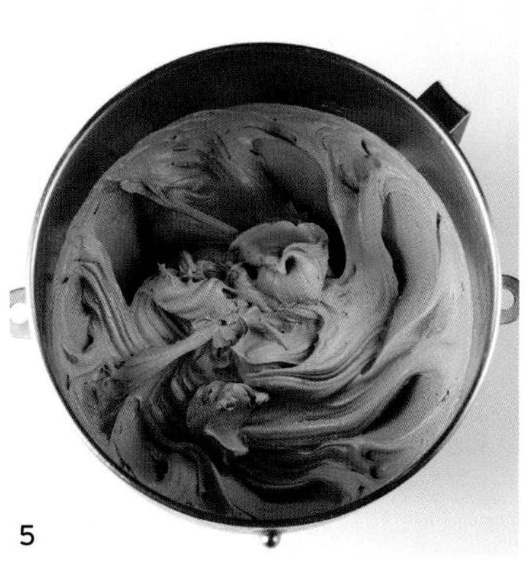
5

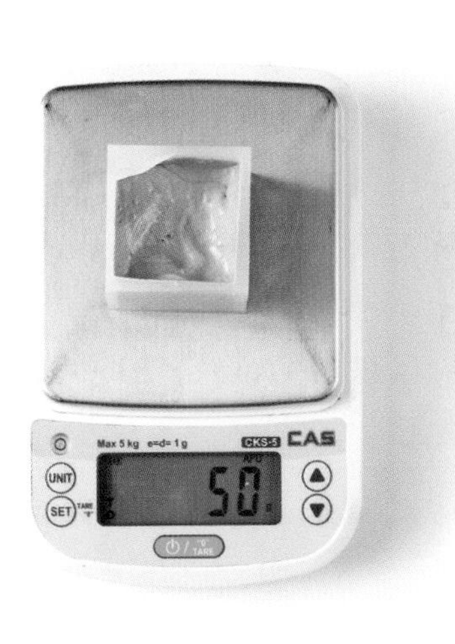
Max 5 kg e=d= 1 g
CKS-5 CAS
UNIT
SET
50
6

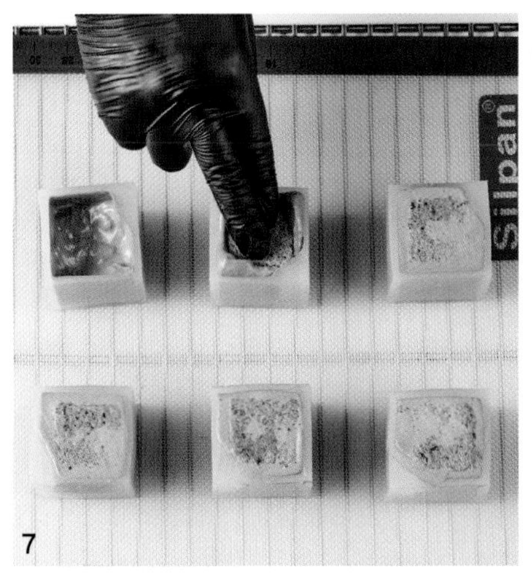
Silpan
7

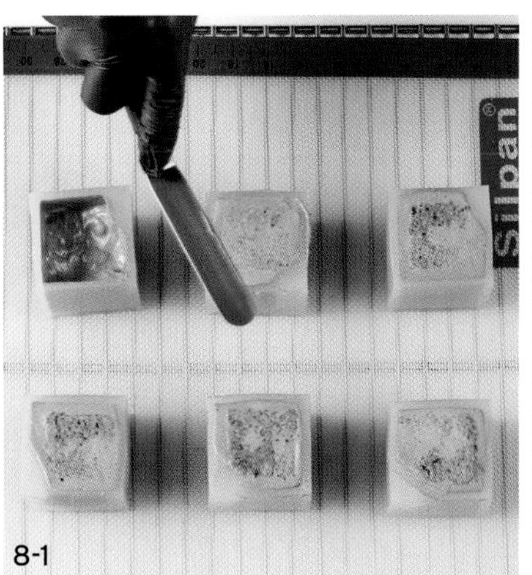
Silpan
8-1

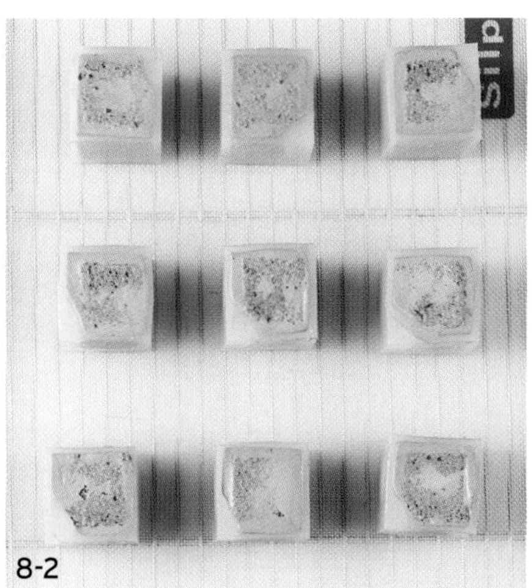
8-2

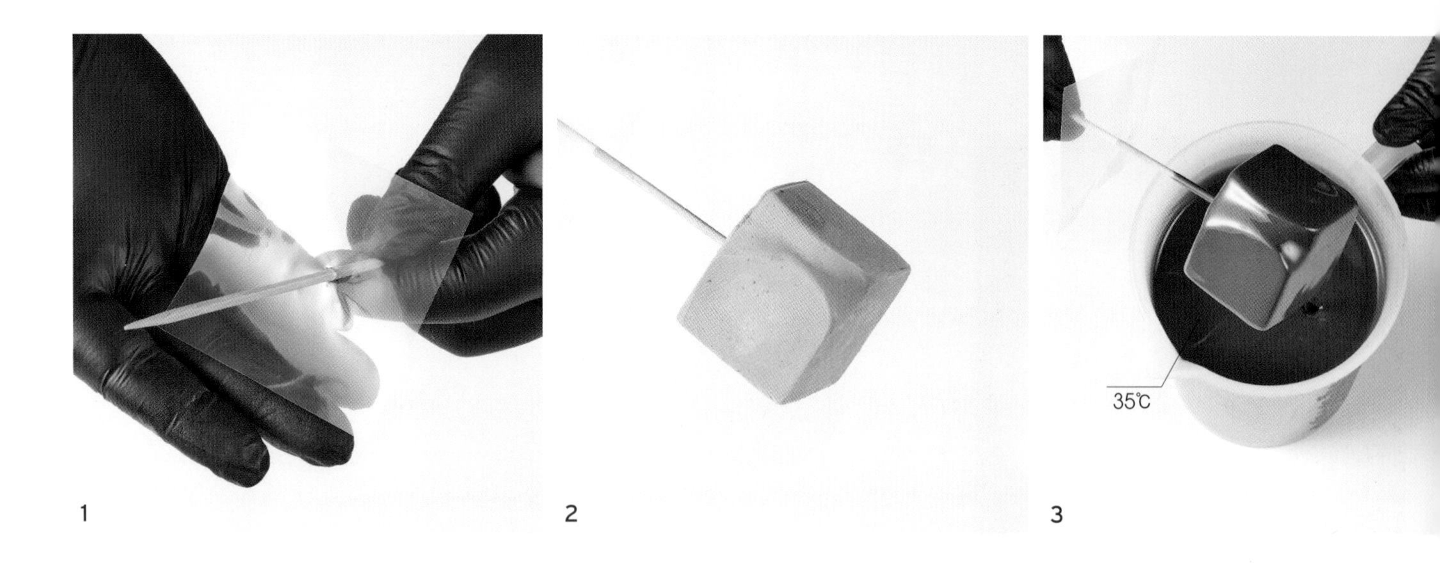

✤ ✤ ✤ ✤

FINISH 마무리

1 튼튼한 대나무 꼬치에 얼린 무스보다 조금 더 크게 자른 초콜릿 필름을 꽂는다.

2 얼린 무스를 꽂는다.

3 얼린 무스에 녹인 밀크 스프레이(35℃)를 코팅한다.

4 타공 팬에 수직으로 내리면서 꼬치를 아래로 뺀다.

5 냉동실로 옮겨 굳힌다.

6 밀크 스프레이와 다크 스프레이로 분사하고, 붓으로 카카오파우더를 골고루 묻혀준 후 열풍기로 열을 가해 명암과 음영을 표현한다. 리얼 초콜릿 사블레 위에 올려 완성한다.

1. Skewer a piece of acetate film cut slightly larger than the frozen mousse on a sturdy bamboo skewer.
2. Insert the skewer into the frozen mousse.
3. Dip the frozen mousse in the melted milk spray (35°C).
4. Pull the skewer perpendicularly through a perforated tray to remove the skewer.
5. Keep in the freezer to harden.
6. Spray with milk and dark spray mixtures, coat cacao powder evenly with a brush, and apply heat with a heat gun to create contrast and shading. Place on the real chocolate sablé to finish.

16

THE SCENT OF SPRING

봄의 향기

PAIRING & TEXTURE
페어링 & 텍스처

벚꽃, 체리, 복숭아, 화이트초콜릿, 요거트 등으로 페어링하여 잘 어우러지도록 연출했다.

Cherry blossom, cherry, peach, white chocolate, and yogurt are paired to create harmony.

TECHNIQUE
테크닉

- 룸 템퍼라처 글레이즈, 펙틴 미러 글레이즈를 이용해 우아한 텍스처를 연출했다.
- 벚꽃 티 젤리 위에 펙틴 미러 글레이즈를 코팅해 유리 같은 느낌을 연출했다.
- 일반적으로 몰드에 넣는 순서를 거꾸로 하여 스펀지부터 채우는 방식의 테크닉을 사용했다.
- 겔 크림 콜드라는 차가운 수분과 반응하는 재료를 사용하여 열을 가하지 않는 프레시한 맛을 연출했다.

- I created a glamorous texture with room temperature glaze and pectin mirror glaze.
- I used a pectin mirror glaze over the cherry blossom tea jelly to give a glass-like look.
- I reversed the standard technique of filling the mold by putting in the sponge first.
- An ingredient that reacts with cold water called Gelcrem Cold was used to create a fresh taste without applying heat.

DESIGN
디자인

- 두 가지 몰드를 사용해 조립하는 방식으로 완성했다.
- 젤리 안에 실제 벚꽃 잎을 넣어 봄에 피어나는 꽃의 느낌을 극대화시켰다.

- Two molds were used to assemble.
- Real cherry blossom petals were placed inside the jelly to maximize the look of a blooming flower in spring.

✦ HOW TO COMPOSE THIS RECIPE ✦

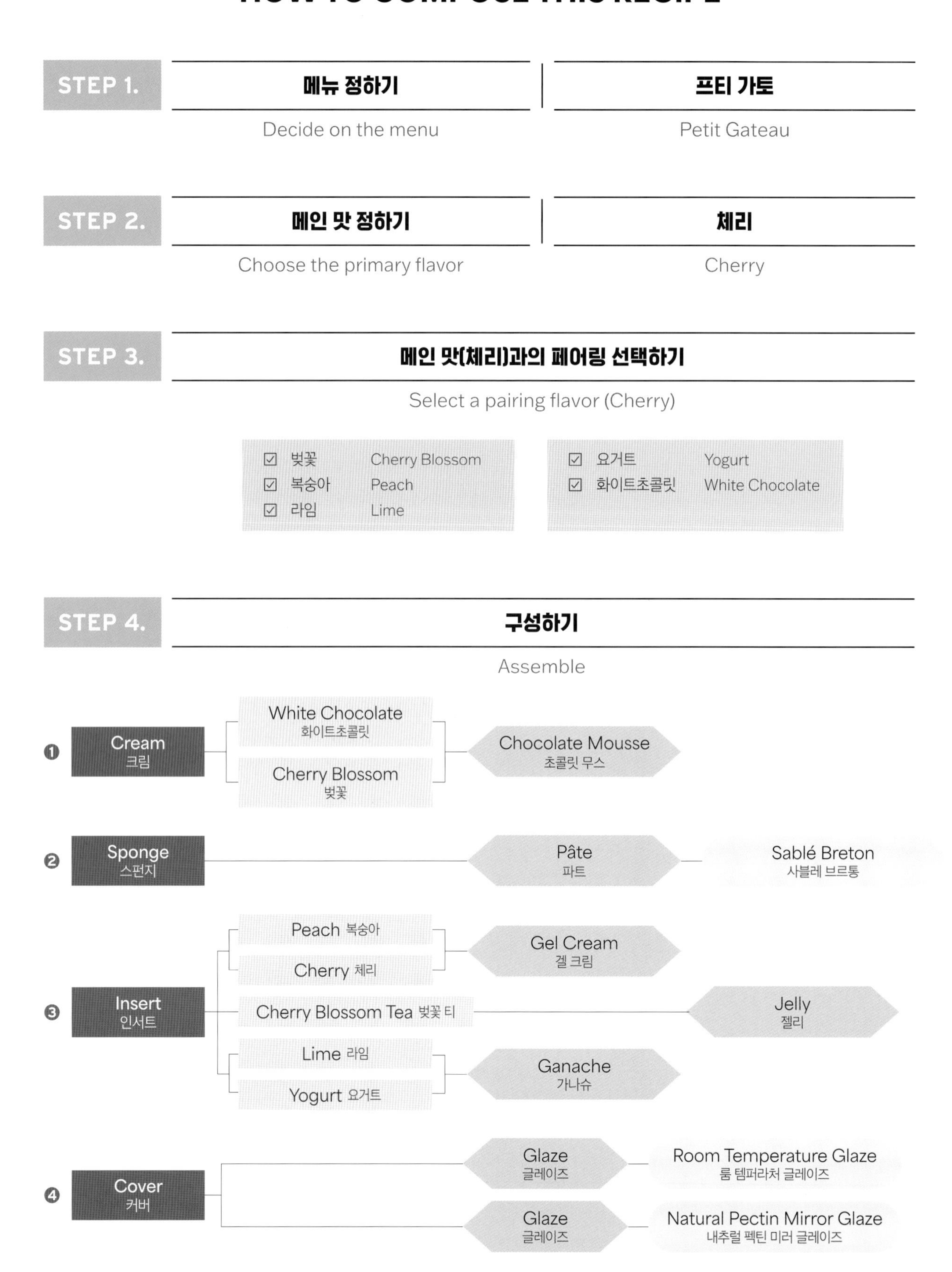

벚꽃 티 젤리
Cherry Blossom Tea Jelly
내추럴 펙틴 미러 글레이즈
Natural Pectin Mirror Glaze
초콜릿 디스크
Chocolate Disk
벚꽃 무스
Cherry Blossom Mousse
라즈베리 룸 템퍼라처 글레이즈
Raspberry Room Temperature Glaze
체리 복숭아 겔 크림
Cherry Peach Gel Cream
요거트 라임 가나슈
Yogurt Lime Ganache
사블레 브르통
Sablé Breton

INGREDIENTS 20개 분량/ 20 cakes

사블레 브르통

Sablé Breton

INGREDIENTS		g
버터 (Bridel)	Butter	260g
설탕	Sugar	200g
소금	Salt	5g
박력분	Cake flour	500g
베이킹파우더	Baking powder	20g
노른자	Egg yolks	100g
바닐라에센스 (Aroma Piu)	Vanilla essence	4g
레몬제스트	Lemon zest	1개 분량/ 1 lemon
레몬에센스 (Aroma Piu)	Lemon essence	1g
TOTAL		**1090g**

체리 복숭아 겔 크림

Cherry Peach Gel Cream

INGREDIENTS		g
체리 퓌레	Cherry purée	130g
화이트 피치 퓌레	White peach purée	100g
TPT 시럽	TPT syrup	60g
겔크림콜드	Gelcrem Cold	13g
벚꽃 향료 (ES)	Cherry blossom flavor (ES)	2방울/ 2 drops
TOTAL		**303g**

- TPT 시럽 = 물과 설탕을 1:1 비율로 끓여 만든 것으로, 식혀 사용한다.
- TPT syrup: A syrup made by boiling water and sugar in a 1:1 ratio. Use after cooling.

요거트 라임 가나슈

Yogurt Lime Ganache

INGREDIENTS		g
우유	Milk	60g
젤라틴매스 (×5)	Gelatin mass (×5)	13g
화이트초콜릿 (Zephyr white 34%)	White chocolate	190g
그릭요거트	Greek yogurt	140g
요거트파우더	Yogurt powder	5g
라임제스트	Lime zest	1/2개 분량/ 1/2 lime
TOTAL		**408g**

벚꽃 티 젤리

Cherry Blossom Tea Jelly

INGREDIENTS		g
물	Water	240g
설탕	Sugar	140g
로거스트빈검	Locust bean gum	1g
젤라틴매스 (×5)	Gelatin mass (×5)	45g
벚꽃 향료 (ES)	Cherry blossom flavor (ES)	1방울/ 1 drop
벚꽃 잎	Cherry blossom petals	1g
TOTAL		**427g**

벚꽃 무스

Cherry Blossom Mousse

INGREDIENTS		g
생크림A	Heavy cream A	200g
젤라틴매스 (×5)	Gelatin mass (×5)	40g
벚꽃 페이스트	Cherry blossom paste	15g
화이트초콜릿 (Zephyr white 34%)	White chocolate	200g
벚꽃 향료 (ES)	Cherry blossom flavor (ES)	1방울/ 1 drop
벚꽃 리큐르 (Dijon cherry blossom)	Cherry blossom liqueur	15g
생크림B	Heavy cream B	400g
TOTAL		**870g**

내추럴 펙틴 미러 글레이즈

Natural Pectin Mirror Glaze

INGREDIENTS		g
물	Water	400g
물엿	Corn syrup	240g
설탕A	Sugar A	80g
NH펙틴	NH pectin	15g
구연산삼나트륨	Trisodium citrate	4g
설탕B	Sugar B	730g
젤라틴매스 (×5)	Gelatin mass (×5)	165g
TOTAL		**1634g**

＊만드는 법

❶ 냄비에 물, 물엿을 넣고 가열하다가 40℃가 되기 전에 미리 섞어둔 [설탕A, NH펙틴, 구연산삼나트륨]을 넣고 주걱으로 저어가며 가열한다.

❷ 끓어오르면 설탕B를 두 번에 나눠 넣으며 섞는다.

❸ 설탕B가 모두 녹으면 젤라틴매스를 넣고 98℃까지 가열한다.

❹ 바트에 붓고 밀착 랩핑한 후 냉장 보관한다.

❺ 사용하기 전날 실온에 꺼내두어 기포가 생기지 않게 녹인 후 바믹서로 블렌딩해 30~32℃로 맞춰 사용한다.

＊Procedure

❶ Heat water and corn syrup. Before it reaches 40°C, add the previously mixed [sugar A, pectin NH, and trisodium citrate] and cook while stirring with a spatula.

❷ Once it boils, stir in sugar B half at a time.

❸ When all the sugar B dissolves, add gelatin mass and cook to 98°C.

❹ Pour onto a stainless-steel tray, cover with plastic wrap, making sure the wrap is in contact with the glaze, and refrigerate.

❺ Leave the glaze out at room temperature the day before use to prevent air bubbles from forming; blend and use at 30~32°C.

라즈베리 룸 템퍼라처 글레이즈

Raspberry Room Temperature Glaze

INGREDIENTS		g
화이트초콜릿 (Zephyr white 34%)	White chocolate	400g
딸기초콜릿 (Callebaut)	Strawberry chocolate	300g
식용유	Vegetable oil	200g
빨간색 색소	Red color	0.1g
하얀 색소 (이산화티타늄)	White coloring (Titanium dioxide)	5g
라즈베리 코팅 크리스피 (Sosa)	Wet-proof raspberry crispy	20g
TOTAL		**925.1g**

- 국제적인 추세를 반영하여 이산화티타늄은 최소량으로 사용했다.
- Titanium dioxide was used in a minimal amount to reflect international trends.

*** 만드는 법**

❶ 녹인 화이트초콜릿과 딸기초콜릿이 담긴 비커에 식용유(50℃)를 넣고 바믹서로 블렌딩한다.

❷ 빨간색 색소, 적당한 크기로 다진 라즈베리 코팅 크리스피를 넣고 블렌딩한 후 30~35℃로 맞춰 사용한다.

*** Procedure**

❶ Add vegetable oil (50°C) into a beaker with melted white chocolate and strawberry chocolate and combine using an immersion blender.

❷ Blend with red food coloring and roughly chopped raspberry crispies. Use at 30~35°C.

초콜릿 디스크

Chocolate Disk

INGREDIENTS		g
화이트초콜릿 (Zephyr white 34%)	White chocolate	1000g
카카오버터	Cacao butter	70g
하얀 색소 (이산화티타늄)	White coloring (Titanium dioxide)	10g
빨간색 색소	Red coloring	0.5g
TOTAL		**1080.5g**

- 국제적인 추세를 반영하여 이산화티타늄은 최소량으로 사용했다.
- Titanium dioxide was used in a minimal amount to reflect international trends.

*** 만드는 법**

❶ 비커에 녹인 화이트초콜릿, 50℃ 이하로 녹인 카카오버터를 넣고 바믹서로 블렌딩한다.

❷ 흰색 색소와 빨간색 색소를 넣고 블렌딩한다.

❸ 템퍼링한 후 지름 5.5cm 원형 커터와 8cm 벚꽃 모양 커터로 자르고 굳혀 사용한다.

*** Procedure**

❶ Blend melted white chocolate and cacao butter melted to below 50°C with an immersion blender.

❷ Blend with white and red food colorings.

❸ Temper the mixture and cut with a 5.5 cm round and an 8 cm cherry blossom-shaped cutter. Use after crystallization.

NOTE.

SABLÉ BRETON

사블레 브르통

1 믹싱볼에 포마드 상태의 버터, 설탕, 소금을 넣고 비터로 가볍게 풀어준다.

2 체 친 박력분과 베이킹파우더 1/2을 넣고 믹싱한다.

3 노른자(30℃), 바닐라에센스, 레몬제스트를 넣고 믹싱한다.

4 남은 가루 재료를 모두 넣고 믹싱한다.

5 볼 안에서 반죽을 치대가며 한 덩어리로 만든다.

6 랩핑한 후 직사각형으로 만들어 냉장고에서 숙성시킨다.

TIP. 0.8cm 두께로 밀어 펴 냉동 보관한 후 자르기 쉬운 단단한 상태가 되면 지름 5cm 원형 커터로 잘라 타공매트를 깐 철판 위에 올리고 타공매트를 덮는다. 170℃로 예열된 오븐에서 뎀퍼를 100% 열고 20분간 굽고, 덮은 타공매트를 제거해 2분간 더 굽는다.

1 Lightly beat softened butter, sugar, and salt with a paddle.

2 Mix with sifted cake flour and half of baking powder.

3 Add egg yolks (30°C), vanilla essence, and lemon zest and mix.

4 Add the remaining powder ingredients and mix.

5 Knead until the dough comes together.

6 Wrap with plastic wrap, shape it into a rectangle, and refrigerate.

TIP. Roll out to 0.8 cm thickness, keep in the freezer, and when it becomes hard enough to cut, cut with a 5 cm round cutter. Place on a baking sheet lined with a perforated mat and cover with another perforated mat. Bake in an oven preheated to 170°C with the damper open at 100% for 20 minutes. Remove the covered perforated mat and bake for another 2 minutes.

CHERRY PEACH GEL CREAM

체리 복숭아 겔 크림

1 비커에 체리 퓌레, 화이트 피치 퓌레, TPT 시럽을 넣고 겔크림콜드를 흘려 넣어가면서 바믹서로 블렌딩한다.

TIP. 모든 재료는 차가운 상태로 사용한다.
차가운 온도에서 반응하는 겔을 사용하므로 가열할 필요가 없고, 결과적으로 더 신선한 맛과 향으로 완성할 수 있다.
다만 일반 겔보다 좀 더 오래 블렌딩해야 한다.

2 벚꽃 향료를 넣고 가볍게 섞는다.

3 몰드(실리코마트 SF044)에 12g씩 채우고 바닥에 쳐 평평하게 만든 후 냉동한다.

1 In a beaker, add cherry purée, white peach purée, and TPT syrup. Combine with an immersion blender while drizzling in the Gelcrem Cold.

TIP. Use all the ingredients cold.
Because the gelling agent reacts at cold temperatures, there is no need to cook, resulting in a fresher taste and aroma.
However, it needs to be blended a bit longer than general gel.

2 Lightly mix with cherry blossom flavor.

3 Fill 12 grams in the mold (Silikomart SF044), tap it on the table to even out the cream, and freeze.

✤

1

2

3

4

5

6

✤ ✤

1

2

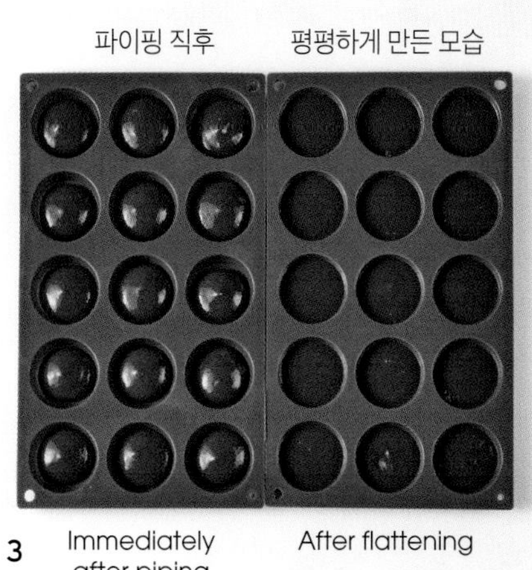

3 Immediately after piping　After flattening

✤ ✤ ✤

YOGURT LIME GANACHE

요거트 라임 가나슈

요거트 라임 가나슈

1 냄비에 우유, 젤라틴매스를 넣고 젤라틴매스가 녹을 때까지 가열한다. (약 60℃)

2 화이트초콜릿이 담긴 비커에 넣고 바믹서로 블렌딩한다.

3 그릭요거트 - 요거트파우더 순서로 넣고 블렌딩해 완전히 유화시킨다.

4 라임제스트를 넣고 섞는다.

5 얼음물이 담긴 볼에 받쳐 26℃로 쿨링한다.

조립

6 얼려둔 체리 복숭아 겔 크림 위에 12g씩 파이핑하고, 바닥에 쳐 평평하게 만들고 급속 냉동(-35 ~ -40℃)한 후 몰드에서 빼내 냉동(-18℃) 보관한다.

Yogurt Lime Ganache

1 Heat milk and gelatin mass until the gelatin melts (about 60°C).

2 Pour into a beaker with white chocolate and combine with an immersion blender.

3 Blend in the order of Greek yogurt – yogurt powder and perfectly emulsify.

4 Mix with lime zest.

5 Cool in an ice bath to 26°C.

Assemble

6 Pipe 12 grams over the frozen cherry peach gel cream, tap it on the table to even out the cream. Blast freeze (-35 ~ -40°C), remove from the mold and freeze (-18°C).

✤ ✤ ✤ ✤

CHERRY BLOSSOM TEA JELLY

벚꽃 티 젤리

1 비커에 물, 미리 섞어둔 설탕과 로거스트빈검을 넣고 바믹서로 블렌딩한다. 냄비로 옮겨 젤라틴매스를 넣고 녹을 때까지 가열한다. (약 60℃)

TIP. 오래 가열해야 하는 쿨리 작업과 다르게 젤라틴매스만 녹으면 불에서 내린다.
로거스트빈검이 제대로 섞이지 않으면 작업할 때마다 다른 상태의 젤리로 완성되므로 항상 일정한 상태로 고르게 블렌딩한다.

2 얼음물이 담긴 볼에 받쳐 30℃로 쿨링한 후 벚꽃 향료를 넣고 섞는다. 다시 25℃로 쿨링한 후 벚꽃 잎을 넣고 섞는다.

TIP. 벚꽃은 줄기와 수술을 제외한 꽃잎 부분만 사용한다.

3 20℃로 맞춰 몰드(DECO RELIEF DR306A)에 20g씩 채운 후 냉장고에 두고 살짝 굳혀준다.

TIP. 20℃보다 낮은 온도로 채우면 벚꽃 잎이 위로 떠올라 완성되었을 때 예쁘지 않다. (중간층에 자리잡아야 완성되었을 때 가장 예쁘다.)

1 Blend water with previously mixed sugar and locust bean gum using an immersion blender. Pour the mixture into a pot with gelatin mass and heat until the gelatin melts (about 60°C).

TIP. Unlike coulis, which takes time to cook, remove from the heat as soon as the gelatin melts.
If the locust bean gum is not mixed properly, the jelly will be different every time you work with it. Therefore, always blend it evenly and consistently.

2 Cool in an ice bath to 30°C and mix with cherry blossom flavor. Cool again to 25°C and mix with cheery blossom petals.

TIP. Use only the petals of cherry blossom, excluding the stems and stamens.

3 Set to 20°C and fill 20 grams in the mold (Deco Relief DR306A). Let it set slightly in a refrigerator.

TIP. If you fill at a temperature lower than 20°C, the petals will float to the top, which will not look nice when finished. (It should settle in the center to look prettiest when set.)

✤ ✤ ✤

60℃
1

2

3

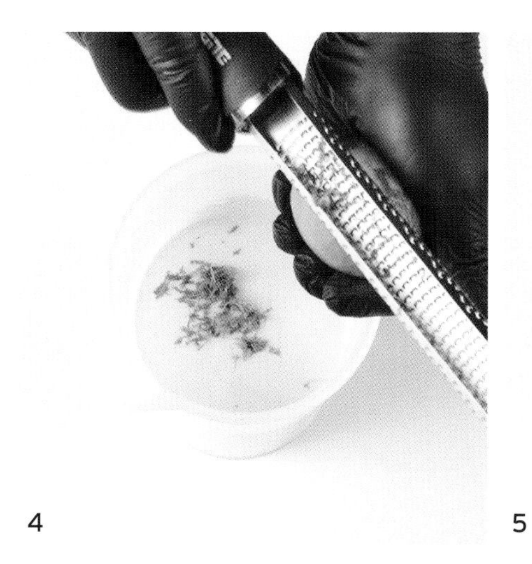
4

5

6

✤ ✤ ✤ ✤

60℃
1

25℃
2

3

✤ ✤ ✤ ✤ ✤

CHERRY BLOSSOM MOUSSE

벚꽃 무스

벚꽃 무스

1 냄비에 생크림A, 젤라틴매스, 벚꽃 페이스트를 넣고 젤라틴매스가 녹을 때까지 주걱으로 저어가며 60℃로 가열한다.

2 화이트초콜릿이 담긴 비커에 붓고 바믹서로 블렌딩한다.

TIP. 수분이 높은 배합이므로 너무 뜨거운 상태로 초콜릿에 붓지 않아도 된다.

3 벚꽃 향료 - 벚꽃 리큐르 순서로 넣고 블렌딩한다.

4 휘퍼를 이용해 손으로 가볍게 휘핑(60~70%)한 생크림B에 **3**을 넣고 주걱으로 가볍게 섞는다.

조립

5 냉장고에서 살짝 굳힌 벚꽃 티 젤리 위에 벚꽃 무스를 10g씩 채운다.

6 지름 5.5cm 원형으로 만든 초콜릿 디스크를 올리고 급속 냉동(-35 ~ -40℃)한 후 냉동(-18℃) 보관한다.

TIP. 초콜릿 디스크를 올리지 않으면 나중에 글레이즈를 씌울 때 무스가 녹아버린다.

7 다른 몰드(실리코마트 SF163)에 벚꽃 무스를 10g씩 파이핑한 후 사블레 브르통을 돌려가면서 바닥까지 넣어준다.

TIP. 보통은 사블레 브르통을 가장 마지막에 넣지만 여기에서는 반대로 인서트 크림으로 마무리했다.

8 사블레 브르통 위에 벚꽃 무스를 18g씩 파이핑한다.

9 얼린 [체리 복숭아 겔 크림 + 요거트 라임 가나슈]를 넣는다.

10 스패출러로 깔끔하게 정리하고 급속 냉동(-35 ~ -40℃)한 후 몰드에서 빼내 냉동(-18℃) 보관한다.

Cherry Blossom Mousse

1 Heat cream A, gelatin mass, and cherry blossom paste in a pot and cook to 60°C while stirring until the gelatin melts.

2 Pour into a beaker with white chocolate and combine with an immersion blender.

TIP. Because this recipe has a high moisture content, pouring it into the chocolate when it's too hot is unnecessary.

3 Blend in the order of cherry blossom flavor - cherry blossom liqueur.

4 Lightly whip (60~70%) cream B with a whisk and gently combine with (**3**) using a spatula.

Assemble

5 Fill 10 grams of cherry blossom mousse over the cherry blossom tea jelly that has been set slightly in the fridge.

6 Top with a 5.5 cm round chocolate disk, blast freeze (-35 ~ -40°C), then keep frozen (-18°C).

TIP. If you don't put the chocolate disk on top, the mousse will melt when you glaze it later.

7 Pipe 10 grams of cherry blossom mousse in another mold (Silikomart SF163). Insert the sablé Breton and push it in while rotating until it reaches the bottom.

TIP. Usually, I place the sablé Breton last, but here, I did the opposite and finished with the insert cream.

8 Pipe 18 grams of cherry blossom mousse over the sablé Breton.

9 Insert the frozen [cherry peach gel cream + yogurt lime ganache].

10 Clean up with a spatula, blast freeze (-35 ~ -40°C), remove from the mold, then keep frozen (-18°C).

60℃
1
2
3

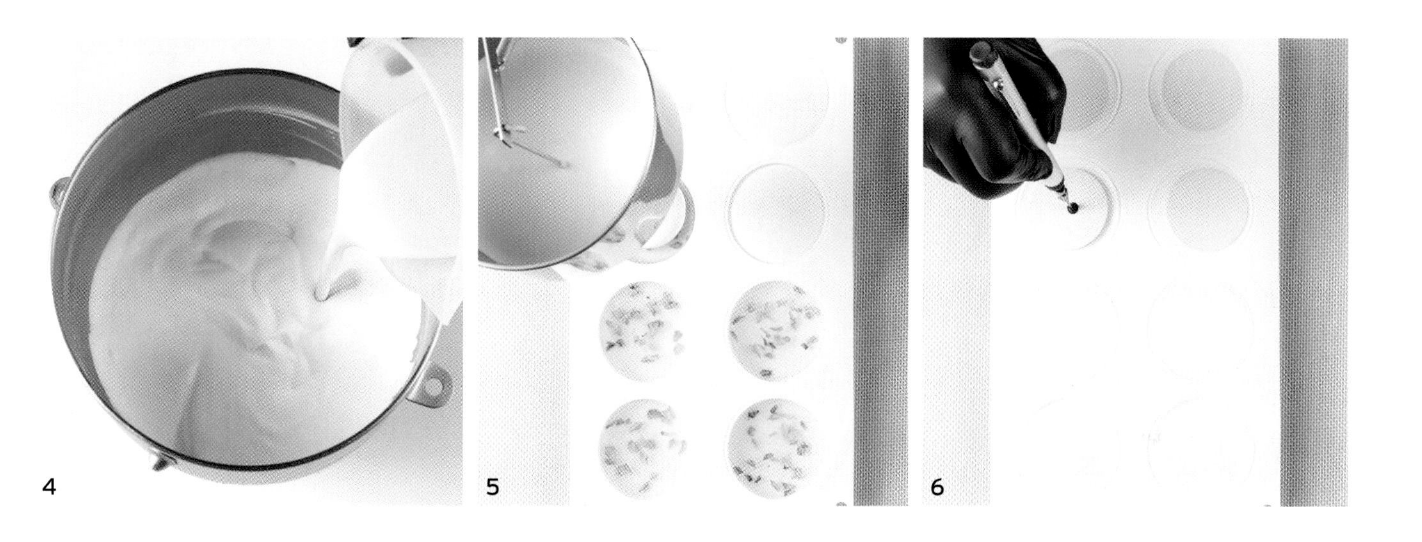
4
5
6

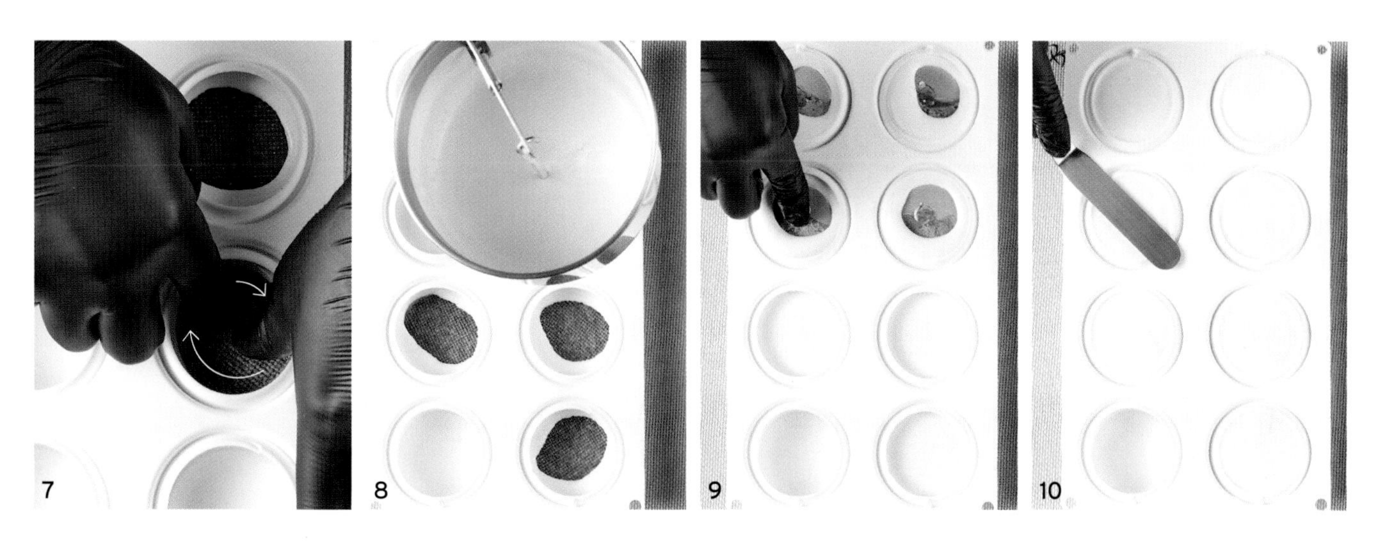
7
8
9
10

✤ ✤ ✤ ✤ ✤ ✤

FINISH 마무리

1 얼린 무스에 라즈베리 룸 템퍼라처 글레이즈(35℃)를 코팅하고 바닥면을 깔끔하게 정리한 후 굳힌다.

TIP. 윗면을 제외한 옆면, 아랫면에 고르게 코팅한다.

2 굳힌 벚꽃 티 젤리에 내추럴 펙틴 미러 글레이즈(30~32℃)를 코팅한다.

3 **1**의 윗면에 여분의 벚꽃 무스를 소량 파이핑한다.

4 벚꽃 모양으로 만든 초콜릿 장식을 올린 후 정중앙에 잘 맞춰 **2**를 올린다.

5 벚꽃 잎, 금박을 올려 마무리한다.

2-2

1 Coat the frozen mousse with raspberry room temperature glaze (35°C), clean the bottom, and let it set.

TIP. Coat evenly on the sides and the bottom, except the top.

2 Coat the hardened cherry blossom tea jelly with the natural pectin mirror glaze (30~32°C).

3 Pipe a small amount of the extra cherry blossom mousse on the top of (**1**).

4 Top with the cherry blossom-shaped chocolate decoration, and place (**2**) in the exact center.

5 Arrange cherry blossom petals and gold leaves to finish.

17

ERIC'S TIRAMISU DOUGHNUT

에릭 티라미수 도넛

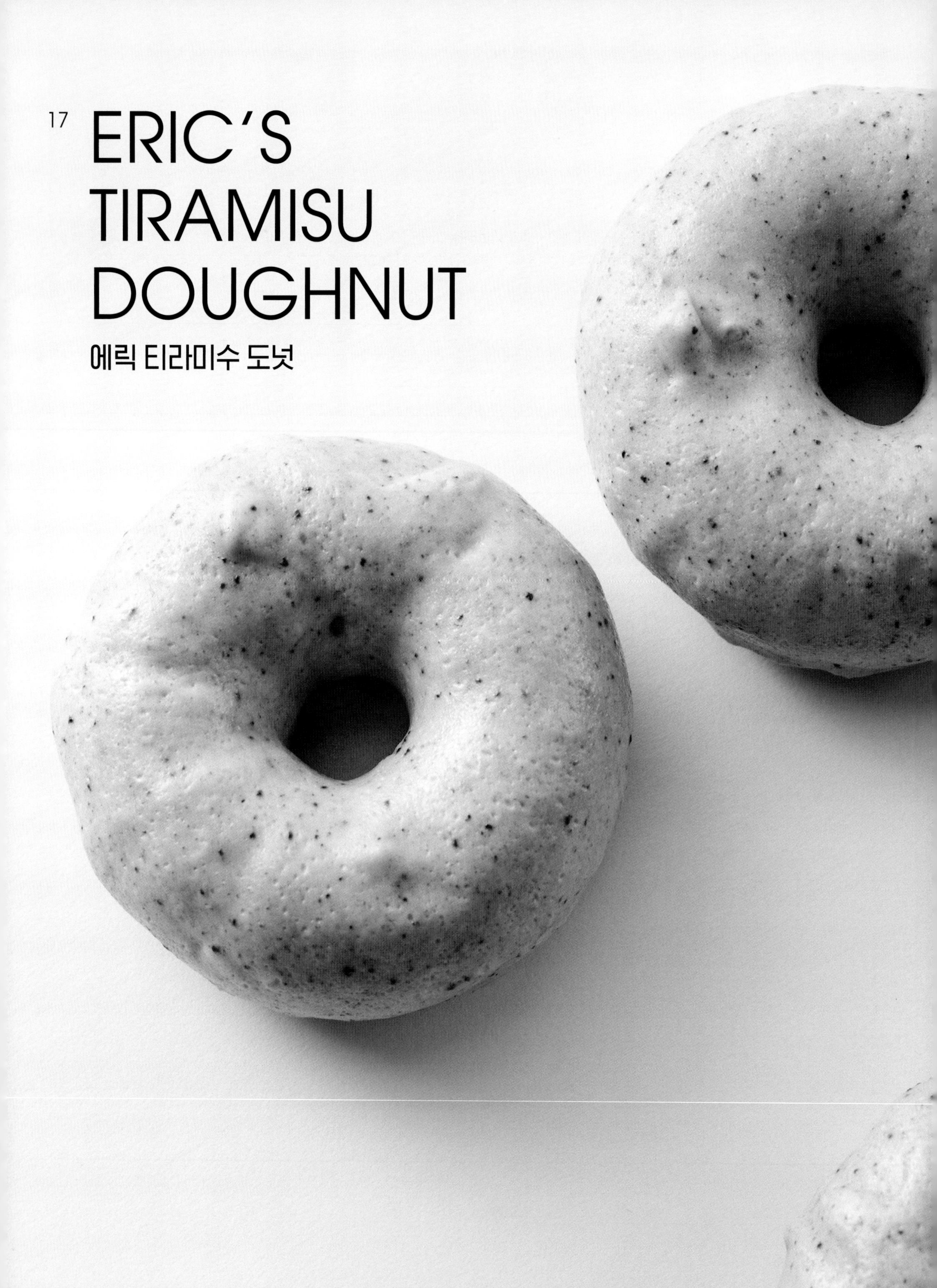

PAIRING & TEXTURE
페어링 & 텍스처

- 파트 사블레의 바삭함, 초콜릿 몽테의 부드러움, 캐러멜의 쫀득함을 느낄 수 있는 제품이다.
- 특히 겉면을 감싼 커피 초콜릿 폼의 식감은 이제까지 느껴보지 못한 새로운 텍스처일 것이다. 입에 넣으면 바스러지듯 부서지는 바삭함과 입 안에서 부드럽게 녹아내리는 상반된 텍스처를 동시에 느낄 수 있다.

- It is a dessert where you can feel the crunchiness of pâte sablé, the smoothness of chocolate montée, and the chewiness of caramel.
- Above all, the texture of the coffee foam on the outside will be unlike anything you've ever experienced before. When you bite into it, you'll feel the contrasting textures of crumbling crunchiness and smoothly melting sensation.

TECHNIQUE
테크닉

초콜릿 폼 글레이즈

- 기존에 사용했던 글레이즈나 피스톨레 기법이 아닌, 커피 초콜릿 폼을 이용해 새로운 커버 텍스처를 개발하였다.

CHOCOLATE FOAM GLAZE

- I developed a new covering texture using coffee chocolate foam rather than the traditional glazing or spraying technique.

DESIGN
디자인

누구에게나 친근하고 익숙한 도넛 모양을 사용했지만, 커피 초콜릿 폼으로 글레이징해 기존에 볼 수 없었던 이색적인 디자인으로 완성했다.

It takes on a doughnut shape that everyone is familiar with, but I glazed it with coffee chocolate foam to create a unique design that has never been seen before.

✦ HOW TO COMPOSE THIS RECIPE ✦

STEP 1.	메뉴 정하기	티라미수
	Decide on the menu	Tiramisu

STEP 2.	메인 맛 정하기	커피
	Choose the primary flavor	Coffee

STEP 3. **메인 맛(커피)과의 페어링 선택하기**

Select a pairing flavor (Coffee)

- ☑ 다크초콜릿 Dark Chocolate
- ☑ 마스카르포네 Mascarpone
- ☑ 크림치즈 Cream Cheese
- ☑ 화이트초콜릿 White Chocolate

STEP 4. **구성하기**

Assemble

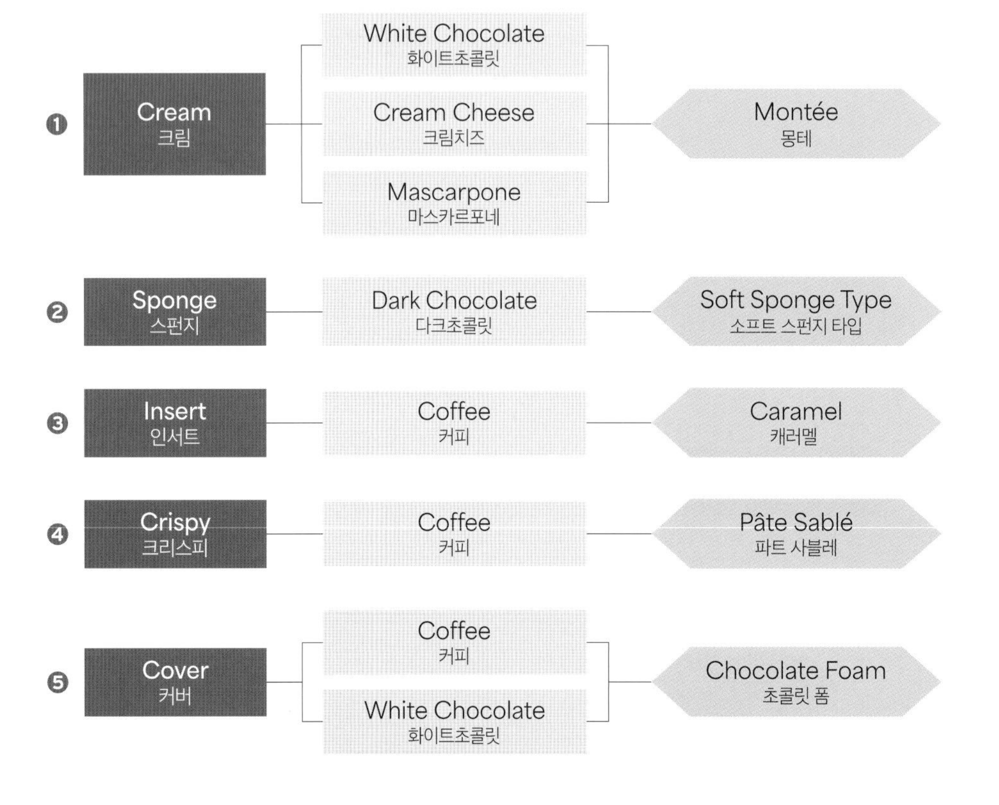

커피 초콜릿 폼
Coffee Chocolate Foam
마스카르포네 크림치즈 몽테
Mascarpone
eam Cheese
Montée
커피 초콜릿 사블레
Coffee Chocolate Sablé
커피 캐러멜
Coffee Caramel
카카오 스펀지
Cacao Sponge

INGREDIENTS 40개 분량/ 40 cakes

카카오 스펀지

Cacao Sponge

(철판(60 × 40cm) 1개 분량)
(1 baking tray (60 × 40 cm))

INGREDIENTS		g
마지팬	Marzipan	180g
설탕A	Sugar A	80g
노른자	Egg yolks	140g
달걀	Eggs	100g
다크초콜릿 (Fleur de cao 70%)	Dark chocolate	70g
버터 (Bridel)	Butter	70g
바닐라에센스	Vanilla essence	10g
박력분	Cake flour	130g
카카오파우더 (Extra Brute)	Cacao powder	30g
흰자	Egg whites	170g
설탕B	Sugar B	90g
TOTAL		**1070g**

커피 초콜릿 사블레

Coffee Chocolate Sablé

INGREDIENTS		g
달걀	Eggs	150g
인스턴트 커피 가루	Instant coffee powder	10g
버터 (Bridel)	Butter	200g
소금	Salt	4g
박력분	Cake flour	400g
아몬드 파우더	Almond powder	40g
옥수수전분	Cornstarch	125g
카카오파우더 (Extra Brute)	Cacao powder	50g
슈거파우더	Sugar powder	200g
TOTAL		**1179g**

*** 만드는 법**

❶ 비커에 달걀(30℃), 인스턴트 커피 가루를 넣고 바믹서로 블렌딩한다.
❷ 믹싱볼에 차가운 상태의 버터를 넣고 가볍게 풀어준 후 슈거파우더, 소금을 넣고 크림화한다.(비터 사용)
❸ 체 친 [박력분, 아몬드파우더, 옥수수전분, 카카오파우더]를 넣고 믹싱한다.
❹ ❶을 조금씩 나눠 넣어가며 믹싱한다.
❺ 반죽을 한 덩어리로 만들어 랩으로 감싸 냉장고에서 숙성시킨다.
❻ 숙성한 반죽을 2.5mm로 밀어 편다.
❼ 지름 6cm와 3cm 원형 커터로 도넛 모양으로 자른 후 굽기 전까지 냉동고에 보관한다.
❽ 철판 - 타공매트 - 사블레 - 타공매트 순서로 올린 후 180℃로 예열된 오븐에서 10분간 굽는다.

*** Procedure**

❶ Blend eggs (30°C) and instant coffee powder in a beaker with an immersion blender.
❷ In a mixing bowl, lightly beat cold butter with a paddle. Cream it with powdered sugar and salt using the paddle.
❸ Mix with sifted [cake flour, almond powder, cornstarch, and cacao powder].
❹ Gradually add ❶ and combine.
❺ When the dough comes together, cover with plastic wrap and refrigerate.
❻ Roll out the refrigerated dough into 2.5 mm thickness.
❼ Cut into doughnut shapes using 6 cm and 3 cm round rings and store in the freezer until use.
❽ Place baking sheet, perforated mat, sablé, and perforated mat in order. Bake in an oven preheated to 180°C for 10 minutes.

마스카르포네 크림치즈 몽테

Mascarpone Cream cheese Montée

INGREDIENTS		g
생크림A	Heavy cream A	250g
노른자	Egg yolks	160g
젤라틴매스 (×5)	Gelatin mass (×5)	28g
화이트초콜릿	White chocolate	120g
연유	Condensed milk	50g
마스카르포네	Mascarpone cheese	250g
크림치즈 (Kiri)	Cream cheese	125g
휘핑크림	Whipping cream	200g
생크림B	Heavy cream B	200g
TOTAL		**1383g**

커피 캐러멜

Coffee Caramel

INGREDIENTS		g
설탕	Sugar	95g
물엿	Corn syrup	155g
생크림	Heavy cream	250g
인스턴트 커피 가루	Instant coffee powder	8g
소금	Salt	2g
TOTAL		**510g**

커피 초콜릿 폼

Coffee Chocolate Foam

INGREDIENTS		g
화이트초콜릿 (Callebaut, Malchoc White 31%)	White chocolate	1000g
식용유	Vegetable oil	200g
인스턴트 커피 가루	Instant coffee powder	28g
TOTAL		**1228g**

커피 시럽

Coffee Syrup

INGREDIENTS		g
TPT 시럽	TPT syrup	700g
인스턴트 커피 가루	Instant coffee powder	50g
TOTAL		**750g**

- TPT 시럽 = 물과 설탕을 1:1 비율로 끓여 만든 것으로, 식혀 사용한다.
- TPT 시럽과 인스턴트커피 가루를 냄비에 넣고 녹을 때까지 가열한 후 식혀 사용한다.
- TPT syrup: A syrup made by boiling water and sugar in a 1:1 ratio. Use after cooling.
- Heat TPT syrup and instant coffee power until the coffee melts. Use after cooling.

NOTE.

✤

CACAO SPONGE

카카오 스펀지

1 써머믹서에 마지팬(30℃), 설탕A, 노른자와 달걀(30℃)을 넣고 갈아준다.

2 **1**이 완전히 섞이면 녹인 다크초콜릿과 버터, 바닐라에센스를 넣고 갈아준다.

3 체 친 박력분, 카카오파우더를 넣고 갈아준다.

4 볼에 흰자(30℃), 설탕B를 넣고 휘핑해 머랭을 만든다.

5 **4**에 **3**을 두 번 나눠 넣어가며 골고루 섞는다.

6 테프론시트를 깐 철판(60 × 40cm)에 고르게 팬닝한 후, 180℃로 예열된 오븐에서 14분간 굽는다.

7 완전히 식으면 5.5cm, 10.5cm 링으로 재단하고 커피 시럽을 전체적으로 적신 후 냉동한다.

TIP. 6cm 링으로 재단한 후 3.4cm 링으로 가운데를 한 번 더 재단해 도넛 모양으로 만든다.

1 Blend marzipan (30°C), sugar A, egg yolks and eggs (30°C) in Thermomix.

2 Once (**1**) is thoroughly combined, add melted chocolate with butter, and vanilla essence and blend.

3 Blend again with sifted cake flour and cacao powder.

4 In a bowl, whip egg whites (30°C) and sugar B to make meringue.

5 Add (**3**) into (**4**) about half at a time; mix thoroughly.

6 Spread on a baking tray (60 × 40 cm) lined with a Teflon sheet. Bake in an oven preheated to 180°C for 14 minutes.

7 When completely cooled, cut with the 5.5 cm and 10.5 cm round rings. Soak in coffee syrup and keep them in the freezer.

TIP. Cut with the 6 cm round ring first, then the center with 3.4 cm ring to make the doughnut shapes.

30℃
1

2

3-1

3-2

4

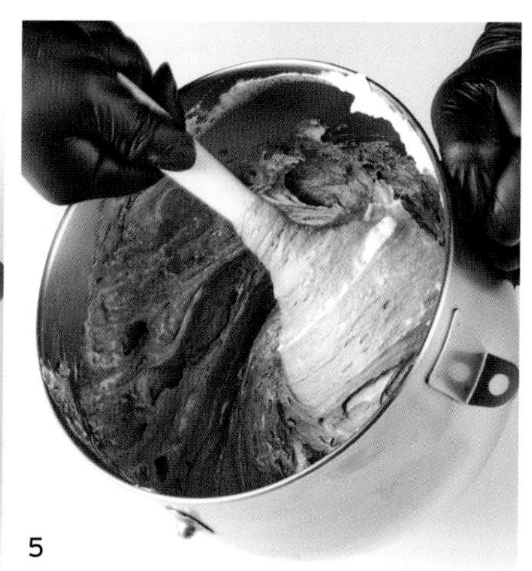
5

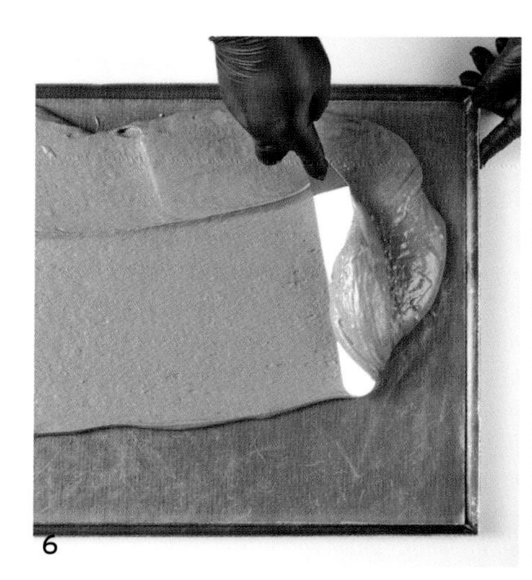
6

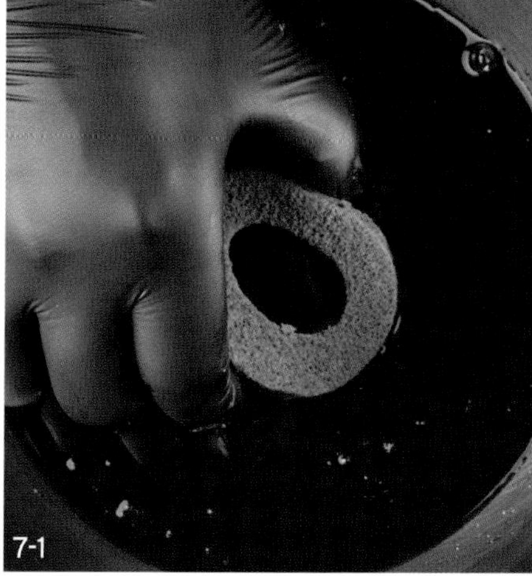
7-1

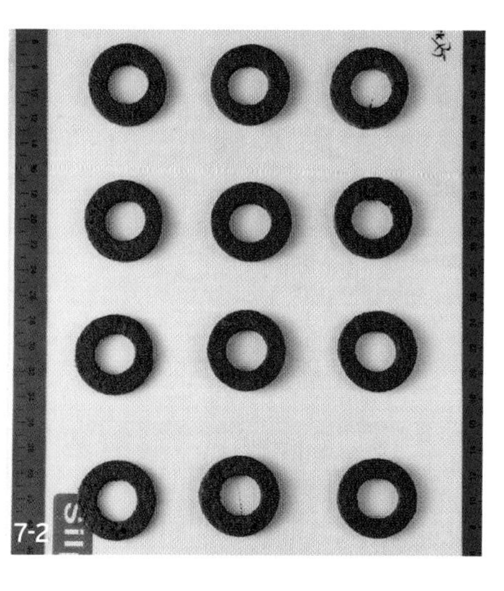
7-2

✤ ✤

MASCARPONE CREAM CHEESE MONTÉE

마스카르포네 크림치즈 몽테

1 냄비에 생크림A를 넣고 따뜻한 정도로 가열한다. (약 70℃)

2 노른자(30℃)가 담긴 볼에 **1**을 넣고 섞는다.

3 다시 냄비로 옮겨 75℃로 가열한다.

4 불에서 내린 후 젤라틴매스를 넣고 녹을 때까지 섞는다.

TIP. 노른자가 익는 현상이 질감으로 나타나지만 바믹서로 갈아주기 때문에 신경쓰지 않아도 된다.

5 화이트초콜릿이 담긴 비커에 **4**와 연유를 넣고 바믹서로 블렌딩한다.

6 마스카르포네, 크림치즈를 넣고 블렌딩한다.

7 차가운 상태의 휘핑크림과 생크림B를 넣고 블렌딩한다.

8 바트에 담아 밀착 랩핑해 냉장 보관한다.

9 사용하기 직전 휘핑(80% 정도)해 사용한다.

1 Heat heavy cream A in a saucepan until warm (about 70°C).

2 Whisk in (**1**) into a bowl with egg yolks (30°C).

3 Pour it back into the saucepan and heat to 75°C.

4 Remove from the heat and stir in gelatin mass until melted.

TIP. You may notice the texture of cooked egg yolks, but you don't have to worry—they'll be blended with an immersion blender.

5 Blend (**4**) and condensed milk in a beaker with white chocolate using an immersion blender.

6 Blend with mascarpone and cream cheese.

7 Blend again with cold whipping cream and heavy cream B.

8 Pour it into a tray and cover it with plastic wrap, making sure the wrap is in contact with the cream, and refrigerate.

9 Whip to about 80% just before use.

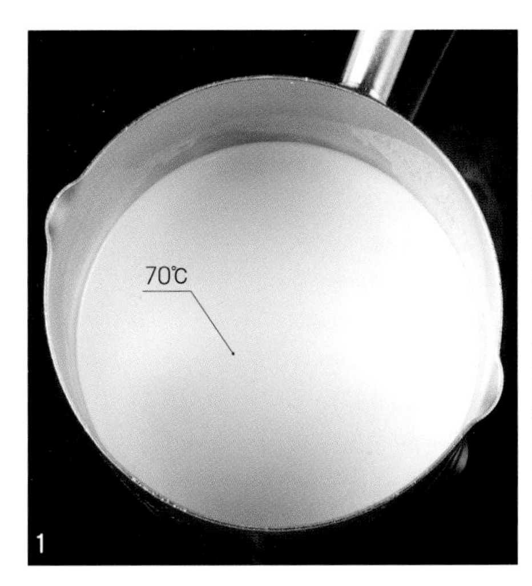

1

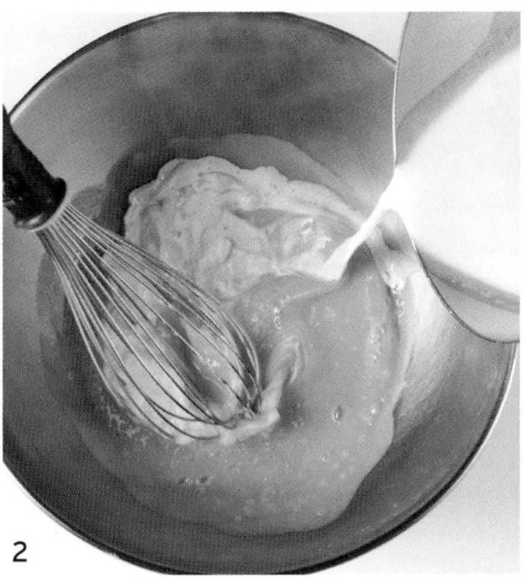
2

3

4

5

6

7

8

9

✤ ✤ ✤

COFFEE CARAMEL

커피 캐러멜

1 냄비에 설탕을 두세 번 나눠 넣어가며 진한 갈색이 될 때까지 캐러멜화시킨다.

TIP. 동시에 따뜻하게 데운 생크림에 인스턴트 커피 가루를 넣고 섞은 후 뜨겁게 데워 준비한다.

2 뜨겁게 데운 물엿을 넣고 휘퍼로 저어가며 섞는다.

3 불을 끈 후 뜨겁게 데워둔 생크림과 인스턴트 커피 가루를 넣고 휘퍼로 저어가며 섞는다.

4 105℃까지 가열한다.

5 불에서 내려 소금을 넣고 섞는다.

6 얼음물이 담긴 볼에 받쳐 30℃까지 쿨링한다.

7 몰드에 12g씩 채운다.

8 스패출러로 깔끔하게 정리하고 급속 냉동(-35 ~ -40℃)한다.

1 In a pot, caramelize the sugar in two or three batches until deep brown.

TIP. At the same time, stir the instant coffee powder into warm whipping cream and heat until hot.

2 Add hot corn syrup and whisk to combine.

3 Remove from the heat, add the hot cream mixed with the coffee powder, and combine with a whisk.

4 Cook to 105°C.

5 Remove from heat and mix with salt.

6 Let it cool to 30°C over an ice bath.

7 Fill the molds with 12 grams of the caramel.

8 Trim with an offset spatula and blast freeze (-35 ~ -40°C).

1

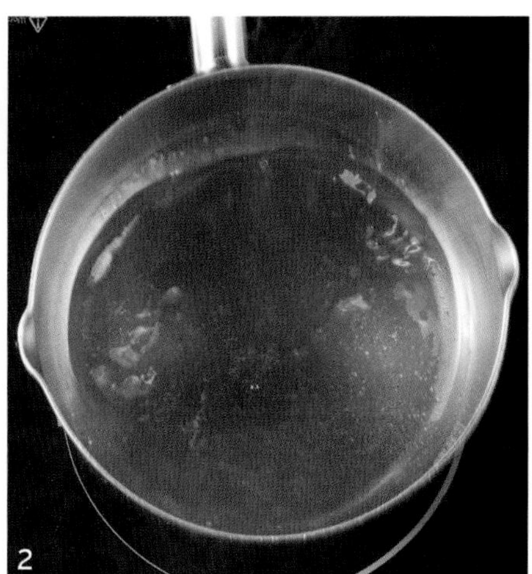
2

3

4

5

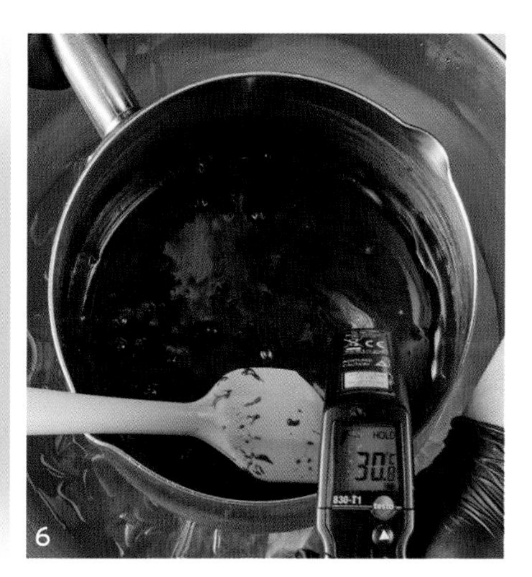
6

7

8

MONTAGE

몽타주

1 도넛 모양 몰드(실리코마트 SF354)에 마스카르포네 크림치즈 몽테를 30g씩 파이핑한다.

2 얼린 커피 캐러멜을 넣는다.

3 마스카르포네 크림치즈 몽테를 6g씩 파이핑한다.

4 얼린 카카오 스펀지를 올린다.

5 급속냉동고(-35 ~ -40℃)에서 얼린다.

1 Pipe 30 grams of mascarpone cream cheese montée into the doughnut-shaped molds (Silikomart SF354).

2 Insert the frozen coffee caramel.

3 Pipe 6 grams of mascarpone cream cheese montée.

4 Top with the frozen cacao sponge.

5 Freeze in the blast freezer (-35 ~ -40°C).

1
2
3

4
5

✤ ✤ ✤ ✤ ✤

COFFEE CHOCOLATE FOAM & FINISH

커피 초콜릿 폼 & 마무리

1 비커에 템퍼링한 화이트초콜릿, 식용유, 인스턴트 커피 가루를 넣고 바믹서로 블렌딩한다.

2 열풍기로 30~35℃로 데운 사이폰 통(500ml)에 담는다.

3 가스를 1차로 12초 충전하고 뒤집어 흔들어준 후, 2차로 10초 충전하고 다시 뒤집어 흔들어준다. (충전용 가스 20초)

TIP. 흔들어준 사이폰 통을 5분 정도 그대로 두어 안정화시킨 후 사용한다.

4 커피 초콜릿 폼을 볼에 분사한다.

5 얼린 무스를 커피 초콜릿 폼에 담갔다 빼내 자연스러운 모양으로 코팅한다.

6 웜 매트 또는 따뜻하게 데운 철판에서 밑면을 깔끔하게 정리한다.

7 커피 사블레 위에 올려 급속냉동고(-35 ~ -40℃)에서 빠르게(약 10분) 얼린 후, 냉장고에 보관한다.

TIP. 작업 직후 급속 냉동하지 않으면 불안정한 커피 초콜릿 폼이 액체 상태로 다시 변하면서 주저앉아버릴 수 있다.

1 Blend tempered white chocolate, vegetable oil, and instant coffee powder in a beaker with an immersion blender.

2 Pour into a siphon bottle (500 ml) warmed to 30~35°C with a hot air gun.

3 Charge with gas for 12 seconds, turn it over and shake, then charge again for 10 seconds, turn it over, and shake again. (20 seconds of gas for charging)

TIP. Leave the shaken siphon for 5 minutes to stabilize before using.

4 Dispense the coffee chocolate foam into a bowl.

5 Dip the frozen mousse into the coffee chocolate foam and remove to coat in its natural shape.

6 Use a warm mat or a warmed baking sheet to neatly clean up the bottom of the cake.

7 Place on the coffee sable, blast freeze (-35 ~ -40°C) quickly for about 10 minutes, and refrigerate.

TIP. If you don't freeze them quickly immediately after glazing, the unstable coffee foam may change back to liquid and collapse.

1

2

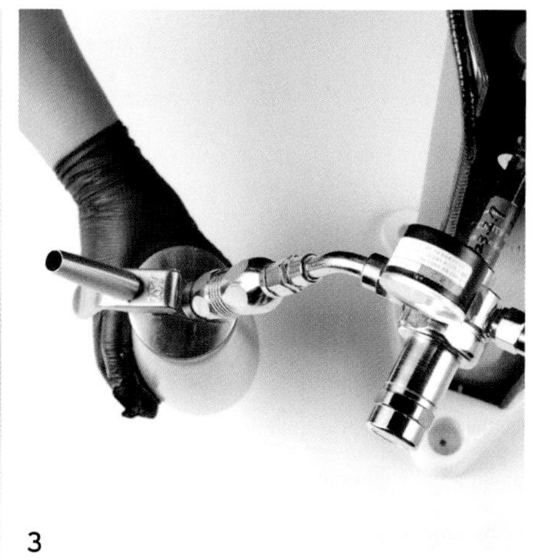

3

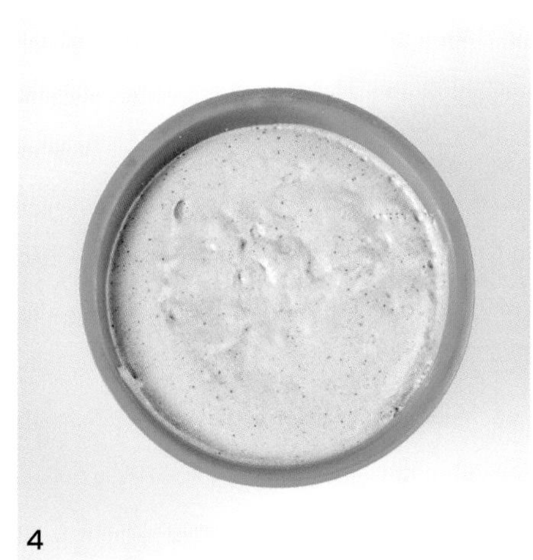

4

5-1

5-2

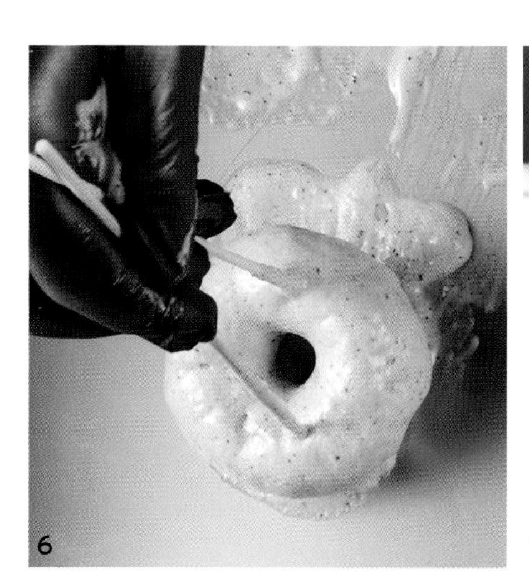

6

7-1

7-2

18

GREEN CACAO
그린 카카오

PAIRING & TEXTURE
페어링 & 텍스처

- 말차와 팥을 사용해 가장 보편적이면서도 기본적인 페어링을 연출했다.
- 말차의 맛을 최대한 극대화시켜 말차 마니아들의 니즈를 충족시킨 제품이다.

- This dessert is made with matcha and red beans, the most common and basic pairings.
- I maximized the flavor of matcha to satisfy the needs of matcha enthusiasts.

TECHNIQUE
테크닉

몰드에 템퍼링한 카카오 컬러로 명암과 음영을 표현해 글로시하고 입체적인 질감으로 완성했다.

The mold is coated with tempered cacao color to create contrast and shading for a glossy, three-dimensional texture.

DESIGN
디자인

카카오 농장을 처음 견학했을 때 직접 수확한 카카오 포드는 초록색과 노란색이 예쁘게 섞여 있었다. 이 기억을 바탕으로 정말 리얼한 카카오 포드를 연출하기 위해 색 분사 작업을 먼저 한 후 크림을 넣는 방식을 선택했다.

When I first visited a cacao farm, the cacao pods I picked were a beautiful mix of green and yellow. Based on this memory, I color-sprayed the mold first and then added the cream to create a genuinely realistic cacao pod.

✦ HOW TO COMPOSE THIS RECIPE ✦

STEP 1.	**메뉴 정하기** Decide on the menu	**프티 가토** Petit Gateau
STEP 2.	**메인 맛 정하기** Choose the primary flavor	**말차** Matcha

STEP 3. **메인 맛(말차)과의 페어링 선택하기**

Select a pairing flavor (Matcha)

- ☑ 팔 — Red Bean
- ☑ 마스카르포네 — Mascarpone
- ☑ 화이트초콜릿 — White Chocolate
- ☑ 바닐라 — Vanilla

STEP 4. **구성하기**

Assemble

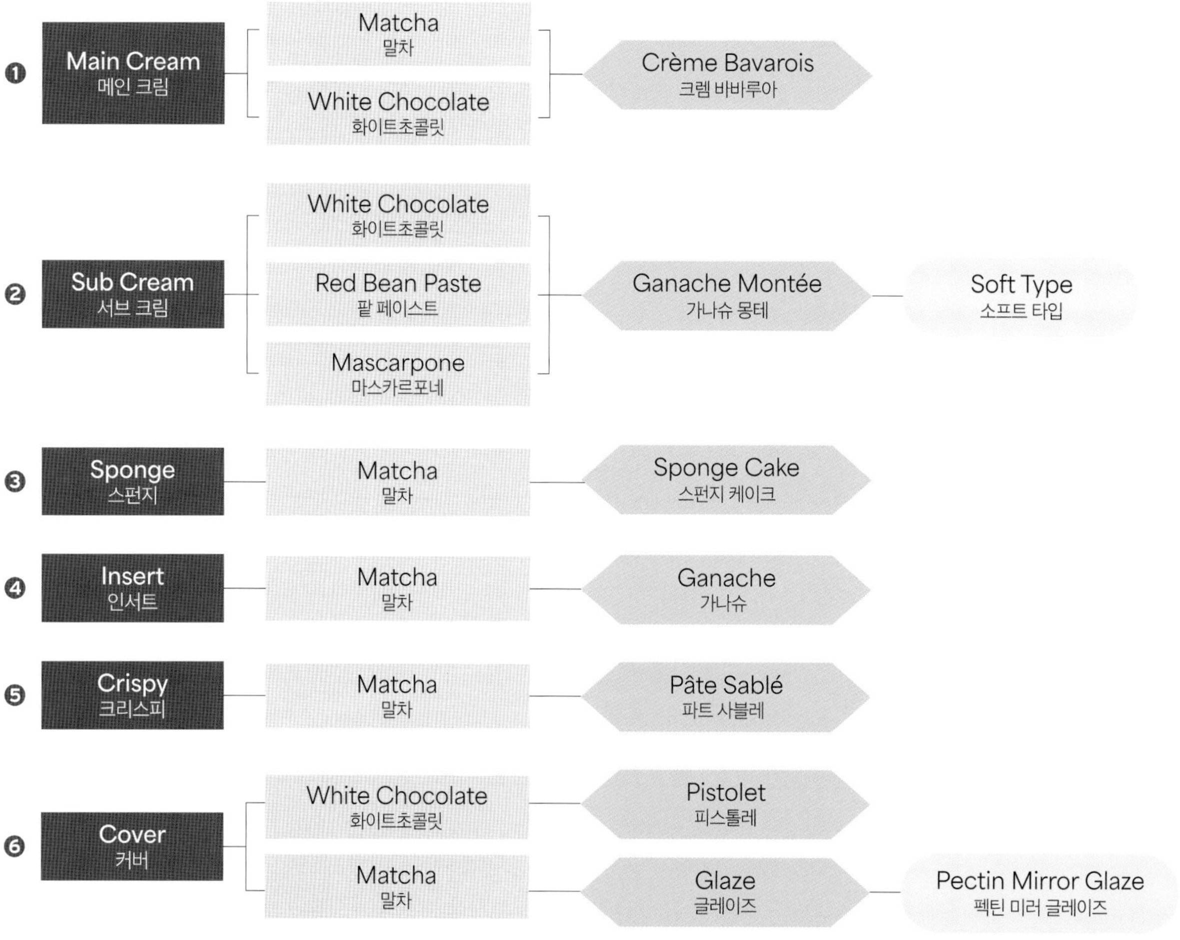

	Component	Flavor	Type	Detail
❶	Main Cream 메인 크림	Matcha 말차 / White Chocolate 화이트초콜릿	Crème Bavarois 크렘 바바루아	
❷	Sub Cream 서브 크림	White Chocolate 화이트초콜릿 / Red Bean Paste 팥 페이스트 / Mascarpone 마스카르포네	Ganache Montée 가나슈 몽테	Soft Type 소프트 타입
❸	Sponge 스펀지	Matcha 말차	Sponge Cake 스펀지 케이크	
❹	Insert 인서트	Matcha 말차	Ganache 가나슈	
❺	Crispy 크리스피	Matcha 말차	Pâte Sablé 파트 사블레	
❻	Cover 커버	White Chocolate 화이트초콜릿	Pistolet 피스톨레	
		Matcha 말차	Glaze 글레이즈	Pectin Mirror Glaze 펙틴 미러 글레이즈

옐로우 & 그린 스프레이
Yellow &
Green Spray
말차 펙틴 미러 글레이즈
Matcha Pectin
Mirror Glaze
팥 크림
Red Bean Cream
말차 바바루아
Matcha Bavarois
말차 가나슈
Matcha Ganache
말차 사블레
Matcha Sablé
말차 스펀지
Matcha Sponge

INGREDIENTS

30개 분량/ 30 cakes

말차 사블레
Matcha Sablé

INGREDIENTS		g
버터 (Bridel)	Butter	180g
슈거파우더	Sugar powder	140g
박력분	Cake flour	340g
아몬드파우더	Almond powder	40g
말차파우더	Matcha powder	15g
소금	Salt	3g
달걀	Eggs	75g
TOTAL		**793g**

말차 스펀지
Matcha Sponge

(철판(60 × 40cm) 1개 분량)
(1 baking tray (60 × 40 cm))

INGREDIENTS		g
우유	Milk	75g
버터 (Bridel)	Butter	50g
말차파우더	Matcha powder	22g
노른자	Egg yolks	120g
달걀	Eggs	300g
물엿	Corn syrup	20g
설탕	Sugar	200g
박력분	Cake flour	130g
옥수수전분	Cornstarch	5g
TOTAL		**922g**

말차 가나슈
Matcha Ganache

INGREDIENTS		g
생크림	Heavy cream	75g
화이트초콜릿 (Zephyr white 34%)	White chocolate	180g
말차파우더	Matcha powder	12g
TOTAL		**267g**

바닐라 마스카르포네 몽테 *

Vanilla Mascarpone Montée

INGREDIENTS		g
휘핑크림A	Whipping cream A	150g
젤라틴매스 (×5)	Gelatin mass (×5)	25g
바닐라빈	Vanilla bean	1/2개 분량/ 1/2 pc
화이트초콜릿 (Zephyr white 34%)	White chocolate	100g
연유	Concentrated milk	40g
휘핑크림B	Whipping cream B	350g
마스카르포네	Mascarpone	100g
골드럼 (PAN RUM)	Gold rum	6g
TOTAL		**771g**

*** 만드는 법**

❶ 냄비에 휘핑크림A, 젤라틴매스, 바닐라빈 껍질을 넣고 70℃로 가열한다.
❷ 화이트초콜릿, 연유, 바닐라빈 씨가 담긴 비커에 넣고 바믹서로 블렌딩한다.
❸ 마스카르포네 - 휘핑크림B - 골드럼 순서로 넣어가며 블렌딩한다.
❹ 바트에 담아 밀착 랩핑한 후 냉장고에서 12시간 동안 숙성시킨다.
❺ 사용할 때는 비터로 부드럽게 풀어 사용한다.

*** Procedure**

❶ Heat whipping cream A, gelatin mass, and vanilla bean pod (without the seeds) in a saucepan to 70°C.
❷ Pour into a beaker with white chocolate, condensed milk, and vanilla bean seeds. Combine with an immersion blender.
❸ Blend in the order of mascarpone – whipping cream B – gold rum.
❹ Pour onto a stainless-steel tray, cover with plastic wrap, making sure the wrap is in contact with the cream, and refrigerate for 12 hours.
❺ Gently beat to loosen with a paddle to use.

팥 크림

Red Bean Cream

INGREDIENTS		g
바닐라 마스카르포네 몽테 *	Vanilla mascarpone montée *	600g
통팥	Red bean paste	250g
TOTAL		**850g**

크렘 앙글레이즈 (되직한 타입) *

Crème Anglaise (cooked thicker)

• 1권 Book 1, p.172

INGREDIENTS		g
우유	Milk	250g
바닐라빈	Vanilla bean	1/2개 분량/ 1/2 pc
설탕	Sugar	50g
노른자	Egg yolks	100g
TOTAL		**400g**

*** 만드는 법**

❶ 냄비에 우유, 바닐라빈 껍질, 설탕 1/2을 넣고 80℃까지 가열한다.

❷ 볼에 노른자(30℃), 남은 설탕을 넣고 거품이 생길 때까지 가볍게 섞는다.

❸ ❷에 ❶을 천천히 넣어가며 섞는다.

❹ 다시 냄비로 옮겨 중불에서 83~85℃까지 가열한다.

❺ 불에서 내려 83~85℃ 온도를 1~2분간 유지한다.
- 균이 사멸하는 온도에서 일정 시간 두어 살균되는 시간을 준다.

❻ 체에 거른 후 바닐라빈 씨를 넣고 블렌더로 섞는다. 완성된 크렘 앙글레이즈는 넓은 바트에 펼쳐 담고 밀착 랩핑해 냉장 보관한다.

*** Procedure**

❶ Heat milk, vanilla bean pod (without the seeds), and half sugar to 80°C.

❷ In a separate bowl, whisk egg yolks (30°C) and the remaining sugar lightly until it starts to get foamy.

❸ Gradually whisk in ❶ into ❷.

❹ Pout it back in the pot and cook to 83~85°C over a medium heat.

❺ Remove from the heat and retain 83~85°C for 1~2 minute.
- Allow it some time for sterilization by leaving it at a temperature for a while where bacteria die.

❻ Strain and blend with vanilla bean seeds. Spread the crème anglaise on a large stainless-steel tray, wrap it tightly, and refrigerate.

말차 바바루아

Matcha Bavarois

INGREDIENTS		g
크렘 앙글레이즈 (되직한 타입) *	Crème anglaise * (cooked thicker)	290g
젤라틴매스 (×5)	Gelatin mass (×5)	60g
화이트초콜릿 (Zephyr white 34%)	White chocolate	280g
말차파우더	Matcha powder	36g
연유	Condenced milk	80g
그린티 리큐르 (Dijon The Vert)	Green tea liqueur	80g
생크림	Heavy cream	750g
TOTAL		**1576g**

말차 펙틴 미러 글레이즈

Matcha Pectin Mirror Glaze

INGREDIENTS		g
말차 우린 물	Matcha-infused water	400g
물엿	Corn syrup	250g
설탕A	Sugar A	90g
NH펙틴	NH pectin	15g
구연산삼나트륨	Trisodium citrate	4g
설탕B	Sugar B	730g
젤라틴매스 (×5)	Gelatin mass (×5)	165g
TOTAL		**1654g**

- 말차 우린 물은 말차파우더와 물을 1:100 비율로 섞어 사용한다.
  (미리 만들어두면 변색될 수 있으므로 사용하기 직전에 만든다.)
- Matcha-infused water: Mix matcha powder and water in a 1:100 ratio to infuse.
  (Make it just before use, as it may discolor if you make it beforehand.)

옐로우 스프레이

Yellow Spray

INGREDIENTS		g
카카오버터	Cacao butter	200g
화이트초콜릿 (Zephyr white 34%)	White chocolate	200g
노란색 색소	Yellow color	20g
TOTAL		**420g**

- 볼에 모든 재료를 넣고 녹여 바믹서로 블렌딩한 후 28~29℃로 템퍼링해 사용한다.
- Melt all the ingredients in a bowl, combine with an immersion blender, and temper to 28~29°C to use.

그린 스프레이

Green Spray

INGREDIENTS		g
카카오버터	Cacao butter	200g
화이트초콜릿 (Zephyr white 34%)	White chocolate	200g
초록색 색소	Green color	8g
말차파우더	Matcha powder	12g
TOTAL		**420g**

- 볼에 모든 재료를 넣고 녹여 바믹서로 블렌딩한 후 28~29℃로 템퍼링해 사용한다.
- Melt all the ingredients in a bowl, combine with an immersion blender, and temper to 28~29°C to use.

✤

MATCHA SABLÉ

말차 사블레

1 믹싱볼에 차가운 상태의 버터, 슈거파우더를 넣고 가볍게 풀어준다.

2 체 친 [박력분, 아몬드파우더, 말차파우더, 소금] 절반을 넣고 섞는다.

3 크럼블 상태가 되면 달걀(30℃)을 넣고 섞는다.

4 남은 가루 재료를 모두 넣고 섞는다.

5 날가루가 보이지 않으면 한 덩어리로 치대 랩핑한 후 사각형으로 만들어 냉장고에서 12시간 휴지시킨다.

6 두께 2.5mm로 밀어 펴 자르기 좋은 상태로 냉동한다. 카카오 포드 모양 커터로 자르고 철판 - 타공매트 - 사블레 - 타공매트 순서로 올린 후 150℃로 예열된 오븐에서 약 12분간 굽는다.

1 Lightly beat cold butter and sugar powder in a mixing bowl.

2 Add half of sifted [cake flour, almond powder, matcha powder, and salt].

3 When it turns crumbly, mix with eggs (30°C).

4 Mix with the remaining powder ingredients.

5 When the powder ingredients are no longer visible, knead the dough into a ball, wrap in plastic wrap, shape it into a rectangle, and refrigerate for 12 hours.

6 Roll out to 2.5 mm and freeze. Cut with a cacao pod-shaped cutter. Place in the order of baking sheet - perforated silicon mat - sablé - perforated silicon mat. Bake in an oven preheated to 150°C for about 12 minutes.

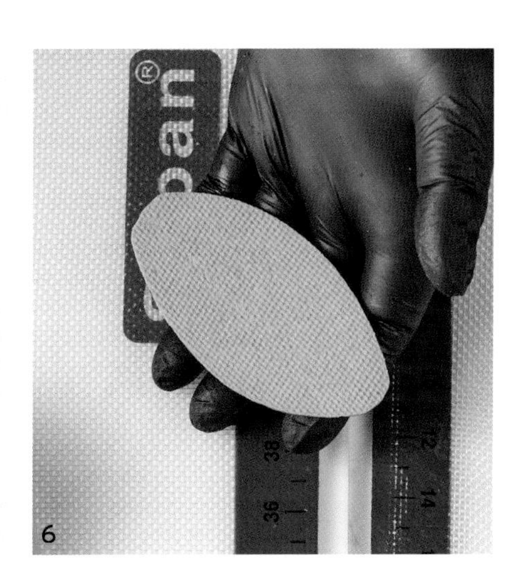

무스에 사용한 실리코마트 Cacao 120 몰드에 들어 있는 커터를 사용했다.

I used the cutter that came with the Silikomart Cacao 120 mold used for the mousse.

✤ ✤

MATCHA SPONGE

말차 스펀지

1 볼에 우유, 녹인 버터(45℃), 말차파우더를 담고 휘퍼로 섞는다.

2 바믹서로 블렌딩한다.

3 믹싱볼에 노른자, 달걀, 물엿, 설탕을 담고 따뜻한 물이 담긴 볼에 받쳐 저어가며 45~50℃로 온도를 올린다.

4 미색의 뽀얀 상태가 되고, 기공이 조밀하고 건강한 상태의 거품이 생길 때까지 휘핑한다.

5 **4**의 1/3을 덜어 **2**와 함께 섞는다.

6 체 친 박력분, 옥수수전분을 넣고 섞는다.

7 남은 **4**를 넣고 섞는다.

8 테프론시트를 깐 철판(60 × 40cm)에 팬닝한다.

9 데크 오븐 기준 윗불 200℃, 아랫불 180℃에서 16분간 굽는다.

1 Mix milk, melted butter (45°C), and matcha powder in a bowl with a whisk.

2 Combine using an immersion blender.

3 In a mixing bowl, add egg yolks, eggs, corn syrup, and sugar. Place it over warm water and whisk until it reaches 45~50°C.

4 Whip until it turns pale and forms a dense and steady foam.

5 Scoop out 1/3 of (**4**) and mix with (**2**).

6 Mix with sifted cake flour and cornstarch.

7 Mix with the remaining (**4**).

8 Spread onto a baking tray (60 × 40 cm) lined with a Teflon sheet.

9 Bake in a deck oven with 200°C on the top and 180°C on the bottom for 16 minutes.

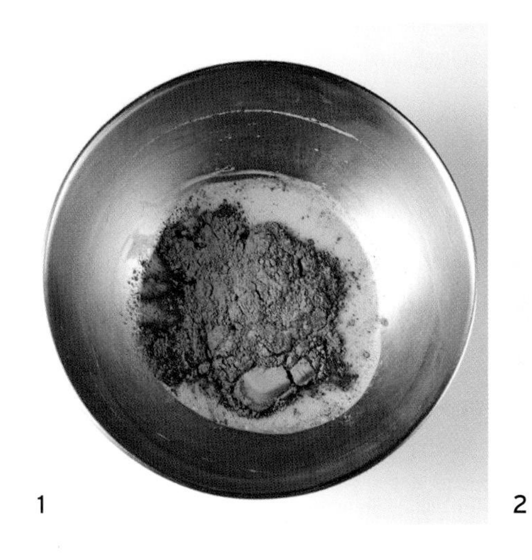
1

2

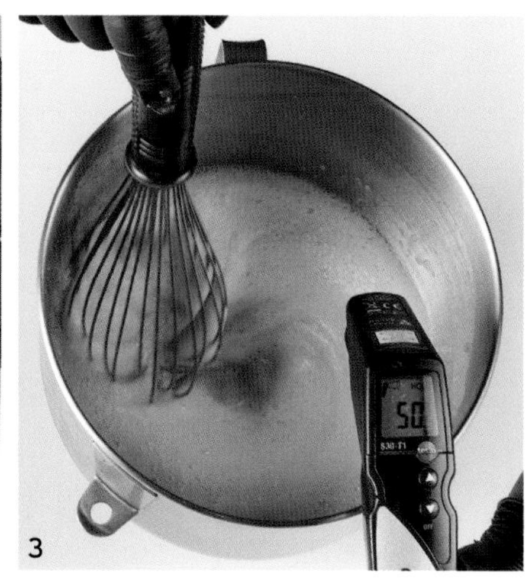

3

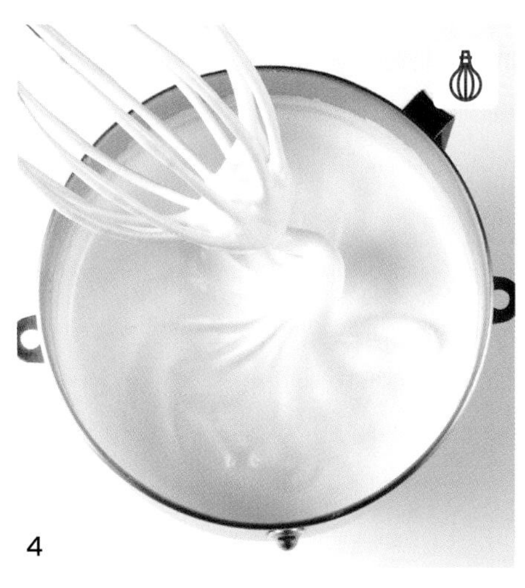
4

5

6

7

8

9

✤ ✤ ✤

MATCHA GANACHE

말차 가나슈

말차 가나슈

1 냄비에 생크림을 담고 70℃까지 가열한다.

2 비커에 녹인 화이트초콜릿(45~50℃)과 말차파우더를 넣고 섞은 후 **1**을 붓고 블렌딩한다.

조립

3 말차 스펀지에 말차 가나슈 240g을 고르게 펴 바른다. 말차 가나슈가 굳으면 카카오 포드 모양 커터로 자른 후 냉동한다.

TIP. 무스에 사용한 실리코마트 Cacao 120 몰드에 들어 있는 커터를 사용했다.

Matcha Ganache

1 Heat cream in a saucepan to 70°C.

2 Blend with white chocolate (45~50°C) and matcha powder in a beaker using an immersion blender. Add (**1**) and blend again.

Assemble

3 Evenly spread 240 grams of matcha ganache on the matcha sponge. When the ganache sets, cut with the cacao pod-shaped cutter and freeze.

TIP. I used the cutter that came with the Silikomart Cacao 120 mold used for the mousse.

✤ ✤ ✤ ✤

RED BEAN CREAM

팥 크림

1 믹싱볼에 모든 재료를 넣고 골고루 섞는다.

2 몰드에 23g씩 채운다.

3 스패출러로 평평하게 정리한 후 냉동한다.

1 Mix all the ingredients evenly in a mixing bowl.

2 Fill 23 grams in the mold.

3 Evenly spread with a spatula and freeze.

✤ ✤ ✤

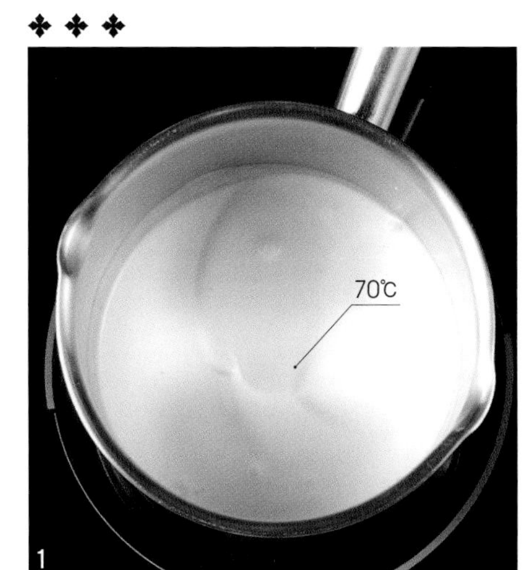

1

2

3

1

2

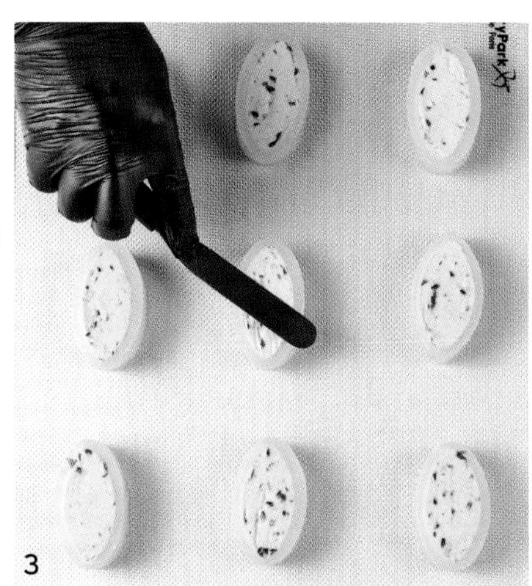
3

✤ ✤ ✤ ✤ ✤

MATCHA BAVAROIS

말차 바바루아

1 냄비에 크렘 앙글레이즈, 젤라틴매스를 넣고 젤라틴매스가 녹을 때까지 가열한다. (약 60℃)

2 녹인 화이트초콜릿(45~50℃)이 담긴 볼에 **1**과 말차파우더를 넣고 바믹서로 블렌딩한다.

3 연유를 넣고 블렌딩한다.

4 그린티 리큐르를 넣고 블렌딩한다.

5 믹싱볼에 생크림을 넣고 60~70% 정도로 휘핑한다.

6 **5**에 두세 번 나눠 넣어가며 섞는다.

7 얼음물이 담긴 볼에 받쳐 23℃로 쿨링해 사용한다.

1 Heat crème Anglaise and gelatin mass in a pot until the gelatin melts (about 60°C).

2 Pour into a bowl with white chocolate (45~50°C) and matcha powder; mix with an immersion blender.

3 Blend with condensed milk.

4 Add green tea liqueur and continue to blend.

5 Whip the heavy cream in a mixing bowl to 60~70%.

6 Mix with (**5**) in two or three batches.

7 Cool in an ice bath to 23°C to use.

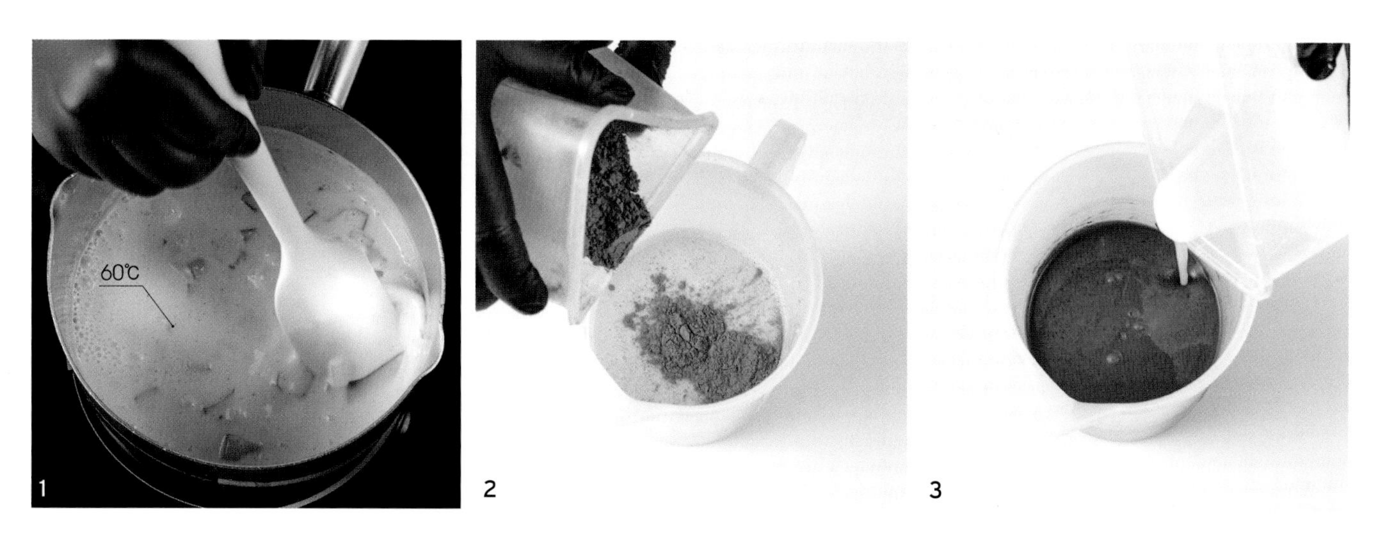
60℃
1
2
3

4
5
6

23℃
7

✤ ✤ ✤ ✤ ✤ ✤

MATCHA PECTIN MIRROR GLAZE

말차 펙틴 미러 글레이즈

1 냄비에 말차 우린 물, 물엿을 넣고 가열한다.

2 45℃가 되기 전에 미리 섞어둔 [설탕A, NH펙틴, 구연산삼나트륨]을 넣고 섞어가며 가열한다.

3 한번 끓어오르면 설탕B를 3~4번 나눠 넣어가며 가열한다.

4 설탕이 다 녹으면 펙틴 반응을 확인한 후 불에서 내려 젤라틴매스를 넣고 섞는다.

5 바믹서로 블렌딩한다.

6 글레이즈 표면에 키친타월을 덮은 후 걷어내 기포를 제거한다.

7 기포가 생기지 않게 주걱 위에 부어가며 바트에 옮긴다.

8 밀착 랩핑해 냉장고에서 보관한다.

TIP. 사용하기 직전 바믹서로 블렌딩해 30~35℃로 맞춰 사용한다.
말차가 변색될 수 있으므로 오래 보관하지 않는다.

1 Heat matcha-infused water and corn syrup in a pot.

2 Before it reaches 45°C, mix with previously mixed [sugar A and pectin NH with trisodium citrate];
heat while stirring.

3 Once it boils, add sugar B in three or four batches while cooking.

4 When all the sugar dissolves, and the pectin activates, remove from the heat and mix with gelatin mass.

5 Combine with an immersion blender.

6 Cover the glaze with a paper towel and remove to eliminate air bubbles.

7 Transfer into a stainless-steel tray while pouring it onto a spatula to avoid forming air bubbles.

8 Cover with plastic wrap, making sure the wrap is in contact with the glaze, and refrigerate.

TIP. Blend right before using, to 30~35°C. Do not store for too long as matcha may discolor.

1
2
3

4
5
6

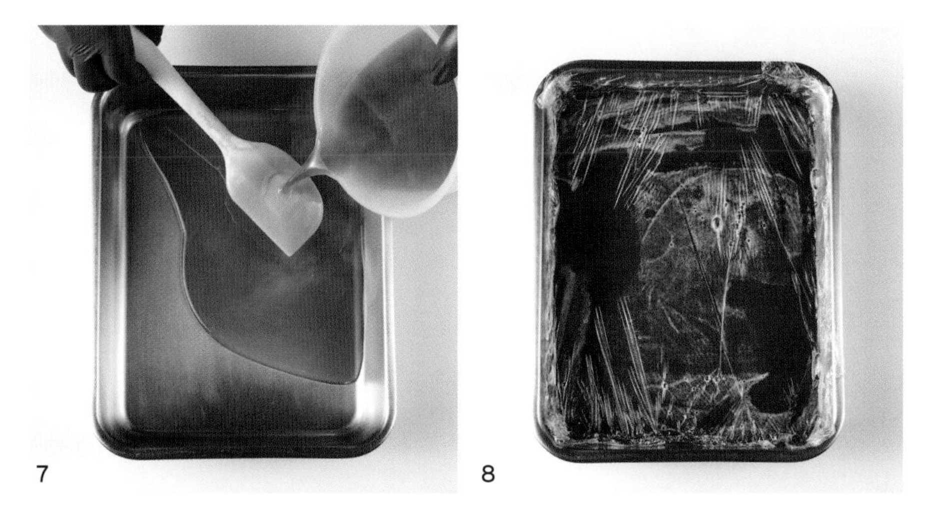
7
8

✤ ✤ ✤ ✤ ✤ ✤ ✤

FINISH 마무리

1 카카오포드 모양 몰드(실리코마트 Cacao 120)에 템퍼링한 옐로우 & 그린 스프레이를 분사한다.

2 스프레이를 분사한 직후 말차 바바루아를 40g씩 채운다.

TIP. 스프레이가 굳은 후 바바루아를 채우면 스프레이에 크랙이 생겨 완성도가 떨어진다.

3 얼린 팥 크림을 넣는다.

4 말차 바바루아를 10g 채운다.

5 얼린 [말차 스펀지 + 말차 가나슈]를 넣는다.

6 스패출러로 깔끔하게 정리한 후 급속냉동고(-30 ~ -40℃)에서 빠르게 얼린다.

7 얼린 무스를 몰드에서 빼내 냉동고(-18℃)에 옮긴 후 냉동고 온도가 되면 말차 펙틴 미러 글레이즈(30~35℃)로 코팅한다.

8 무스 바닥을 깔끔하게 정리한 후 말차 사블레 위에 올려 마무리한다.

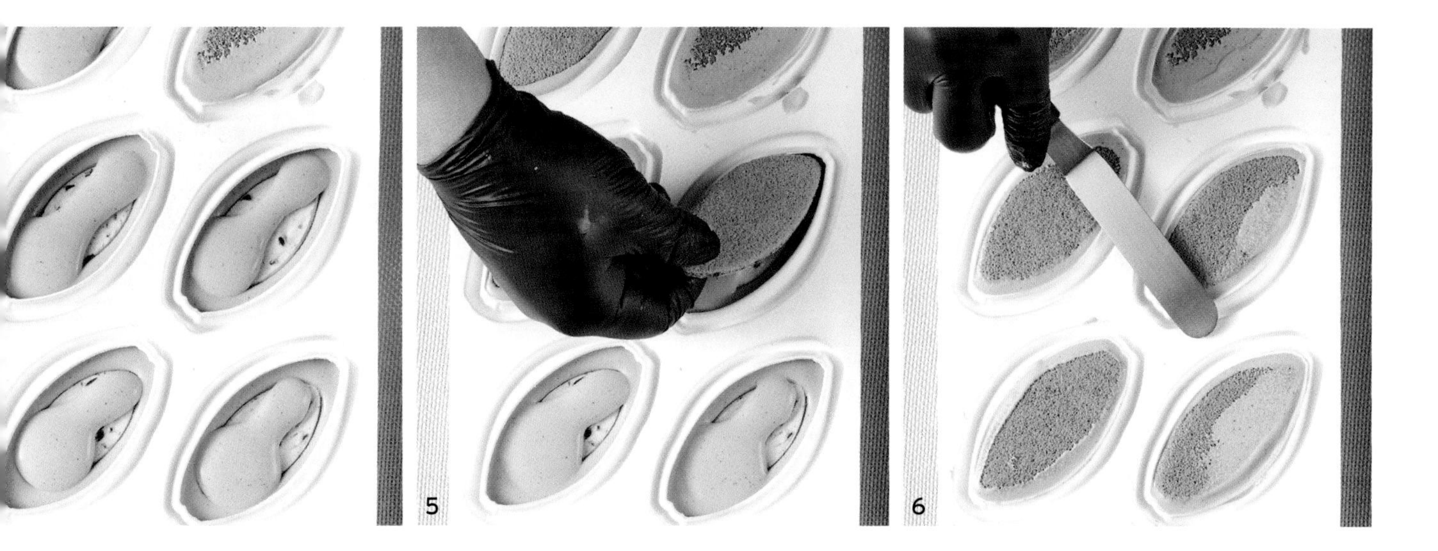

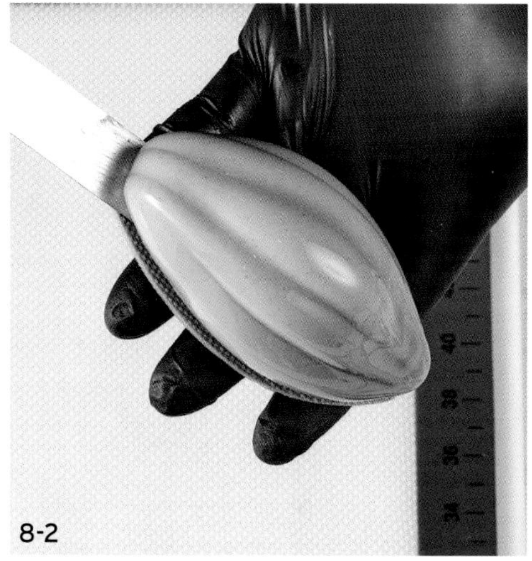

1 Spray yellow and green sprays in the cacao pod-shaped mold (Silikomart Cacao 120).

2 Fill 40 grams of matcha bavarois immediately after spraying.

TIP. If you fill the bavarois after the sprayed mixture sets, it will crack and lower the level of completion.

3 Insert the frozen red bean cream.

4 Fill 10 grams of matcha bavarois.

5 Insert the frozen [matcha sponge + matcha ganache].

6 Organize neatly with an offset spatula and freeze rapidly in a blast freezer (-30 ~ -40°C).

7 Remove the frozen mousse from the mold and store in the freezer (-18°C). When it reaches the freezer temperature, coat with the matcha pectin mirror glaze (30~35°C).

8 Neatly organize the bottom of the mousse and place on the matcha sablé to finish.

19

TROPICAL ROSEMARY

트로피컬 로즈마리

PAIRING & TEXTURE

페어링 & 텍스처

코코넛, 패션푸르트, 로즈마리와 페어링한 여름 디저트를 표현했다.

It is a dessert depicting summer, which I paired with coconut, passion fruit, and rosemary.

TECHNIQUE

테크닉

몰드의 형태에 맞춰 데커레이션 크림의 모양을 잡아 그대로 얼려 오차 없이 완벽하게 밀착되도록 만들었다.

The decoration cream was prepared according to the shape of the mold and frozen for a perfect fit.

DESIGN

디자인

- 정형화된 모양의 시판 몰드를 사용했지만 여기에 개성 있는 형태의 크림을 만들어 올려 독특한 디자인으로 만들었다.
- 케이크는 글로시하게, 데커레이션 크림은 매트하게 마무리해 상반되는 질감을 동시에 느낄 수 있게 연출했다.

- I used a commercially available mold with a standard shape, but the unique form of the cream created an exclusive design.
- The cake is glossy, and the decoration cream is matte to display the contrasting textures in one dessert.

✤ HOW TO COMPOSE THIS RECIPE ✤

STEP 1.	메뉴 정하기	앙트르메
	Decide on the menu	Entremets

STEP 2.	메인 맛 정하기	파인애플
	Choose the primary flavor	Pineapple

STEP 3. **메인 맛(파인애플)과의 페어링 선택하기**

Select a pairing flavor (Pineapple)

- ☑ 코코넛 Coconut
- ☑ 로즈마리 Rosemary
- ☑ 사워크림 Sour Cream
- ☑ 아몬드 Almond
- ☑ 패션푸르트 Passion Fruit

STEP 4. **구성하기**

Assemble

	Component	Flavor	Type	Detail
❶	Main Cream 메인 크림	Coconut 코코넛 Rosemary 로즈마리	Chocolate Mousse 초콜릿 무스	
❷	Sub Cream 서브 크림	Sour Cream 사워크림 Mascarpone 마스카르포네	Ganache Montée 가나슈 몽테	Thicker Type 되직한 타입
❸	Sponge 스펀지	Coconut 코코넛	Madeleine 마들렌	
❹	Insert 인서트	Pineapple 파인애플 Passion Fruit 패션푸르트	Compote 콩포트	
❺	Crispy 크리스피	Almond 아몬드	Crunch 크런치	
❻	Cover 커버	Mint 민트	Glaze 글레이즈	Mirror Glaze 미러 글레이즈
		White Chocolate 화이트초콜릿	Pistolet 피스톨레	

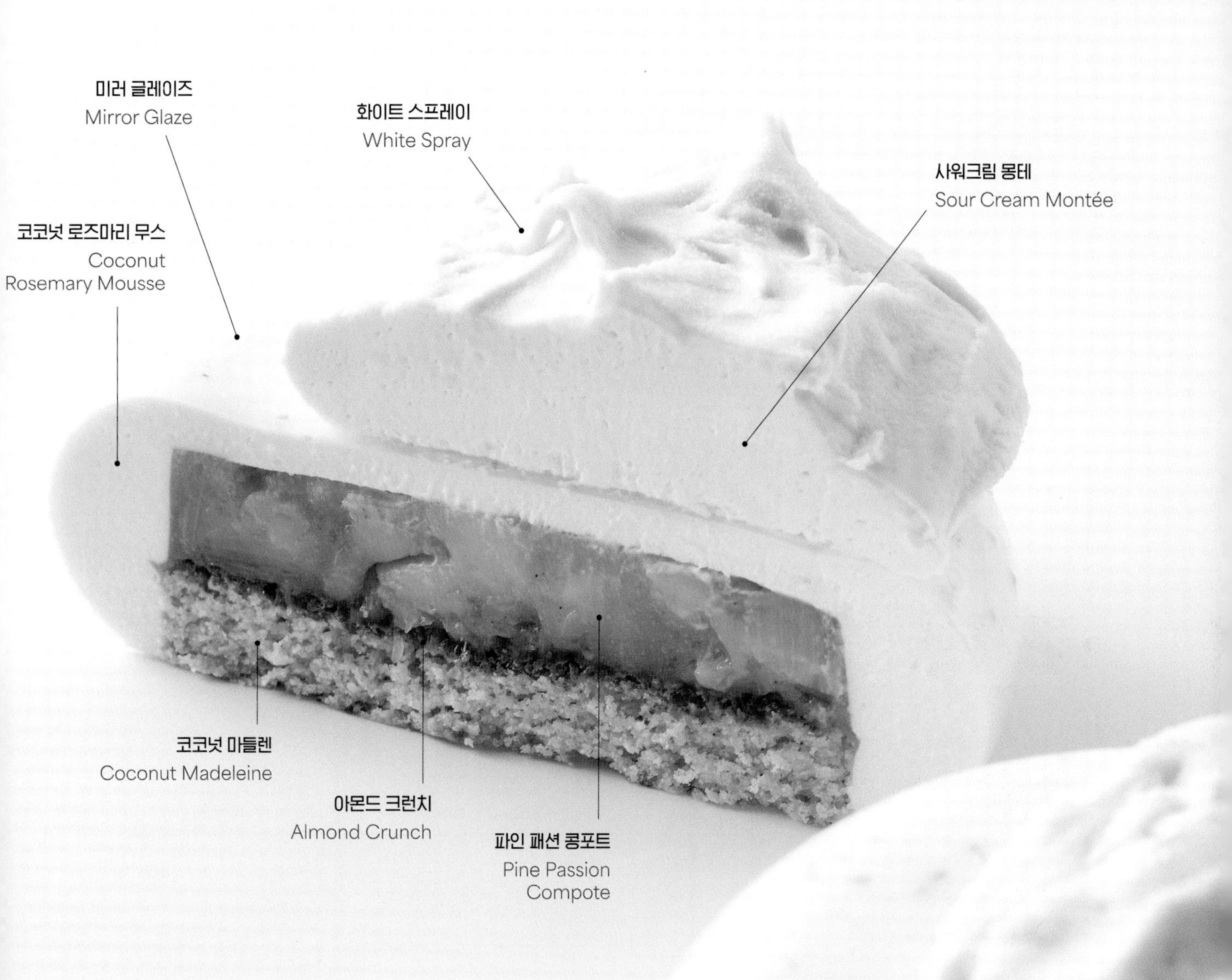
미러 글레이즈
Mirror Glaze
화이트 스프레이
White Spray
사워크림 몽테
Sour Cream Montée
코코넛 로즈마리 무스
Coconut
Rosemary Mousse
코코넛 마들렌
Coconut Madeleine
아몬드 크런치
Almond Crunch
파인 패션 콩포트
Pine Passion
Compote

INGREDIENTS
5개 분량/ 5 cakes

코코넛 마들렌
Coconut Madeleine

(철판(60 × 40cm) 1개 분량)
(1 baking tray (60 × 40 cm))

INGREDIENTS		g
달걀	Eggs	450g
코코넛밀크	Coconut milk	150g
꿀	Honey	50g
코코넛롱	Shredded dried coconut	120g
설탕	Sugar	400g
T55밀가루	T55 flour	200g
아몬드파우더	Almond powder	100g
레몬제스트	Lemon zest	2g
식용유	Vegetable oil	30g
버터 (Bridel)	Butter	450g
바닐라 페이스트	Vanilla paste	20g
소금	Salt	7.5g
베이킹파우더	Baking powder	20g
TOTAL		**1999.5g**

아몬드 크런치
Almond Crunch

INGREDIENTS		g
밀크초콜릿 (Alunga 41%)	Milk chocolate	30g
정제버터	Clarified butter	20g
아몬드 페이스트	Almond paste	120g
파에테포요틴	Paillete feuilletine	50g
TOTAL		**220g**

파인 패션 콩포트
Pine Passion Compote

INGREDIENTS		g
파인애플 퓌레	Pineapple purée	450g
패션프루트 퓌레	Passion fruit purée	100g
물	Water	100g
설탕	Sugar	250g
NH펙틴	NH pectin	12g
로거스트빈검	Locust bean gum	1g
바닐라빈 껍질	Vanilla bean pod	1개 분량/ 1 pc
파인애플	Pineapple	1개/ 1 ea
TOTAL		**913g**

코코넛 로즈마리 베이스 *

Coconut Rosemary Base

INGREDIENTS		g
코코넛밀크	Coconut milk	200g
코코넛 퓌레	Coconut purée	120g
코코넛롱	Shredded dried coconut	50g
로즈마리 잎	Rosemary leaves	3g
생크림	Heavy cream	(중량 부족 시 사용) Use when needed
TOTAL		**373g**

코코넛 로즈마리 무스

Coconut Rosemary Mousse

INGREDIENTS		g
코코넛 로즈마리 베이스 *	Coconut rosemary base *	250g
전화당	Inverted sugar	70g
젤라틴매스 (×5)	Gelatin mass (×5)	54g
화이트초콜릿 (Zephyr white 34%)	White chocolate	150g
코코넛 리큐르 (Malibu)	Coconut liqueur	40g
소금	Salt	2g
생크림	Heavy cream	500g
TOTAL		**1066g**

사워크림 몽테

Sour Cream Montée

INGREDIENTS		g
우유	Milk	60g
전화당	Inverted sugar	20g
젤라틴매스 (×5)	Gelatin mass (×5)	50g
화이트초콜릿 (Zephyr white 34%)	White chocolate	150g
마스카르포네	Mascarpone cheese	30g
사워크림	Sour cream	150g
휘핑크림	Whipping cream	450g
레몬 리큐르 (Dijon Lemon)	Dijon lemon liqueur	15g
라임 퓌레	Lime purée	15g
TOTAL		**940g**

미러 글레이즈
Mirror Glaze

INGREDIENTS		g
물	Water	150g
설탕	Sugar	300g
물엿	Corn syrup	300g
연유	Condensed milk	200g
젤라틴매스 (×5)	Gelatin mass (×5)	120g
초록색 색소	Green color	0.5g
노란색 색소	Yellow color	적당량/ QS
다진 애플민트	Apple mint, chopped	2g
TOTAL		**1072.5g**

＊만드는 법

❶ 냄비에 물, 설탕, 물엿을 넣고 103℃로 가열한다.
❷ 연유, 젤라틴매스를 넣고 섞는다.
❸ 비커에 옮겨 바믹서로 블렌딩한다.
❹ 바트에 담아 밀착 랩핑한 후 냉장고에 보관한다.
❺ 사용하기 전날 실온에 꺼내 50℃로 녹여 쿨링한 후 다진 애플민트 잎을 넣고 블렌딩해 30℃로 맞춰 사용한다

＊Procedure

❶ Heat water, sugar, and corn syrup to 103°C.
❷ Mix with condensed milk and gelatin mass.
❸ Pour into a beaker and combine with an immersion blender.
❹ Pour onto a tray, cover with plastic wrap, making sure the wrap is in contact with the glaze, and refrigerate.
❺ The day before use, leave the glaze at room temperature and melt it to 50°C. Blend with chopped apple mint leaves and use at 30°C.

화이트 스프레이
White Spray

INGREDIENTS		g
화이트초콜릿 (Zephyr white 34%)	White chocolate	200g
카카오버터	Cacao butter	200g
TOTAL		**400g**

＊만드는 법

❶ 비커에 녹인 화이트초콜릿(45~50℃), 50℃ 이하로 녹인 카카오버터를 담고 바믹서로 블렌딩한다.
❷ 50℃로 맞춰 사용한다.

＊Procedure

❶ Blend melted white chocolate (45~50°C) and cacao butter melted to below 50°C in a beaker and combine with an immersion blender.
❷ Use at 50°C.

NOTE.

✤

COCONUT MADELEINE

코코넛 마들렌

1 비커에 달걀(30℃), 코코넛밀크, 꿀을 담고 바믹서로 블렌딩한 후 냉장고에서 12시간 숙성시킨다.

TIP. 사용할 때는 30~35℃로 맞춰 사용한다.

2 로보쿱에 코코넛롱, 설탕을 넣고 곱게 갈아준다.

TIP. 코코넛롱은 오븐에서 살짝 로스팅해 사용한다.

3 볼에 30℃로 맞춘 **1**과 **2**, 체 친 T55밀가루, 아몬드파우더를 넣고 고르게 섞는다.

4 레몬제스트를 넣고 섞는다.

5 식용유, 녹인 버터(45℃), 바닐라 페이스트를 넣고 섞는다.

6 소금, 베이킹파우더를 넣고 섞는다.

7 테프론시트를 깐 철판(60 × 40cm)에 고르게 팬닝한 후 190℃로 예열된 오븐에서 15분간 굽고 식힌 뒤 냉동 보관한다.

1 Blend eggs (30°C), coconut milk, and honey in a beaker with an immersion blender. Refrigerate for 12 hours.

TIP. Use at 30~35°C.

2 Finely grind shredded dried coconut, and sugar in the Robot Coupe.

TIP. Lightly roast the shredded coconut before use.

3 Combine (**1**) at 30°C, (**2**), sifted T55 flour, and almond powder in a bowl.

4 Mix with lemon zest.

5 Mix with vegetable oil, melted butter (45°C), and vanilla paste.

6 Mix with salt and baking powder.

7 Spread evenly on a baking tray (60 × 40 cm) lined with a Teflon sheet and bake in an oven preheated to 190°C for 15 minutes. Cool at room temperature and freeze.

✤ ✤

ALMOND CRUNCH

아몬드 크런치

아몬드 크런치

1 볼에 40℃로 녹인 밀크초콜릿과 정제버터, 아몬드 페이스트를 넣고 섞은 후 파에테포요틴을 넣고 섞는다.

조립

2 코코넛 마들렌을 지름 10.5cm 원형 커터로 자르고 아몬드 크런치 20g을 바른 후 냉동한다.

Almond Crunch

1 Combine milk chocolate melted to 40°C, clarified butter, and almond paste in a bowl, then mix with paillete feuilletine.

Assemble

2 Cut coconut madeleine with a 10.5 cm round cutter, spread 20 grams of almond crunch, and freeze.

1
2
3
4
5
6
7
1
Silpan
2

✤ ✤ ✤

PINE PASSION COMPOTE

파인 패션 콩포트

파인애플 쿨리

1 냄비에 파인애플 퓌레, 패션푸르트 퓌레, 물, 바닐라빈 껍질, 미리 섞어둔 [설탕, NH펙틴, 로거스트빈검]을 넣고 휘퍼로 저어가며 펙틴 반응이 일어날 때까지 가열한다.

2 2/3를 덜어 바트에 붓고 밀착 랩핑해 냉장 보관한다.

파인애플 패션푸르트 콩포트

3 냄비에 남은 파인애플 쿨리 전량(1/3), 1cm 두께로 길게 자른 파인애플을 넣고 가열한다.

4 충분히 끓어오르면 바트에 붓고 밀착 랩핑해 냉장 보관한다.

5 **4**의 파인애플 300g을 사방 1cm 크기로 깍둑썰어 준비한다.

6 만들어둔 파인애플 쿨리 200g을 바믹서로 부드럽게 풀어준다.

7 **5**와 **6**을 섞는다.

조립

8 몰드(실리코마트 SF042 - 지름 10.3cm, 높이 2cm)에 160g씩 채운 후 스패출러로 평평하게 정리한다.

9 얼린 [코코넛 마들렌 + 아몬드 크런치]를 뒤집어 올린 후 냉동한다.

Pineapple Coulis

1 Add pineapple purée, passion fruit purée, water, vanilla bean pod, and previously mixed [sugar, pectin NH, and locust bean gum] in a pot. Heat while stirring until the pectin activates.

2 Pour 2/3 onto a tray, cover with plastic wrap, making sure the wrap is in contact with the coulis, and refrigerate.

Pineapple Passion Fruit Compote

3 Cook the remaining pineapple coulis (1/3), pineapple cut into 1 cm wide strips.

4 When it boils, pour onto a tray, cover with plastic wrap, making sure the wrap is in contact with the compote, and refrigerate.

5 Cut 300 grams of pineapple from (**4**) into 1 cm cubes.

6 Gently loosen the prepared 200 grams of pineapple coulis with an immersion blender.

7 Combine (**5**) and (**6**).

Assemble

8 Fill 160 grams in the mold (Silikomart SF042, 10.3 cm in diameter and 2 cm in height) and organize with a spatula.

9 Insert the frozen [coconut madeleine + almond crunch] upside down and freeze.

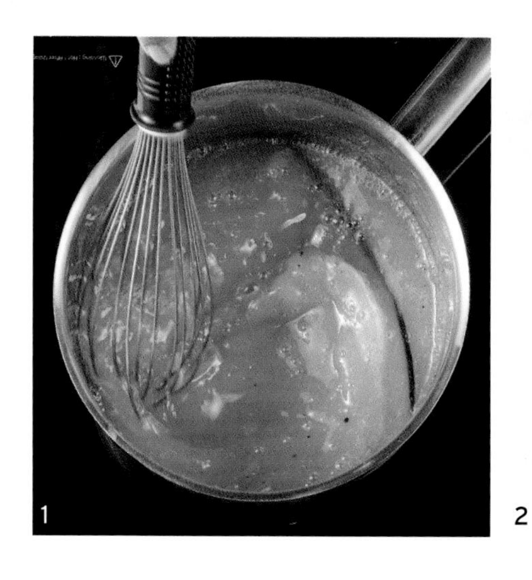
1

2

3

4

5

6

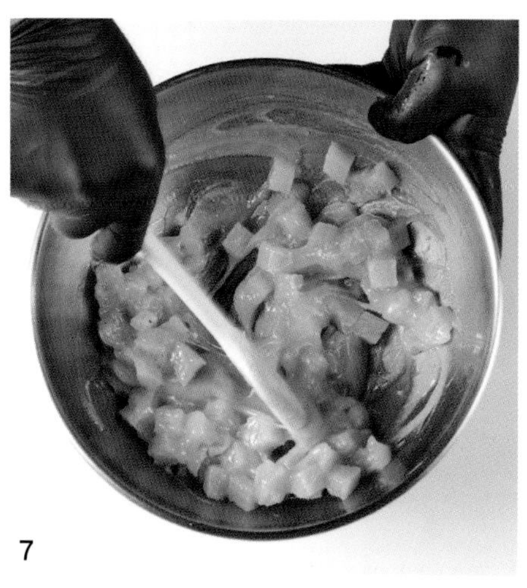
7

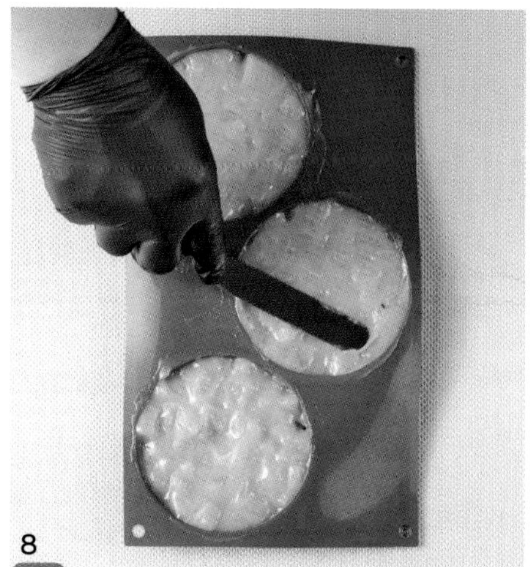
8

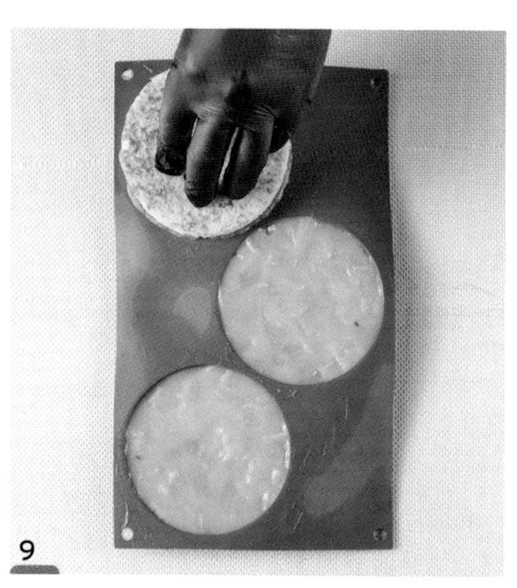
9

✤ ✤ ✤ ✤

COCONUT ROSEMARY MOUSSE

코코넛 로즈마리 무스

코코넛 로즈마리 베이스

1 냄비에 코코넛밀크, 코코넛 퓌레, 데운 코코넛롱, 로즈마리 잎을 넣고 가열하다가 끓어오르면 불에서 내려 냄비 입구를 랩핑해 10분간 인퓨징한다.

TIP. 코코넛롱은 전자레인지에서 1분 정도 데워 따뜻한 상태로 사용한다.

2 체에 거른다.

TIP. 걸러진 액체를 사용한다.

3 중량을 체크한 후 총 250g이 되도록 생크림(분량 외)을 추가해 준비한다.

코코넛 로즈마리 무스

4 냄비에 코코넛 로즈마리 베이스 250g, 전화당, 젤라틴매스를 넣고 젤라틴매스가 녹을 때까지 가열한다. (약 60℃)

5 화이트초콜릿이 담긴 볼에 붓고 바믹서로 블렌딩한다.

6 코코넛 리큐르, 소금을 넣고 블렌딩한다.

7 볼에 차가운 상태의 생크림을 넣고 60~70% 정도로 휘핑한 후 **6**을 두 번 나눠 넣어가며 섞는다.

8 얼음물이 담긴 볼에 받쳐 15℃까지 쿨링한다.

Coconut Rosemary Base

1 Heat coconut milk, coconut purée, warmed shredded dried coconut, and rosemary leaves in a pot. When it boils, remove from the heat, wrap to cover the opening of the pot, and infuse for 10 minutes.

TIP. Warm the shredded dried coconut for about 1 minute in the microwave to use.

2 Stain over the sieve.

TIP. Use the strained liquid.

3 Check the weight and add the cream (other than requested) to make a total of 250 grams.

Coconut Rosemary Mousse

4 Heat 250 grams of coconut rosemary base, inverted sugar, and gelatin mass in a saucepan to about 60°C until the gelatin melts.

5 Blend with white chocolate in a beaker with an immersion blender.

6 Blend with coconut liqueur and salt.

7 In a mixing bowl, whip the cold heavy cream to 60~70%. Add (**6**) in two batches to mix.

8 Cool in an ice bath to 15°C.

1

2

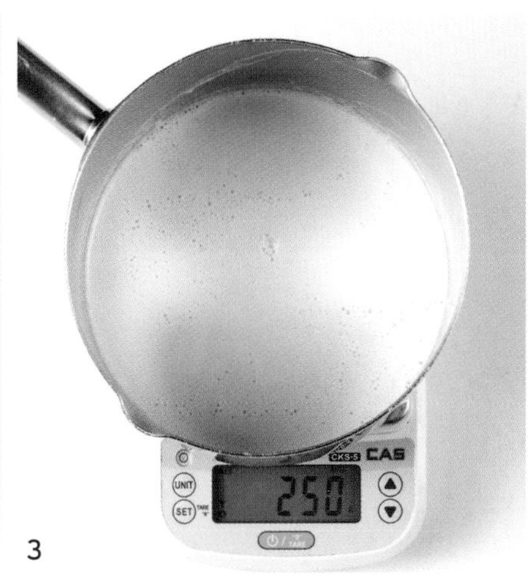

3

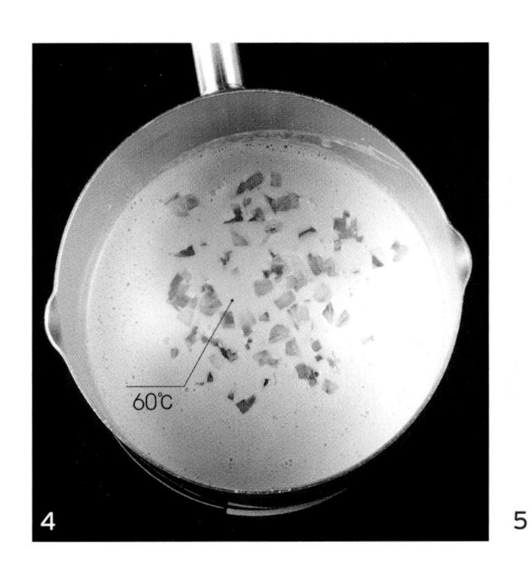

4

5

6

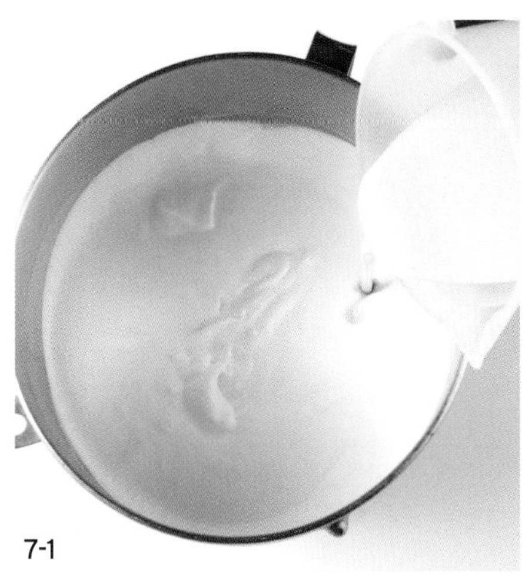

7-1

7-2

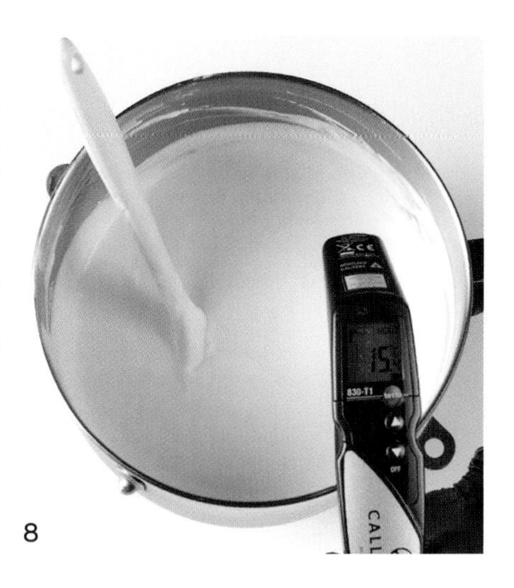

8

조립

9 몰드(실리코마트 SFT376)에 로즈마리 무스를 200g 채운다.

10 공기가 차지 않도록 얼린 [코코넛 마들렌 + 아몬드 크런치 + 파인 패션 콩포트]를 돌려가면서 넣어준다.

11 스패출러로 깔끔하게 정리해 급속 냉동(-35 ~ -40℃)한 후 몰드에서 빼내 냉동(-18℃) 보관한다.

Assemble

9 Fill 200 grams of rosemary mousse in the mold (Silikomart SFT376).

10 Insert the frozen [coconut madeleine + almond crunch + pine passion compote] while rotating to prevent air from getting in.

11 Clean up with an offset spatula. Blast freeze (-35 ~ -40°C), remove from the mold, and freeze (-18°C).

✤ ✤ ✤ ✤ ✤

SOUR CREAM MONTÉE

사워크림 몽테

1 냄비에 우유, 젤라틴매스를 넣고 젤라틴매스가 녹을 때까지 가열한다. (65~70℃)

2 화이트초콜릿이 담긴 볼에 붓고 바믹서로 블렌딩한다.

3 마스카르포네 - 사워크림 - 휘핑크림 순서로 넣어가며 블렌딩한다.

4 레몬 리큐르, 라임 퓌레를 넣고 블렌딩한다.

5 바트에 담아 밀착 랩핑해 냉장고에서 12시간 동안 숙성시킨다.

6 숙성시킨 사워크림 몽테는 사용하기 직전 부드럽게 휘핑해 사용한다.

1 Heat milk and gelatin mass to 65~70°C until the gelatin melts.

2 Blend with white chocolate in a beaker with an immersion blender.

3 Blend in the order of mascarpone – sour cream – whipping cream.

4 Add lemon liqueur and lime purée and continue to blend.

5 Pour onto a tray, cover with plastic wrap, making sure the wrap is in contact with the cream, and refrigerate for 12 hours.

6 Gently whip the refrigerated sour cream montée just before use.

9

10

11

1

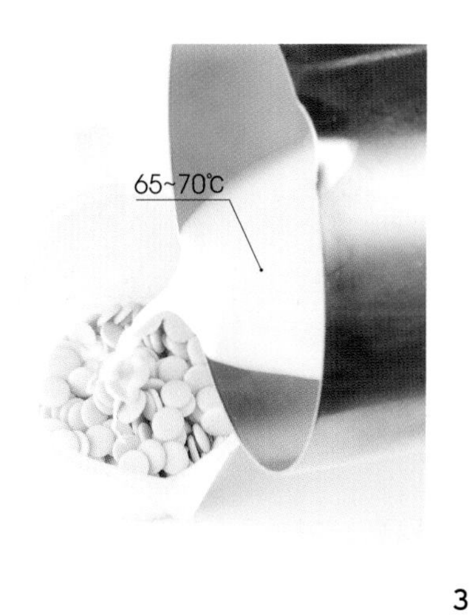

2

3

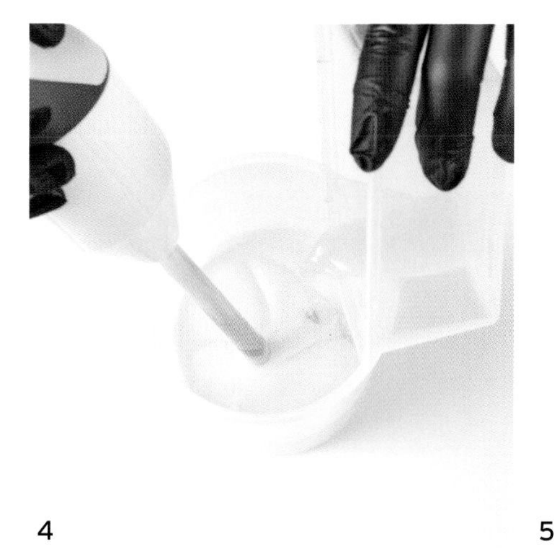

4

5

6

1

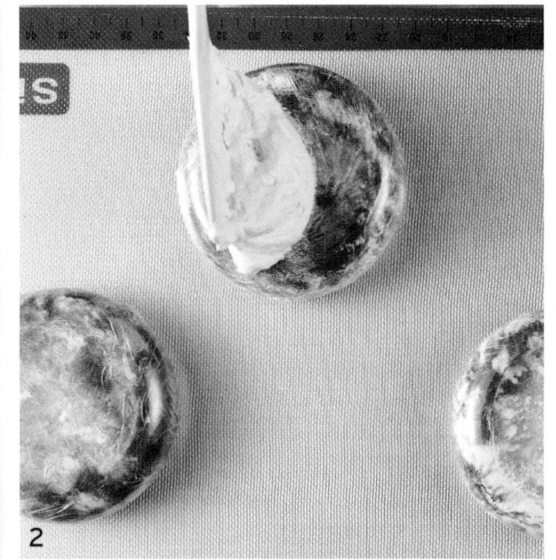
2

3

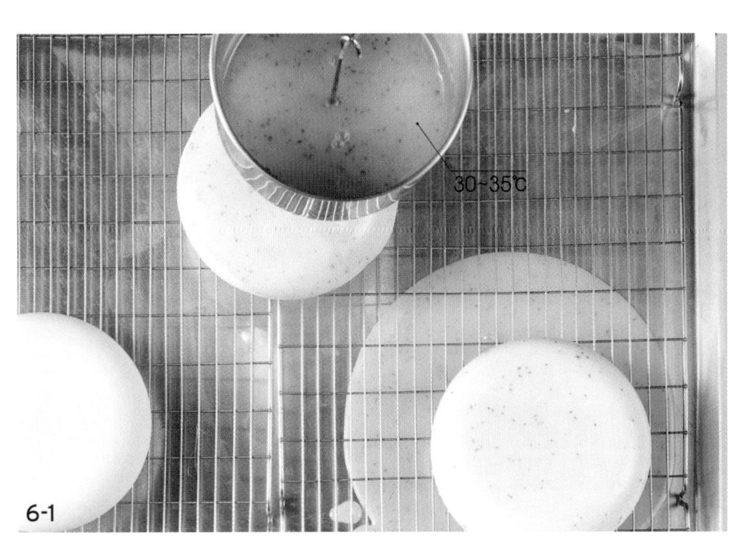

6-1

6-2

✤ ✤ ✤ ✤ ✤ ✤

FINISH 마무리

1　사용한 몰드와 동일한 크기의 모형을 랩핑해 준비한다.

TIP. 최종적으로 조립한 후에도 곡선의 옆면에 들뜸 없이 밀착되도록 하기 위해 동일한 모양과 크기의 초콜릿 모형을 만들어 사용했다.

2　휘핑한 사워크림 몽테를 한 주걱 떠 가장자리에 올린다.

TIP. 자연스러운 모양으로 연출되도록 주걱으로 툭 떨어트리는 느낌으로 올린다.

3　손으로 지저분하게 묻어 있는 부분을 깔끔하게 정리한다.

4　냉동실에서 얼린다.

5　얼린 사워크림 몽테에 화이트 스프레이를 분사하고 냉동실에서 굳힌다.

6　얼린 무스에 글레이즈(30~35℃)를 코팅한다.

7　얼린 사워크림 몽테를 올린다. 여분의 파인애플 패션 쿨리를 군데군데 파이핑하고 허브를 올려 마무리한다.

5

7-2

1 Prepare a mock-up, wrapped with plastic wrap, of the same size as the mold.

TIP. To ensure that it adheres tightly to the curved side without any gap, I made a chocolate mock-up of the same shape and size.

2 Place a spoonful of whipped sour cream montée on the edge.

TIP. Place it as if to drop to create a natural look.

3 Clean up any messy areas with your hands.

4 Freeze completely.

5 Spray the sour cream montée with the white spray mixture and freeze.

6 Coat the frozen mousse with mirror glaze (30~35°C).

7 Top with the frozen sour cream montée. Pipe extra pineapple passion coulis on different spots and decorate with herbs to finish.

20

CAFÉ SACHER TORTE

카페 자허 토르테

PAIRING & TEXTURE

페어링 & 텍스처

클래식 디저트인 자허 토르테를 앙트르메로 재해석한 메뉴. 초콜릿과 살구, 커피로 페어링했다.

A reinterpreted classic dessert, Sacher Torte, as an entremet. I paired the chocolate with apricot and coffee.

TECHNIQUE

테크닉

- 정형화된 모양의 시판 몰드에 무늬를 추가해 자칫 밋밋해보일 수 있는 디자인을 입체적이면서도 새로운 느낌으로 재탄생시켰다.
- 무늬가 있는 부분에 피스톨레 작업을 한 다음 글레이즈를 씌워 상반되는 질감을 극명하게 표현했다.
- 살구 콩포트를 얼리지 않은 상태로 채워 넣어 과일의 신선함이 강조되도록 표현했다.

- I recreated a three-dimensional, new look by adding patterns to a commercially available mold with a formatted shape that could have otherwise been plain.
- The patterned area was sprayed and then glazed to show the contrasting textures clearly.
- The compote was filled unfrozen to emphasize the freshness of the fruit.

DESIGN

디자인

- 기후 변화로 인한 메마른 대지의 느낌을 표현하고자 한 디자인이다.
- 2개의 몰드를 사용하고, 기성품 몰드에 새로운 문양을 추가한 새로운 방식의 디자인이다.

- This design is intended to express the look of dry land caused by climate change.
- It is a new method of design in which I used two molds and added a new pattern to a ready-made mold.

HOW TO COMPOSE THIS RECIPE

STEP 1.

메뉴 정하기	앙트르메
Decide on the menu	Entremet

STEP 2.

메인 맛 정하기	살구
Choose the primary flavor	Apricot

STEP 3.

메인 맛(살구)과의 페어링 선택하기

Select a pairing flavor (Apricot)

- ☑ 오렌지 Orange
- ☑ 커피 Coffee
- ☑ 캐러멜 Caramel
- ☑ 밀크초콜릿 Milk Chocolate
- ☑ 헤이즐넛 Hazelnut
- ☑ 코코넛 Coconut

STEP 4.

구성하기

Assemble

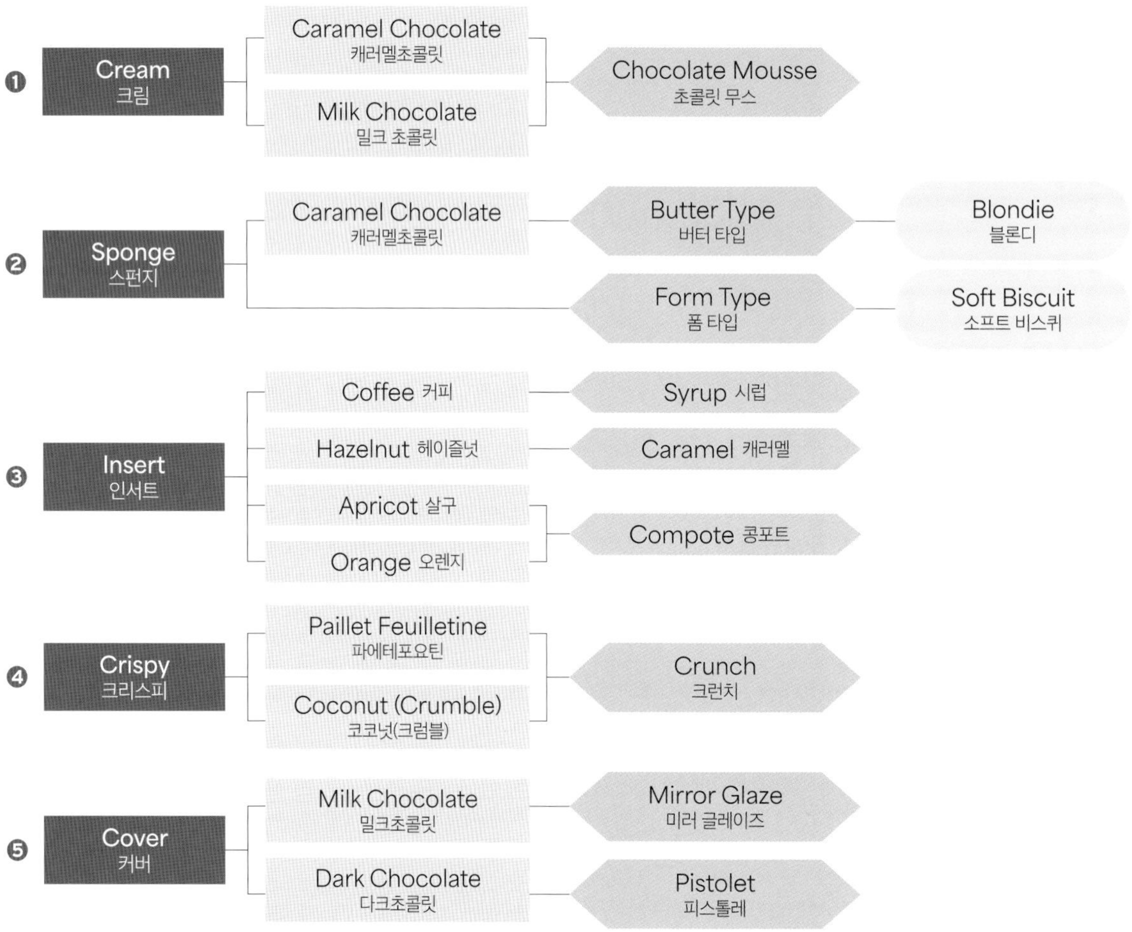

❶ Cream 크림
- Caramel Chocolate 캐러멜초콜릿 / Milk Chocolate 밀크 초콜릿 → Chocolate Mousse 초콜릿 무스

❷ Sponge 스펀지
- Caramel Chocolate 캐러멜초콜릿 → Butter Type 버터 타입 → Blondie 블론디
- Form Type 폼 타입 → Soft Biscuit 소프트 비스퀴

❸ Insert 인서트
- Coffee 커피 → Syrup 시럽
- Hazelnut 헤이즐넛 → Caramel 캐러멜
- Apricot 살구 / Orange 오렌지 → Compote 콩포트

❹ Crispy 크리스피
- Paillet Feuilletine 파에테포요틴 / Coconut (Crumble) 코코넛(크럼블) → Crunch 크런치

❺ Cover 커버
- Milk Chocolate 밀크초콜릿 → Mirror Glaze 미러 글레이즈
- Dark Chocolate 다크초콜릿 → Pistolet 피스톨레

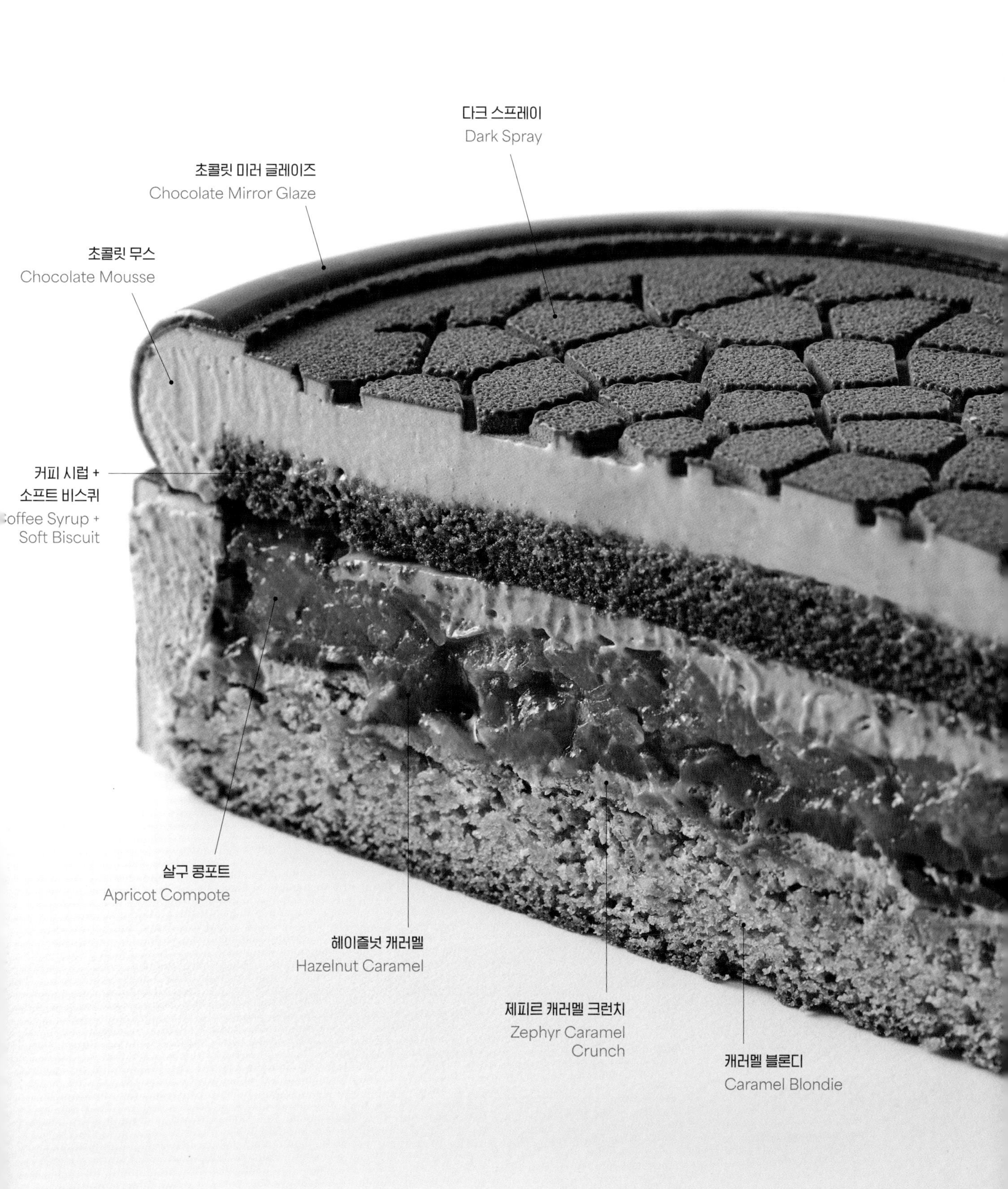

다크 스프레이
Dark Spray
초콜릿 미러 글레이즈
Chocolate Mirror Glaze
초콜릿 무스
Chocolate Mousse
커피 시럽 +
소프트 비스퀴
offee Syrup +
Soft Biscuit
살구 콩포트
Apricot Compote
헤이즐넛 캐러멜
Hazelnut Caramel
제피르 캐러멜 크런치
Zephyr Caramel
Crunch
캐러멜 블론디
Caramel Blondie

INGREDIENTS 4개 분량/ 4 cakes

소프트 비스퀴
Soft Biscuit

(철판(60 × 40cm) 1개 분량)
(1 baking tray (60 × 40 cm))

INGREDIENTS		g
달걀	Eggs	260g
노른자	Egg yolks	55g
설탕A	Sugar A	200g
흰자	Egg whites	120g
설탕B	Sugar B	60g
박력분	Cake flour	230g
옥수수전분	Cornstarch	30g
버터 (Bridel)	Butter	13g
바닐라에센스 (Aroma Piu)	Vanilla essence	7g
TOTAL		**975g**

*** 만드는 법**

❶ 믹싱볼에 달걀, 노른자, 설탕A를 넣고 뜨거운 물이 담긴 볼에 받쳐 휘퍼로 저어가며 50℃로 온도를 올린 다음 고속으로 휘핑한다.
❷ 26℃가 되면 저속으로 낮춰 기공을 정리해 마무리한다.
❸ 다른 볼에 흰자를 휘핑하면서 설탕B를 넣어가며 휘핑한다.
- ❶의 온도가 28℃가 되었을 때 머랭 작업을 시작하면 완성되는 속도가 비슷해 섞는 작업이 수월하다.

❹ 기공이 조밀하고 단단한 상태가 되면 저속으로 마무리한다.
❺ 볼에 ❷의 절반과 체 친 박력분, 옥수수전분을 넣고 섞는다.
❻ 녹인 버터(45℃), 바닐라에센스를 넣고 섞는다.
❼ 남은 가루 재료를 넣고 섞는다.
❽ ❹를 두 번 나눠 넣어가며 섞는다.
❾ 테프론시트를 깐 철판에 0.7cm 두께로 팬닝한다.
❿ 데크 오븐 기준 윗불 220℃, 아랫불 200℃에서 10분간 굽고 식힌다.
⓫ 13cm 원형 링으로 재단한다.

*** Procedure**

❶ Add eggs, egg yolks, and sugar A in a mixing bowl. Set it over a double boiler, whisk until it reaches 50°C, then whip on high speed.
❷ When it reaches 26°C, reduce to low speed to regulate the bubbles and finish.
❸ Whip egg whites in a separate bowl while gradually adding sugar B.
- Start working on the meringue when the temperature of ❶ reaches 28°C to finish at about the same time for easier preparation.

❹ When it forms dense and stiff peaks, reduce to low speed and finish.
❺ Combine half of ❷ with sifted cake flour and cornstarch in a bowl.
❻ Mix with melted butter (45°C) and vanilla essence.
❼ Mix with the remaining powder ingredients.
❽ Add ❹ to mix in two batches.
❾ Spread onto a baking tray lined with a Teflon sheet to 0.7 cm thickness.
❿ Bake in a deck oven with 220°C on the top and 200°C on the bottom for 10 minutes, and cool.
⓫ Cut it with a 13 cm round ring.

커피 시럽
Coffee Syrup

INGREDIENTS		g
TPT 시럽	TPT syrup	600g
인스턴트 커피가루	Instant coffee powder	45g
TOTAL		**645g**

- TPT 시럽 = 물과 설탕을 1:1 비율로 끓여 만든 것으로, 식혀 사용한다.
- 냄비에 모든 재료를 넣고 가열한 후 식혀 사용한다.
- TPT syrup: A syrup made by boiling water and sugar in a 1:1 ratio. Use after cooling.
- Boil all the ingredients and cool to use.

캐러멜 블론디

Caramel Blondie

(철판(42 × 32cm) 1개 분량)
(1 baking tray (42 × 32 cm))

INGREDIENTS		g
달걀	Eggs	290g
설탕	Sugar	320g
전화당	Inverted sugar	50g
캐러멜초콜릿 (Zephyr Caramel 35%)	Caramel chocolate	290g
버터 (Bridel)	Butter	330g
오렌지제스트	Orange zest	1개 분량/ 1 orange
박력분	Cake flour	150g
아몬드파우더	Almond powder	70g
베이킹파우더	Baking powder	2g
소금	Salt	4g
TOTAL		**1506g**

살구 콩포트

Apricot Compote

INGREDIENTS		g
살구 퓌레	Apricot purée	110g
오렌지 퓌레	Orange purée	110g
설탕	Sugar	70g
NH펙틴	NH pectin	8g
꿀	Honey	50g
당적살구	Apricots in syrup	250g
젤라틴매스 (×5)	Gelatin mass (×5)	40g
TOTAL		**638g**

- 당적살구는 냉동 살구를 해동해 물기를 제거한 후 건조기에서 말려 사용해도 좋다.
- For the apricots in syrup, you can also can also thaw the frozen apricots, drain them, and dry them in a dryer before use.

코코넛 크럼블 *

Coconut Crumble

INGREDIENTS		g
버터 (Bridel)	Butter	90g
코코넛슈거	Coconut sugar	50g
설탕	Sugar	25g
소금	Salt	2g
박력분	Cake flour	150g
코코넛롱	Shredded dried coconut	30g
옥수수전분	Cornstarch	15g
물	Water	20g
TOTAL		382g

제피르 캐러멜 크런치

Zephyr Caramel Crunch

INGREDIENTS		g
코코넛 크럼블 *	Coconut crumble *	100g
파에테포요틴	Paillete feuilletine	110g
캐러멜초콜릿 (Zephyr Caramel 35%)	Caramel chocolate	100g
레몬제스트	Lemon zest	1/2개 분량/ 1/2 lemon
TOTAL		310g

초콜릿 무스

Chocolate Mousse

INGREDIENTS		g
휘핑크림A	Whipping cream A	250g
생크림A	Heavy cream A	200g
젤라틴매스 (×5)	Gelatin mass (×5)	48g
연유	Condensed milk	80g
밀크초콜릿 (Alunga 41%)	Milk chocolate	240g
캐러멜초콜릿 (Zephyr Caramel 35%)	Caramel chocolate	280g
커피 리큐르 (Kahlua)	Coffee liqueur	60g
휘핑크림B	Whipping cream B	400g
생크림B	Heavy cream B	400g
TOTAL		1958g

헤이즐넛 캐러멜

Hazelnut Caramel

INGREDIENTS		g
설탕	Sugar	80g
물엿	Corn syrup	130g
생크림	Heavy cream	210g
헤이즐넛 페이스트	Hazelnut paste	24g
소금	Salt	2g
TOTAL		**446g**

*** 만드는 법**

❶ 냄비에 설탕, 물엿을 넣고 캐러멜화시킨다.
❷ 불에서 내려 뜨겁게 데운 생크림을 넣고 휘퍼로 섞는다.
❸ 다시 불에 올려 106℃까지 가열한 후 불에서 내려 헤이즐넛 페이스트, 소금을 넣고 섞는다.
❹ 얼음물이 담긴 볼에 받쳐 쿨링한 후 짤주머니에 담아 사용한다.

*** Procedure**

❶ Caramelize sugar and corn syrup in a pot.
❷ Remove from the heat, add hot cream, and stir with a whisk.
❸ Cook to 106°C, remove from the heat, and mix with hazelnut paste and salt.
❹ Cool in an ice bath and put in a piping bag to use.

초콜릿 글레이즈

Chocolate Glaze

INGREDIENTS		g
물	Water	90g
물엿	Corn syrup	170g
설탕	Sugar	170g
연유	Condensed milk	120g
젤라틴매스 (×5)	Gelatin mass (×5)	80g
밀크초콜릿 (Alunga 41%)	Milk chocolate	220g
TOTAL		**850g**

*** 만드는 법**

❶ 냄비에 물, 물엿, 설탕을 넣고 가열한다.
❷ 104℃가 되면 불에서 내려 연유, 젤라틴매스를 넣고 섞는다.
❸ 밀크초콜릿이 담긴 비커에 붓고 바믹서로 블렌딩한다.
❹ 얼음물이 담긴 볼에 받쳐 쿨링한 후 냉장 보관한다.
❺ 사용할 때는 45℃로 온도를 올린 후 28℃로 맞춰 사용한다.

*** Procedure**

❶ Heat water, corn syrup, and sugar.
❷ When it reaches 104°C, remove from the heat and mix with condensed milk and gelatin mass.
❸ Blend with milk chocolate in a beaker.
❹ Cool in an ice bath and refrigerate.
❺ To use, warm it to 45°C then reduce to 28°C.

다크 스프레이

Dark Spray

INGREDIENTS		g
다크초콜릿 (Mi-Amère 58%)	Dark chocolate	250g
카카오버터	Cacao butter	250g
TOTAL		**500g**

- 볼에 모든 재료를 넣고 녹여 바믹서로 블렌딩한 후 50℃로 맞춰 사용한다.
- Melt all the ingredients in a bowl, combine using an immersion blender, and use at 50°C.

✤

CARAMEL BLONDIE

캐러멜 블론디

1 볼에 달걀, 설탕, 전화당을 넣고 중탕볼에서 45℃까지 온도를 올린다.

TIP. 거품이 많이 생기지 않도록 주의한다.

2 녹인 캐러멜초콜릿(40~45℃)과 녹인 버터(45℃)를 넣고 섞는다.

3 오렌지제스트를 넣고 섞는다.

4 체 친 [박력분, 아몬드파우더, 베이킹파우더], 소금을 넣고 섞는다. (최종 온도 35~40℃)

5 테프론시트를 깐 철판(42 × 32cm)에 팬닝한다.

6 160℃로 예열된 오븐에서 뎀퍼를 100% 열고 25분간 굽다가 140℃로 낮춰 15분간 굽는다.
구운 후 냉동해 지름 15cm 원형 링으로 자른다.

1 Warm eggs, sugar, and inverted sugar over a double boiler to 45°C.

TIP. Be careful not to create too many bubbles.

2 Mix with melted caramel chocolate (40~45°C) and melted butter (45°C).

3 Mix with orange zest.

4 Mix with sifted [cake flour, almond powder, baking powder], and salt. (Final temperature: 35~40°C)

5 Spread onto a baking tray (42 × 32 cm) lined with a Teflon sheet.

6 Bake in an oven preheated to 160°C for 25 minutes with the damper open at 100%, reduce to 140°C, and bake for another 15 minutes. Freeze and cut with a 15 cm round ring.

1 2 3

4-1 4-2

5 6

✤ ✤

APRICOT COMPOTE

살구 콩포트

1 냄비에 살구 퓌레, 오렌지 퓌레, 미리 섞어둔 설탕과 NH펙틴, 꿀을 넣고 휘퍼로 섞어준 후 가열한다.

2 펙틴 반응을 확인한 후 당적살구를 넣고 펙틴 반응이 확인될 때까지 가열한다.

TIP. 당적살구의 수분을 날려주기 위해 재가열한다.

3 불에서 내린 후 젤라틴매스를 넣고 섞는다.

4 얼음물이 담긴 볼에 받쳐 쿨링한다.

5 밀착 랩핑해 냉장고에 보관한다.

1 Whisk apricot purée, orange purée, honey, and previously mixed sugar and pectin NH; cook.

2 After the pectin activates, add the apricots and cook until the pectin reacts again.

TIP. Reheat to cook off the moisture from the apricots.

3 Remove from the heat and stir in the gelatin mass.

4 Cool in an ice bath.

5 Adhere the plastic wrap to the compote and refrigerate.

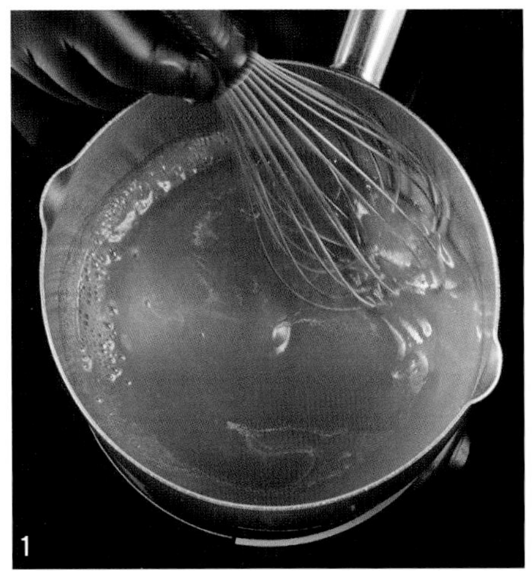
1

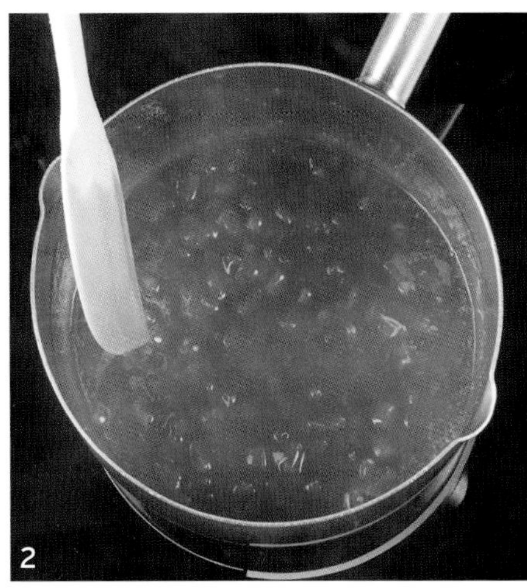
2

3

4

5

✤ ✤ ✤

ZEPHYR CARAMEL CRUNCH

제피르 캐러멜 크런치

제피르 캐러멜 크런치

1 볼에 모든 재료를 넣고 고르게 섞는다.

TIP. 캐러멜초콜릿은 녹인 후 35~40℃로 맞춰 사용한다.

조립

2 캐러멜 블론디에 55g씩 올린 후 평평하게 정리한다.

Zephyr Caramel Crunchy

1 Combine all the ingredients in a bowl.

TIP. Melt the caramel hocolate and use at 35~40°C.

Assemble

2 Spread 55 grams evenly on the caramel blondie.

✤ ✤ ✤ ✤

CHOCOLATE MOUSSE

초콜릿 무스

1 냄비에 휘핑크림A, 생크림A, 젤라틴매스를 넣고 젤라틴매스가 녹을 때까지 가열한다. (약 60℃)

TIP. 연유를 함께 넣고 가열하면 바닥에 눌어붙으므로 함께 가열하지 않는다.

2 불에서 내려 연유를 넣고 섞는다.

3 밀크초콜릿, 캐러멜초콜릿이 담긴 비커에 넣고 바믹서로 블렌딩한다.

4 커피 리큐르를 넣고 블렌딩한다.

5 볼에 휘핑크림B와 생크림B를 넣고 손으로 가볍게 휘핑(60~70%)한 후 **4**를 두 번 나눠 넣어가며 주걱으로 고르게 섞는다.

1 Heat whipping cream A, heavy cream A, and gelatin mass to about 60°C until the gelatin melts.

TIP. Do not boil with condensed milk, as it will stick to the bottom.

2 Remove from the heat and mix with condensed milk.

3 Blend with milk chocolate and caramel chocolate in a beaker using an immersion blender.

4 Continue to blend with coffee liqueur.

5 Gently whip whipping cream B and heavy cream B with a whisk to 60~70%. Add (**4**) in two batches and combine evenly with a spatula.

✤ ✤ ✤

1

2

✤ ✤ ✤ ✤

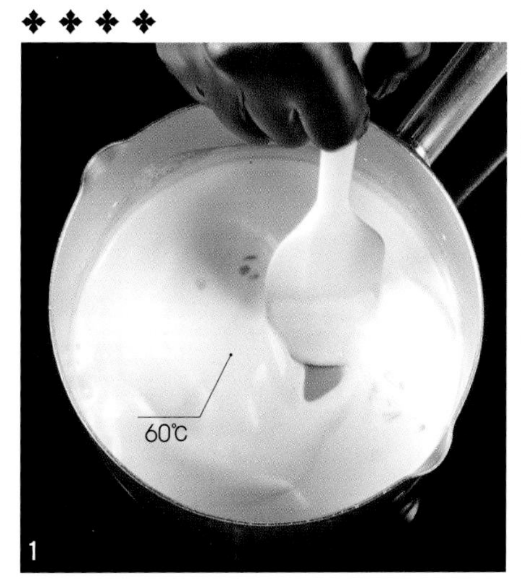

1

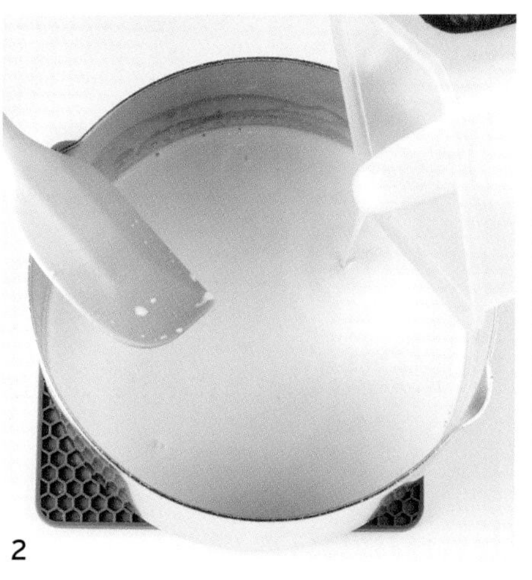
2

3

4

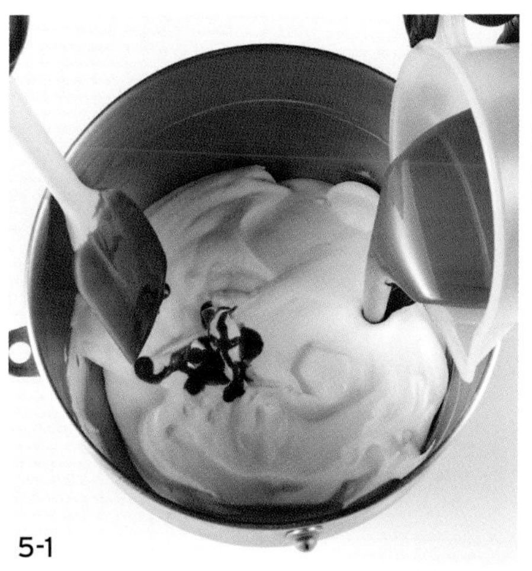
5-1

5-2

1

2

3

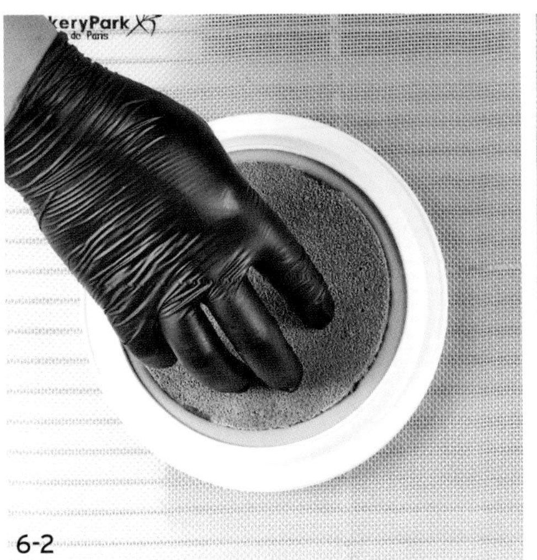
6-2

7

8

✤ ✤ ✤ ✤ ✤

FINISH 마무리

1 제품의 상단이 될 몰드(실리코마트 SFT369)를 준비한다.

TIP. 여기에서는 자체 제작한 무늬 실리콘을 사용했지만, 원하는 무늬나 브랜드 로고로 변형해도 좋다.

2 몰드 안에 자체 제작 몰드를 넣는다.

3 몰드에 초콜릿 무스를 적당량 채운다.

4 무늬가 있는 몰드이므로 무늬 사이사이에 초콜릿 무스가 잘 들어갈 수 있게 스패출러로 펴 바른다.

5 초콜릿 무스를 160g 채운다.

6 얼린 소프트 비스퀴를 넣는다.

TIP. 소프트 비스퀴는 지름 13cm로 자르고 전면에 커피 시럽을 묻혀 냉동실에서 얼려 사용한다.

7 급속 냉동(-35 ~ -40℃)한 후 몰드에서 빼내 냉동(-18℃) 보관한다.

8 다크 스프레이를 분사한다.

9 지름 13cm 원형 링을 정중앙에 올린다.

10 원형 링 바깥쪽에 초콜릿 미러 글레이즈를 코팅한다.

11 원형 링을 제거하고 밑면을 깔끔하게 정리한 후 냉동 보관한다.

5

6-1

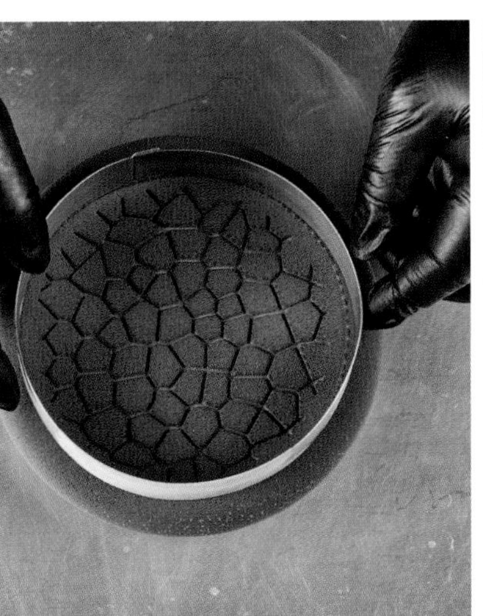

10

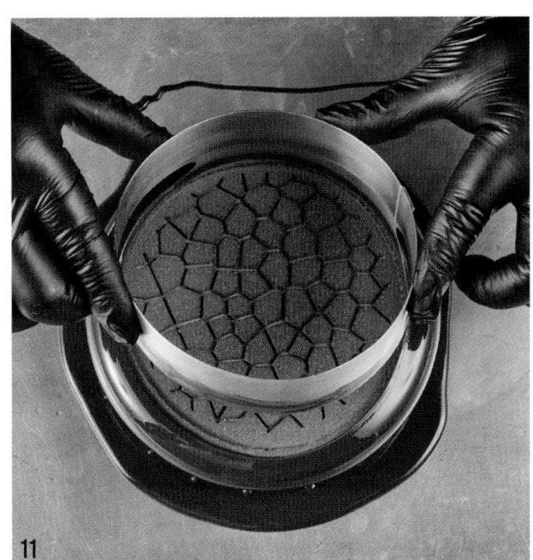

11

1 Prepare the mold (Silikomart SFT369), which will be the top part of the cake.

TIP. I used the silicon I designed myself, but you can change to any pattern or your own logo.

2 Place the custom pattern in the mold.

3 Fill with a moderate amount of chocolate mousse.

4 Spread the mousse with an offset spatula to fill between the patterns.

5 Fill with 160 grams of chocolate mousse.

6 Insert the frozen soft biscuit.

TIP. Cut the soft biscuit to 13 cm in diameter, brush coffee syrup on all sides, and freeze to use.

7 Blast freeze (-35 ~ -40°C), remove from the mold, and store frozen (-18°C).

8 Spray with the dark spray mixture.

9 Place a 13 cm round ring in the center.

10 Coat the outside of the round ring with the chocolate mirror glaze.

11 Remove the ring, clean the bottom, and freeze.

12-1

12-2

13

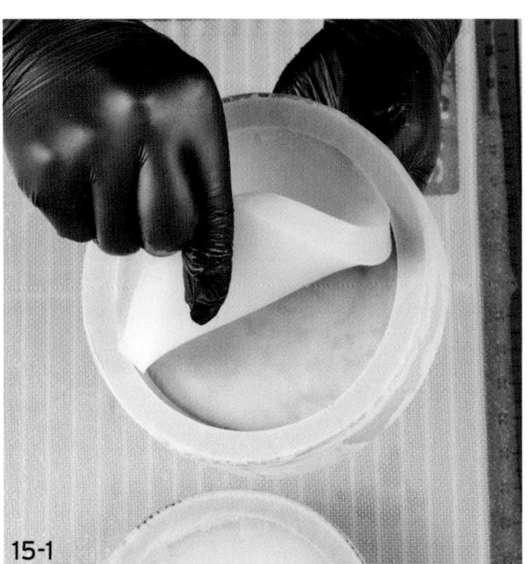
15-1

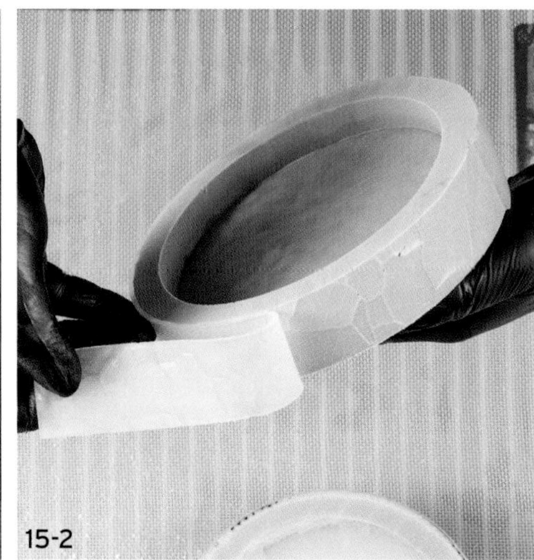
15-2

16

12 제품의 하단이 될 원형 무스 링(지름 17cm, 높이 3.5cm)과 안쪽에 넣을 인서트용 실리콘 몰드(지름 14cm, 높이 1.5cm)를 준비한다.

TIP. 원형 무스 링은 바닥을 랩핑하고 옆면에 링 높이와 동일한 필름을 두른 후 실리콘 무늬 몰드를 둘러 준비한다. 여기에서는 옆면이 나뭇잎 모양인 것과 줄무늬 모양인 것 두 가지를 사용했다.

13 초콜릿 무스 210g을 채운 후 빈 공간까지 잘 들어가도록 스패출러로 펴 바른다.

14 얼린 [캐러멜 블론디 + 제피르 캐러멜 크런치]를 넣는다.

15 급속 냉동(-35 ~ -40℃)한 후 몰드에서 빼낸다.

16 케이크 하단이 되는 부분을 냉동(-18℃) 보관한다.

17 다크 스프레이를 분사한 후 굳으면 헤이즐넛 캐러멜 50g - 살구 콩포트 150g - 초콜릿 무스 30g 순서로 채우고 스패출러로 평평하게 정리한다.

18 하단과 상단을 합쳐 완성한다.

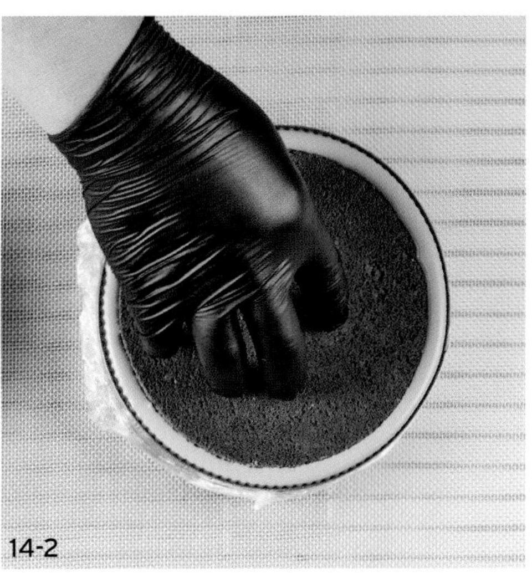

14-2

18-1

18-2

12 Prepare the round mousse ring (Ø 17 cm, H 3.5 cm), which will be the bottom part of the cake, and the silicon mold for the insert (Ø 14 cm, H 1.5 cm).

TIP. To use, wrap the sides and the bottom of the round mousse ring tightly. Line a cake collar equal to the height of the ring, and then place the patterned silicon mold.
I used two silicon designs for the sides: one with a leaf pattern and one with a stripe pattern.

13 Fill with 210 grams of chocolate mousse, and press the mousse with a spatula to fill between the patterns.

14 Insert the frozen [caramel blondie + Zephyr caramel crunch].

15 Blast freeze (-35 ~ -40°C) and remove from the mold.

16 Freeze the sides of the cake (-18°C).

17 Spray with the dark spray mixture. When set, fill in the order of 50 grams of hazelnut caramel, 150 grams of apricot compote, 30 grams of chocolate mousse, and flatten evenly with a spatula.

18 Stack the top on the bottom part to finish.

21

WHOLE FRUIT FIG

홀 프루트 피그

PAIRING & TEXTURE

페어링 & 텍스처

무화과 잎 오일을 이용해 무화과의 프레시한 맛을 최대한 끌어올린 메뉴. 무화과, 치즈, 와인, 발사믹 식초, 초콜릿을 페어링했다.

A menu that maximized the fresh taste of figs using fig leaf oil. I paired figs, cheese, wine, balsamic vinegar, and chocolate.

TECHNIQUE

테크닉

- 일반적인 무스 케이크에서 얼린 인서트를 사용하는 반면, 얼리지 않은 인서트를 넣어 과일의 신선함을 그대로 느낄 수 있게 했다.
- 캐러멜 컬러와 밀크 컬러 두 가지로 분사해 그라데이션 효과를 주었다.

- While the frozen inserts are filled in typical mousse cakes, I used unfrozen inserts to taste the freshness of the fruit.
- I sprayed with two colors, caramel and milk, to give it a gradient effect.

DESIGN

디자인

잘 구워진 브리오슈 번에서 영감을 얻어 브리오슈 형태로 디자인을 하였고, 잘 익은 무화과로 장식해 색감을 살려 더욱 먹음직스럽게 연출했다.

Inspired by a well-baked brioche bun, I designed it as a brioche and decorated it with ripe figs to bring out the color and make it look more appetizing.

✦ HOW TO COMPOSE THIS RECIPE ✦

STEP 1.	메뉴 정하기	프티 가토
	Decide on the menu	Petit Gateau

STEP 2.	메인 맛 정하기	무화과
	Choose the primary flavor	Fig

STEP 3. 메인 맛(무화과)과의 페어링 선택하기

Select a pairing flavor (Fig)

- ☑ 크림치즈 Cream Cheese
- ☑ 마스카르포네 Mascarpone
- ☑ 고르곤졸라 Gorgonzola

- ☑ 무화과 잎 Fig Leaf
- ☑ 바질 Basil

- ☑ 발사믹 식초 Balsamic Vinegar
- ☑ 바닐라 Vanilla
- ☑ 라즈베리 Raspberry

STEP 4. 구성하기

Assemble

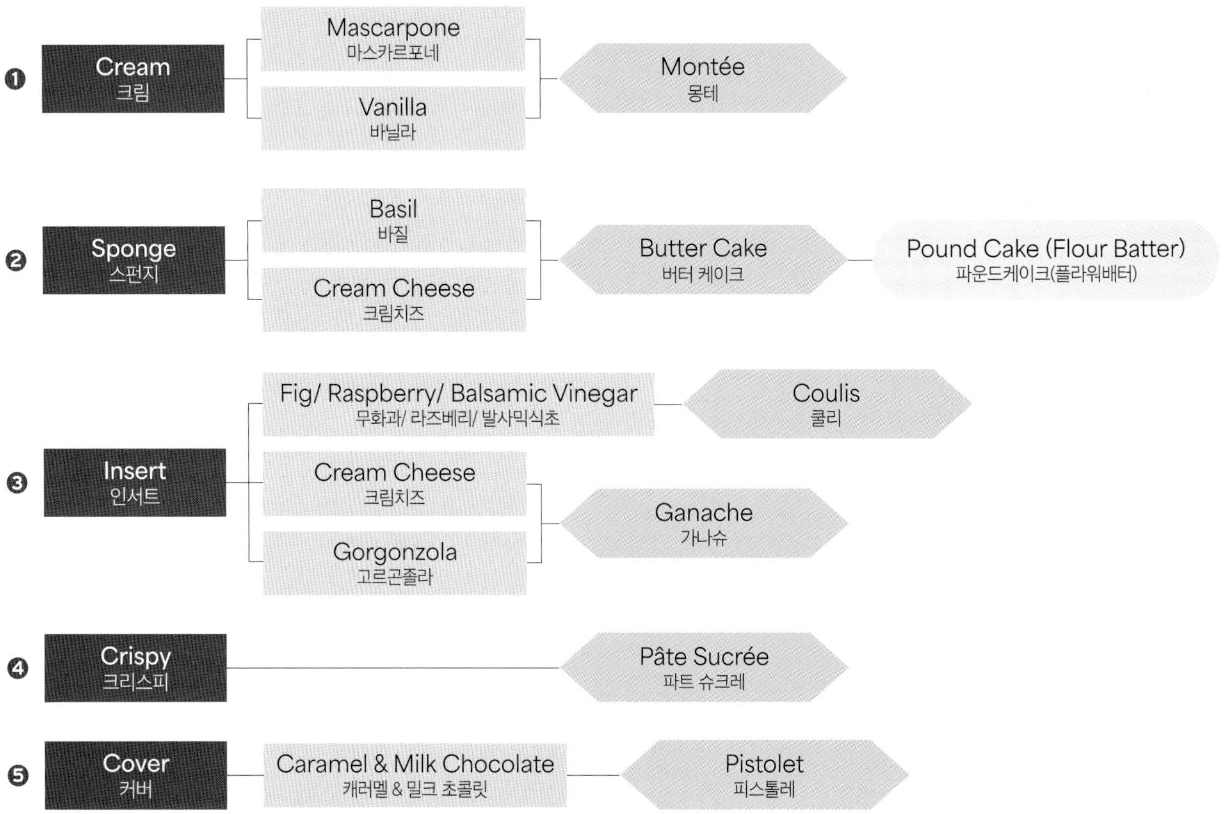

❶ Cream 크림
- Mascarpone 마스카르포네 / Vanilla 바닐라 → Montée 몽테

❷ Sponge 스펀지
- Basil 바질 / Cream Cheese 크림치즈 → Butter Cake 버터 케이크 — Pound Cake (Flour Batter) 파운드케이크(플라워배터)

❸ Insert 인서트
- Fig/ Raspberry/ Balsamic Vinegar 무화과/ 라즈베리/ 발사믹식초 → Coulis 쿨리
- Cream Cheese 크림치즈 / Gorgonzola 고르곤졸라 → Ganache 가나슈

❹ Crispy 크리스피
- → Pâte Sucrée 파트 슈크레

❺ Cover 커버
- Caramel & Milk Chocolate 캐러멜 & 밀크 초콜릿 → Pistolet 피스톨레

무화과 잎 데코젤
Fig Leaf Decorgel
무화과 쿨리 B
Fig Coulis B
캐러멜 & 밀크 스프레이
Caramel & Milk Spray
바닐라 마스카르포네 몽테
Vanilla Mascarpone Montée
크림치즈 고르곤졸라 가나슈
Cream Cheese Gorgonzola Ganache
파트 슈크레
Pâte Sucrée
바질 크림치즈 스펀지
Basil Cream Cheese Sponge
무화과 쿨리 A
Fig Coulis A

INGREDIENTS

12개 분량/ 12 cakes

바질 크림치즈 스펀지

Basil Cream Cheese Sponge

(철판(60 × 40cm) 1개 분량)
(1 baking tray (60 × 40 cm))

INGREDIENTS		g
버터 (Bridel)	Butter	125g
크림치즈 (kiri)	Cream cheese	125g
설탕	Sugar	200g
레몬제스트	Lemon zest	1/2개 분량/ 1/2 lemon
레몬에센스 (Aroma Piu)	Lemon essence	5g
바닐라에센스 (Aroma Piu)	Vanilla essence	8g
박력분	Cake flour	200g
아몬드파우더	Almond powder	50g
베이킹파우더	Baking powder	7.5g
달걀	Eggs	180g
바질	Basil	2g
TOTAL		**902.5g**

파트 슈크레

Pâte Sucrée

INGREDIENTS		g
버터 (Bridel)	Butter	165g
슈거파우더	Sugar powder	165g
소금	Salt	2g
박력분	Cake flour	350g
아몬드파우더	Almond powder	65g
레몬제스트	Lemon zest	1/2개 분량/ 1/2 lemon
우유	Milk	65g
TOTAL		**812g**

- 우유는 단단한 질감을 원할 때 사용한다. (달걀로 대체 가능)
- Use the milk if you want a firm texture. (Can be replaced with eggs)

*** 만드는 법**

❶ 볼에 포마드 상태의 버터, 슈거파우더, 소금을 넣고 비터로 믹싱해 크림화한다.
❷ 체 친 박력분과 아몬드파우더, 레몬제스트를 넣고 믹싱한다.
❸ 우유를 나눠 넣어가며 믹싱한다.
❹ 반죽이 한 덩어리가 될 때까지 치댄다.
❺ 완성된 반죽은 랩으로 감싸 냉장고에서 휴지시킨다.
❻ 휴지시킨 반죽을 3mm 두께로 밀어 편 후 5.5cm 원형 커터로 자른다.
❼ 170℃로 예열된 오븐에서 15분간 굽는다.

*** Procedure**

❶ Cream softened butter, sugar powder, and salt using a paddle attachment.
❷ Mix with sifted cake flour, almond powder, and lemon zest.
❸ Gradually add milk and mix.
❹ Mix until the dough comes together.
❺ Wrap the dough with plastic wrap and refrigerate.
❻ Roll out the refrigerated dough to 3 mm thickness and cut with a 5.5 cm round cutter.
❼ Bake in an oven preheated to 170°C for 15 minutes.

크림치즈 고르곤졸라 가나슈

Cream Cheese Gorgonzola Ganache

INGREDIENTS		g
생크림	Heavy cream	100g
화이트초콜릿 (Zephyr white 34%)	White chocolate	100g
크림치즈	Cream cheese	60g
고르곤졸라	Gorgonzola cheese	6g
버터 (Bridel)	Butter	20g
TOTAL		**286g**

바닐라 마스카르포네 몽테

Vanilla Mascarpone Montée

• p.293

INGREDIENTS		g
휘핑크림A	Whipping cream A	155g
연유	Condensed milk	42g
젤라틴매스 (×5)	Gelatin mass (×5)	30g
화이트초콜릿 (Zephyr white 34%)	White chocolate	100g
마스카르포네	Mascarpone	100g
바닐라빈	Vanilla bean	1/2개 분량/ 1/2 pc
휘핑크림B	Whipping cream B	350g
골드럼 (PAN RUM)	Gold rum	6g
TOTAL		**783g**

무화과 잎 오일 *

Fig Leaf Oil

INGREDIENTS		g
무화과 잎	Fig leaves	60g
식용유	Vegetable oil	140g
TOTAL		**200g**

- 필요한 양에 따라 무화과 잎과 식용유를 3:7 비율로 계량해 사용한다.
- Measure out the fig leaves and vegetable oil in a 3:7 ratio, depending on how much you need.

무화과 잎 데코젤

Fig Leaf Decorgel

INGREDIENTS		g
무화과 잎 오일 *	Fig leaf oil *	30g
내추럴 데코젤	Natural decorgel	100g
TOTAL		**130g**

무화과 쿨리
Fig Coulis

INGREDIENTS		g
무화과(생과)A	Fresh figs A	300g
냉동 라즈베리	Frozen raspberry	200g
설탕	Sugar	200g
NH펙틴	NH pectin	12g
로거스트빈검	Locust bean gum	1g
포트와인	Port wine	70g
라임 퓌레	Lime purée	18g
발사믹식초	Balsamic vinegar	20g
무화과(생과)B	Fresh figs B	100g
TOTAL		**921g**

캐러멜 스프레이
Caramel Spray

INGREDIENTS		g
캐러멜초콜릿 (Zephyr Caramel 35%)	Caramel chocolate	200g
카카오버터	Cacao butter	200g
TOTAL		**400g**

- 볼에 모든 재료를 넣고 녹여 바믹서로 블렌딩한 후 50℃로 맞춰 사용한다.
- Melt all the ingredients in a bowl, combine using an immersion blender, and use at 50°C.

밀크 스프레이
Milk Spray

INGREDIENTS		g
화이트초콜릿 (Zephyr white 34%)	White chocolate	200g
카카오버터	Cacao butter	200g
TOTAL		**400g**

- 볼에 모든 재료를 넣고 녹여 바믹서로 블렌딩한 후 50℃로 맞춰 사용한다.
- Melt all the ingredients in a bowl, combine using an immersion blender, and use at 50°C.

NOTE.

✤

BASIL CREAM CHEESE SPONGE

바질 크림치즈 스펀지

1 믹싱볼에 포마드 상태의 버터를 넣고 비터로 풀어준다.

2 크림치즈를 넣고 믹싱한다.

3 설탕을 넣고 믹싱한다.

4 레몬제스트, 레몬에센스, 바닐라에센스를 넣고 믹싱한다.

5 체 친 [박력분, 아몬드파우더, 베이킹파우더] 절반을 넣고 믹싱한다.

6 가루가 섞이면 달걀(30℃)을 두세 번 나눠 넣어가며 믹싱한다.

7 달걀이 섞이면 남은 가루 재료를 넣고 믹싱한다.

8 다진 바질을 넣고 믹싱한다.

9 테프론시트를 깐 철판(60 × 40cm)에 0.5cm 두께로 팬닝한 후 170℃로 예열된 오븐에서 12분간 굽는다.

TIP. 여기에서는 일정함을 위해 팬닝기로 팬닝했다.
구워져 나온 스펀지의 높이는 1cm 미만이 적당하다.

1 Loosen the softened butter in a mixing bowl using a paddle attachment.

2 Mix with cream cheese.

3 Mix with sugar.

4 Add lemon zest, lemon essence, and vanilla essence.

5 Add half of sifted [cake flour, almond powder, and baking powder].

6 When the powder ingredients are mixed in, add eggs (30°C) in two or three batches.

7 When the eggs are mixed in, add the remaining powder ingredients.

8 Mix with chopped basil.

9 Spread to 0.5 cm on a baking tray (60 × 40 cm) lined with a Teflon sheet. Bake in an oven preheated to 170°C for 12 minutes.

TIP. I used a cake depositor machine for consistency.
The appropriate thickness of the baked sponge should be less than 1 cm.

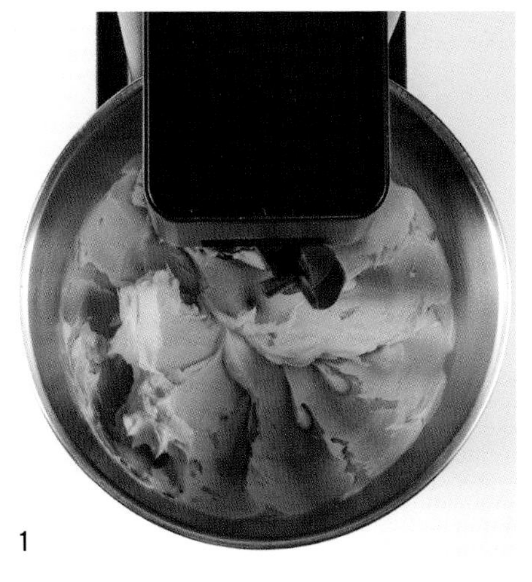
1

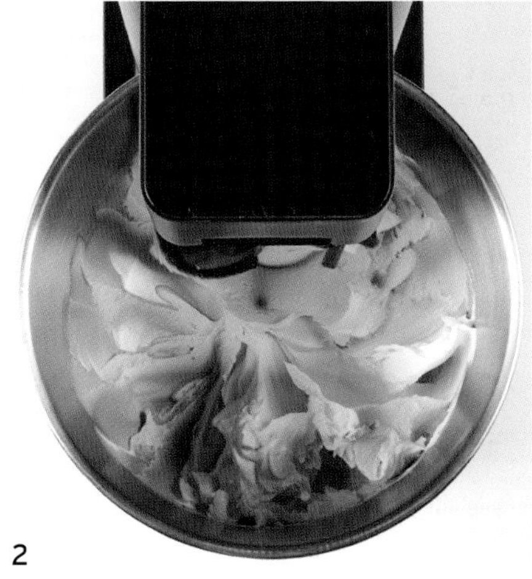
2

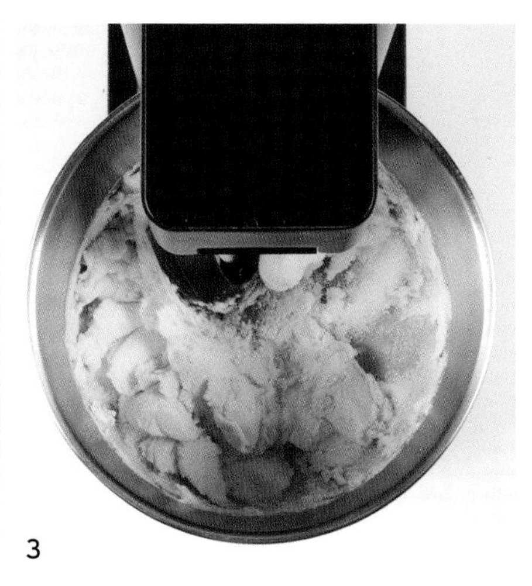
3

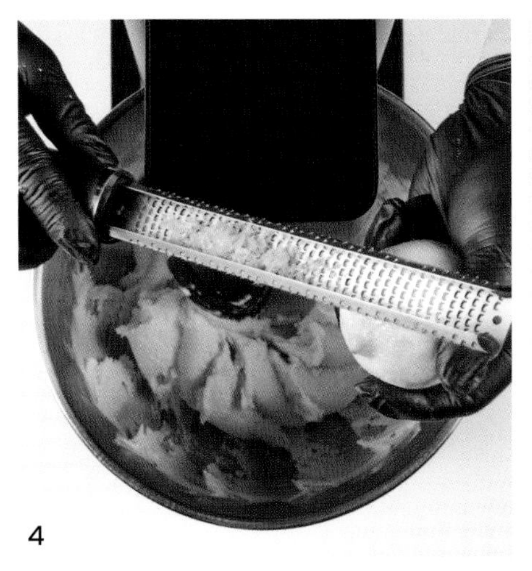
4

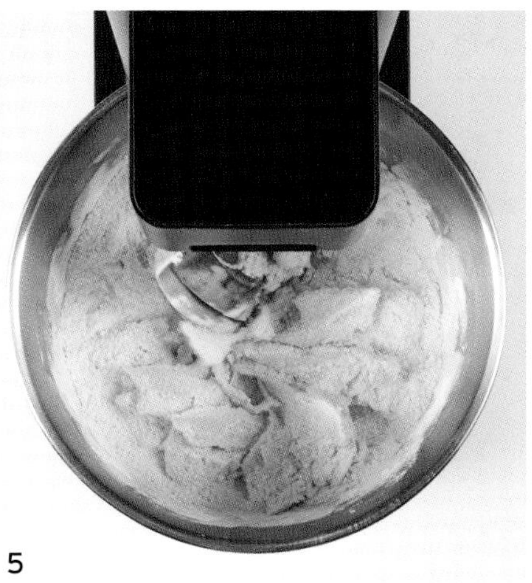
5

6

7

8

9

CREAM CHEESE GORGONZOLA GANACHE

크림치즈 고르곤졸라 가나슈

1 냄비에 생크림을 넣고 뜨겁게 가열한다. (약 60℃)

2 화이트초콜릿이 담긴 비커에 붓고 바믹서로 블렌딩한다.

3 크림치즈, 고르곤졸라를 넣고 블렌딩한다.

4 포마드 상태의 버터를 넣고 블렌딩한다.

5 얼음물이 담긴 볼에 받쳐 25℃로 쿨링한다.

6 바질 크림치즈 스펀지에 고르게 펴 바르고 냉장고에서 굳힌 후 지름 5.5cm 원형 커터로 잘라 냉동한다.

1 Heat the cream until hot (about 60°C).

2 Blend with white chocolate in a beaker using an immersion blender.

3 Blend with cream cheese and gorgonzola.

4 Add softened butter and continue to blend.

5 Cool in an ice bath to 25°C.

6 Spread evenly on the basil cream cheese sponge, set it in the fridge, cut with a 5.5 cm round cutter, and freeze.

FIG LEAF DECORGEL

무화과 잎 데코젤

무화과 잎 오일

1 써머믹서에 깨끗이 씻어 물기를 제거한 무화과 잎, 식용유를 넣고 갈아준다.

TIP. 써머믹서의 온도는 70℃로 설정한다. (무화과 잎이 가지고 있는 특유의 향이 식용유에 잘 흡수되는 온도가 70℃이다.
70℃보다 온도가 높아지면 무화과 잎이 익어버려 너무 어두운 초록빛으로 변한다.)
완성한 무화과 잎 오일은 고운 체에 거르고 최대한 산소의 마찰을 줄일 수 있는 밀폐용기에 담아 보관하며 사용한다.
(가장 좋은 방법은 휴롬 같은 기계를 이용해 엽록소까지 추출하는 것이다.)

무화과 잎 데코젤

2 필요한 양에 맞춰 무화과 잎 오일과 데코젤 뉴트럴을 섞어 완성한다.

Fig Leaf Oil

1 Wash the fig leaves thoroughly and dry them. Grind in the ThermoMixer with vegetable oil.

TIP. Set the temperature of the ThermoMixer to 70°C. (The temperature at which the unique aroma of fig leaves is best absorbed into the oil is 70°C. If it's above 70°C, the leaves will overcook and turn very dark.)
Strain the finished oil through a fine sieve and store it in an airtight container to reduce oxidation as much as possible.
(The best method is to extract chlorophyll using a machine like the Hurom Juicer.)

Fig Leaf Decorgel

2 Combine fig leaf oil and Decorgel Neutral according to the amount needed.

✤ ✤

1

2

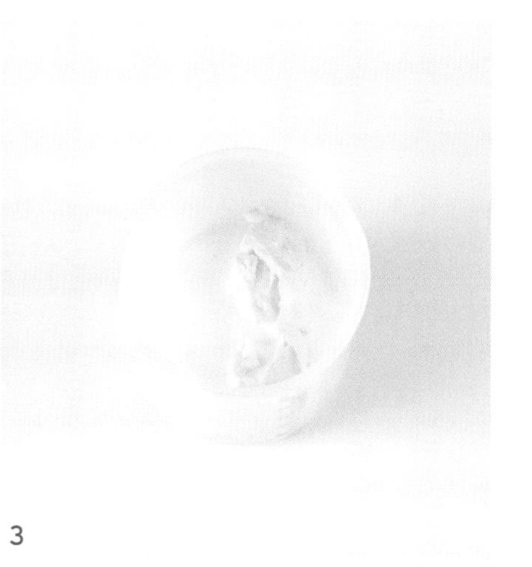
3

4

5

6

✤ ✤ ✤

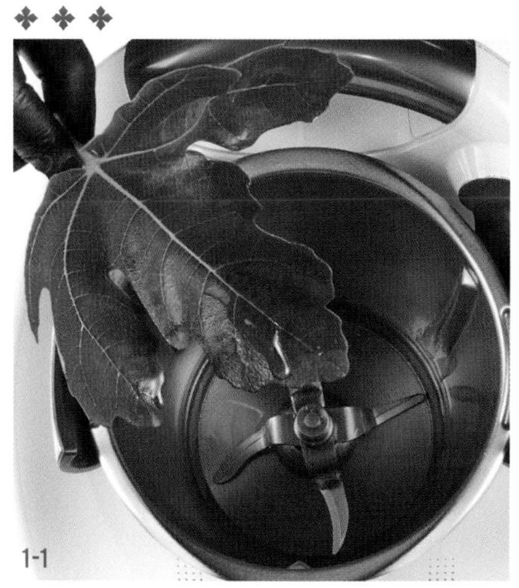
1-1

1-2

2

FIG COULIS

무화과 쿨리

무화과 쿨리

1 비커에 갈기 좋은 크기로 자른 무화과A, 냉동 라즈베리를 넣고 바믹서로 갈아준다.

2 미리 섞어둔 설탕과 NH펙틴과 로거스트빈검을 넣고 블렌딩한다.

3 포트와인 - 라임퓌레 순서로 넣고 블렌딩한다.

4 냄비로 옮겨 펙틴 반응이 일어날 때까지 가열한 후, 얼음물이 담긴 볼에 받쳐 15℃로 쿨링하고 밀착 랩핑해 냉장 보관한다.

5 로보쿱으로 가볍게 갈아준 후 발사믹식초를 넣고 갈아준다.

6 무화과B를 넣고 갈아준다.

TIP. 무화과 특유의 풋내를 살짝 더해주기 위해 마지막 과정에서 조금 더 추가했다.

조립

7 몰드(실리코마트 SF029)에 35g씩 채우고 평평하게 정리한다.

8 얼린 [바질 크림치즈 스펀지 + 크림치즈 고르곤졸라 가나슈]를 올린 후 냉동한다.

Fig Coulis

1 Grind fresh figs A, cut into pieces suitable for grinding, and frozen raspberries in a beaker with an immersion blender.

2 Blend with previously mixed sugar, pectin NH, and locust bean gum.

3 Blend with port wine and lime purée in order.

4 Heat until the pectin activates, cool in an ice bath to 15°C. Wrap tightly and refrigerate.

5 Briefly grind with the Robot Coupe and grind again with balsamic vinegar.

6 Grind with fresh figs B.

TIP. I added a little more at the end to give it a little extra fresh green scent.

Assemble

7 Fill 35 grams in the mold (Silikomart SF029) and flatten evenly.

8 Top with the frozen [basil cream cheese sponge + cream cheese gorgonzola ganache] and freeze.

1 2 3

4

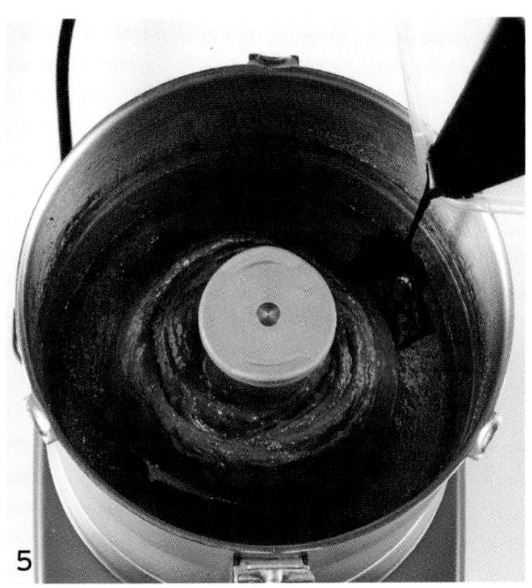

5

6

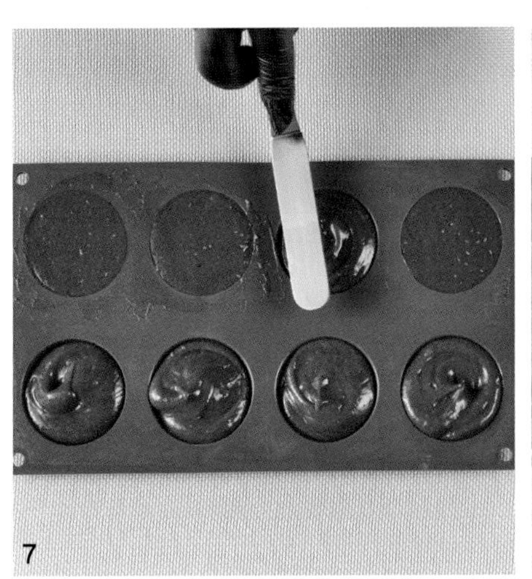

7

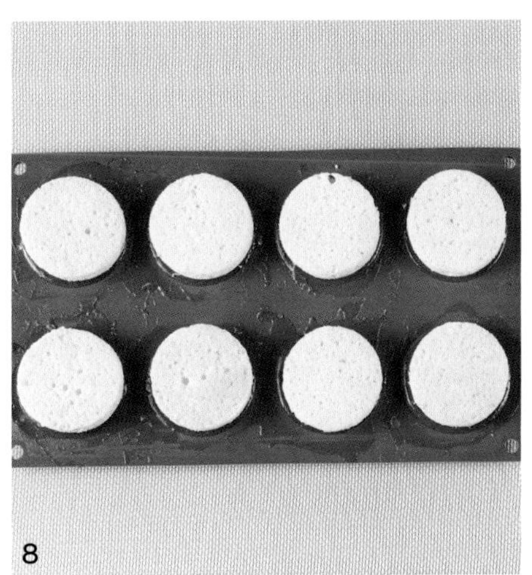

8

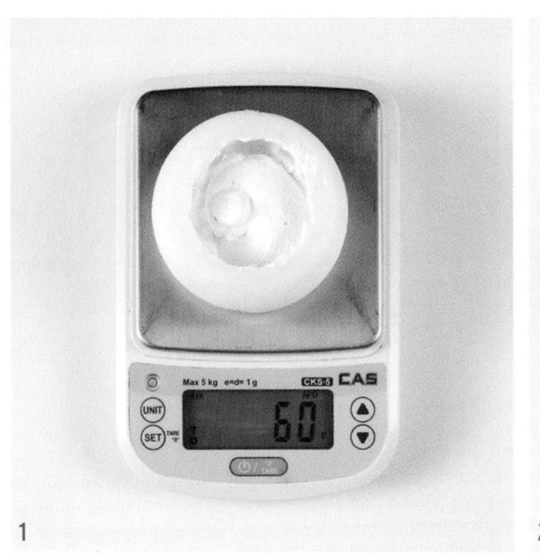

1

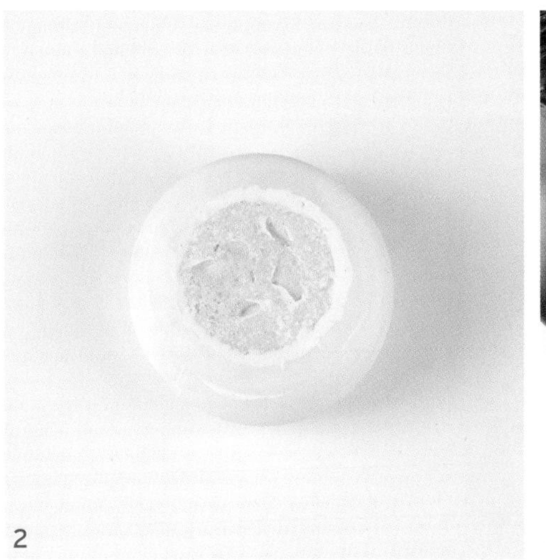
2

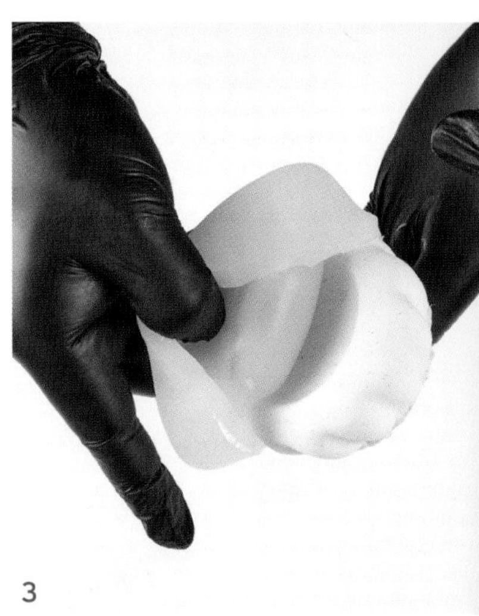
3

✤ ✤ ✤ ✤ ✤

FINISH 마무리

1 몰드에 휘핑한 바닐라 마스카르포네 몽테를 60g씩 채운다.

2 얼린 [바질 크림치즈 스펀지 + 크림치즈 고르곤졸라 가나슈 + 무화과 쿨리]를 넣고 스패출러로 깔끔하게 정리한다.

3 급속 냉동(-35 ~ -40℃)한 후 몰드에서 빼내 냉동(-18℃) 보관한다.

4 캐러멜 스프레이, 밀크 스프레이를 분사해 명암과 음영을 표현하고 파트 슈크레 위에 올린다.

5 무화과 쿨리 14g을 채우고 손질한 무화과를 올린 후, 무화과 잎 데코젤을 파이핑하고 금박으로 장식해 마무리한다.

1. Fill 60 grams of whipped vanilla mascarpone montée in the mold.
2. Insert the frozen [basil cream cheese sponge + cream cheese gorgonzola ganache + fig coulis] and clean up with an offset spatula.
3. Blast freeze (-35 ~ -40°C), remove from the mold, and store frozen (-18°C).
4. Spray with caramel and milk spray mixtures to create contrast and shading. Place it on the pâte sucrée.
5. Fill with 14 grams of fig coulis, top it with trimmed figs, pipe fig leaf decorgel, and garnish with gold leaves to finish.

22

CRISPY BASALT

크리스피 바솔트

PAIRING & TEXTURE
페어링 & 텍스처

바나나를 메인으로 다크 캐러멜 바바루아 크림과 헤이즐넛 캐러멜을 페어링한 메뉴. 크리미한 인서트 안에 크런치함을 추가해 대비되는 식감을 느낄 수 있다.

A menu featuring banana as the main ingredient, paired with dark caramel bavarois cream and hazelnut caramel. I added crunchiness to the creamy insert to create a contrasting texture.

TECHNIQUE
테크닉

- 초콜릿 색소를 이용해 점사, 분사 기법으로 현무암 특유의 거친 질감을 표현했다.
- 압력 조절 장치를 조절하여 분사하는 입자의 크기를 다르게 할 수 있다.
- 케이크 스펀지를 사용하지 않고, 머랭과 라이스 크런치로 파블로바를 연상하게 하는 테크닉을 사용했다.

- The rough texture of basalt was expressed with chocolate coloring using fine and coarse spraying techniques.
- You can adjust the pressure regulator to vary the size of the particles when spraying.
- Instead of using cakes, I used a technique reminiscent of a pavlova with meringue and rice crunchies.

DESIGN
디자인

제주도 현무암에서 영감을 받아 만들게 되었다. 현무암 특유의 리얼한 기공을 표현하기 위해 실제 현무암을 본떠 몰드를 만들었다.

This dessert was inspired by the basalt of Jeju Island. To express the realistic pores unique to basalt, I modeled the basalt to make a mold.

*** 현무암**
현무암은 화산과 마그마의 활동으로 만들어진 암석으로, 표면이 거칠고 크고 작은 구멍이 있는 것이 특징이다. 한국의 화산섬인 제주도에서 쉽게 볼 수 있는 돌이다.

*** Basalt**
Basalt is a rock formed by volcanic and magmatic activity, characterized by a rough surface with large and small pores. You can find it on Jeju Island, a volcanic island in South Korea.

✦ HOW TO COMPOSE THIS RECIPE ✦

STEP 1.	메뉴 정하기	프티 가토
	Decide on the menu	Petit Gateau

STEP 2.	메인 맛 정하기	바나나
	Choose the primary flavor	Banana

STEP 3. 메인 맛(바나나)과의 페어링 선택하기

Select a pairing flavor (Banana)

- ☑ 헤이즐넛 Hazelnut
- ☑ 캐러멜 Caramel
- ☑ 밀크초콜릿 Milk Chocolate
- ☑ 다크초콜릿 Dark Chocolate
- ☑ 캐러멜초콜릿 Caramel Chocolate

STEP 4. 구성하기

Assemble

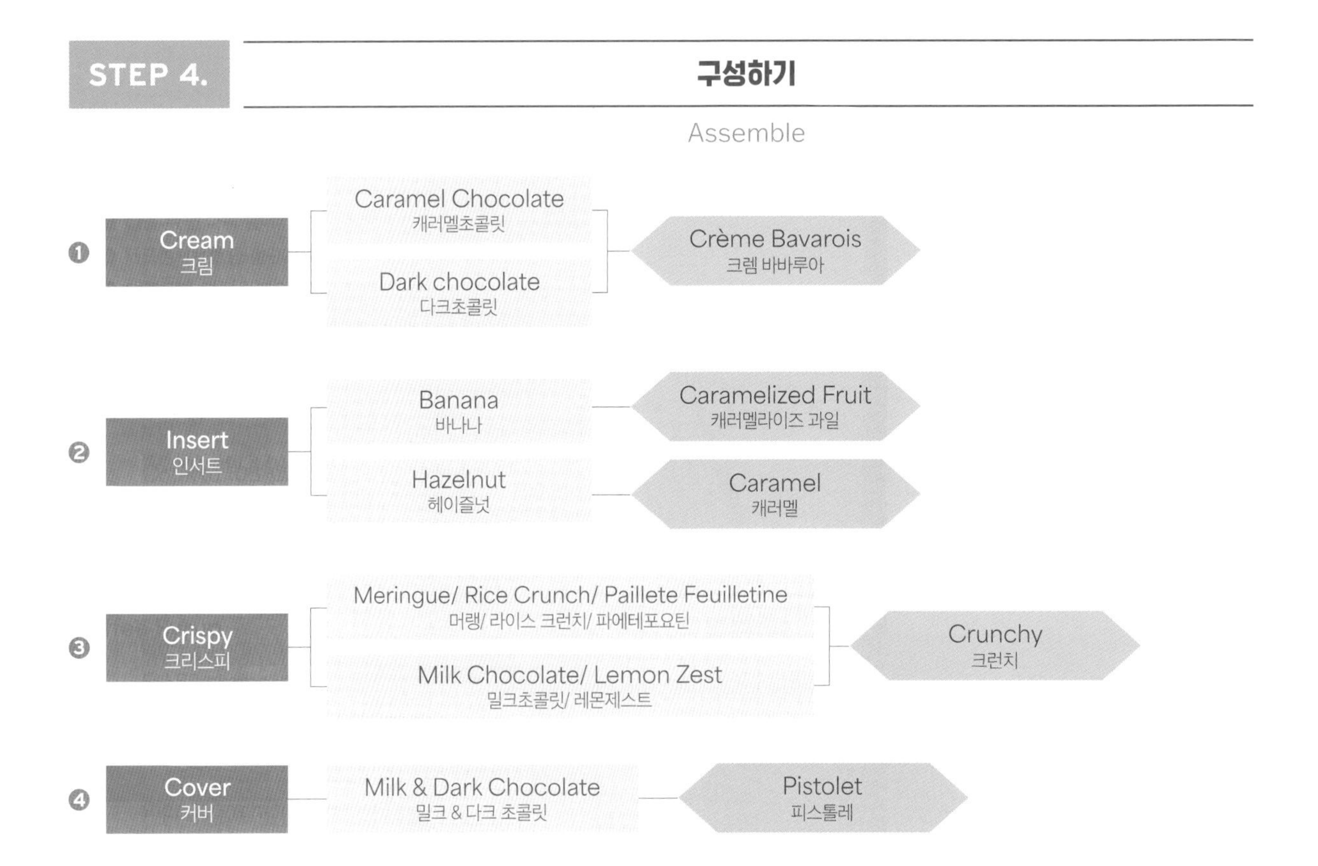

밀크 & 다크 스프레이
Milk & Dark Spray
헤이즐넛 캐러멜
Hazelnut Caramel
다크 캐러멜 바바루아
Dark Caramel Bavarois
캐러멜라이즈 바나나
Caramelized Banana
크런치
Crunchy

INGREDIENTS
10개 분량/ 10 cakes

캐러멜라이즈 바나나
Caramelized Banana

INGREDIENTS		g
황설탕	Brown sugar	60g
버터 (Bridel)	Butter	40g
바나나	Banana	400g
TOTAL		**500g**

머랭 베이스 *
Meringue Base

• p.458

INGREDIENTS		g
흰자	Egg whites	180g
이소말트파우더	Isomalt powder	100g
설탕	Sugar	80g
알부민파우더	Albumin powder	6g
슈거파우더	Sugar powder	120g
탈지분유	Skim milk powder	18g
옥수수전분	Cornstarch	17g
TOTAL		**521g**

크런치
Crunchy

INGREDIENTS		g
밀크초콜릿 (Alunga 41%)	Milk chocolate	120g
카카오버터	Cacao butter	25g
레몬제스트	Lemon zest	1/2개 분량/ 1/2 lemon
라이스크런치	Rice crunchies	70g
머랭 베이스 *	Meringue base *	25g
파에테포요틴	Paillete feuilletine	25g
TOTAL		**265g**

다크 캐러멜 바바루아

Dark Caramel Bavarois

INGREDIENTS		g
크렘 앙글레이즈 (되직한 타입) (p.294)	Crème anglaise (cooked thicker) (p.294)	300g
젤라틴매스 (×5)	Gelatin mass (×5)	28g
다크초콜릿 (Mi-Amère 58%)	Dark chocolate	200g
캐러멜초콜릿 (Zephyr Caramel 35%)	Caramel chocolate	80g
생크림	Heavy cream	400g
바나나 리큐르 (Orchid Cream de Banana)	Banana liqueur	20g
TOTAL		**1028g**

헤이즐넛 캐러멜

Hazelnut Caramel

• p.446

INGREDIENTS		g
생크림	Heavy cream	210g
설탕	Sugar	80g
물엿	Corn syrup	130g
헤이즐넛 페이스트	Hazelnut paste	24g
소금	Salt	2g
TOTAL		**446g**

밀크 스프레이

Milk Spray

INGREDIENTS		g
밀크초콜릿 (Alunga 41%)	Milk chocolate	300g
카카오버터	Cacao butter	300g
TOTAL		**600g**

- 볼에 모든 재료를 넣고 녹여 바믹서로 블렌딩한 후 50℃로 맞춰 사용한다.
- Melt all the ingredients in a bowl, combine using an immersion blender, and use at 50°C.

다크 스프레이

Dark Spray

INGREDIENTS		g
다크초콜릿 (Mi-Amère 58%)	Dark chocolate	300g
카카오버터	Cacao butter	300g
TOTAL		**600g**

- 볼에 모든 재료를 넣고 녹여 바믹서로 블렌딩한 후 50℃로 맞춰 사용한다.
- Melt all the ingredients in a bowl, combine using an immersion blender, and use at 50°C.

✤

CARAMELIZED BANANA

캐러멜라이즈 바나나

1 팬에 황설탕을 넣고 가열한다.

2 진한 갈색으로 캐러멜화되면 포마드 상태의 버터를 넣고 가열한다.

TIP. 황설탕은 백설탕과 다르게 캐러멜화되는 시점을 눈으로 확인하기 어려우므로, 사진처럼 중앙에 기포가 올라오는 시점을 확인하고 버터를 넣는다.

3 버터가 녹으면 사방 1.5cm 크기로 깍둑썬 바나나를 넣고 주걱으로 섞어가며 가열한 후 바트에 담고 밀착 랩핑해 냉장고에 보관한다.

1 Heat brown sugar in a pan.

2 When it caramelizes to a dark brown color, add softened butter and cook.

TIP. Unlike the white sugar, it's harder to tell when the brown sugar is caramelized. So, look for the bubbles in the center, as shown in the photo, and add the butter.

3 When the butter melts, stir in the banana cut into 1.5 cm cubes, and cook. Pour onto a tray, adhere the plastic wrap to the preparation to cover, and refrigerate.

CRUNCHY

크런치

크런치

1 볼에 녹인 밀크초콜릿(45~50℃), 50℃ 이하로 녹인 카카오버터, 레몬제스트를 넣고 섞는다.

2 라이스크런치, 머랭 베이스, 파에테포요틴을 넣고 골고루 섞는다.

3 몰드(DECO RELIEF DR-283)에 20g씩 채우고 스패출러로 평평하게 정리한다. 냉동고에서 10분 정도 굳힌 후 몰드에서 빼내 밀폐용기에 담아 실온 보관한다.

조립

4 가운데 홈에 헤이즐넛 캐러멜을 10g씩 채운다.

5 캐러멜라이즈 바나나를 36g씩 올리고 스패출러로 봉긋하게 모양을 다듬은 후 냉동한다.

Crunchy

1 Combine melted milk chocolate (45~50°C), cacao butter melted to below 50°C, and lemon zest.

2 Mix with rice crunchies, meringue base, and paillete feuilletine.

3 Fill 20 grams in the mold (Deco Relief DR-283) and flatten with a spatula. Freeze for about 10 minutes, remove from the mold and keep at room temperature in an airtight container.

Assemble

4 Fill 10 grams of hazelnut caramel in the center indentation.

5 Top with 36 grams of caramelized banana, trim it into a mound with a spatula and freeze.

✤

1

2

3

✤ ✤

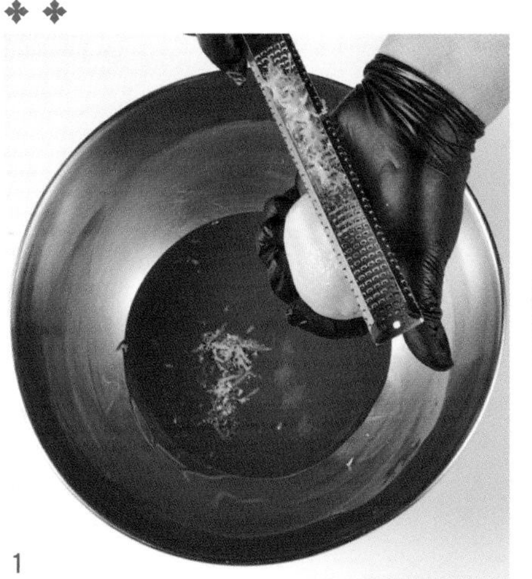
1

2-1

2-2

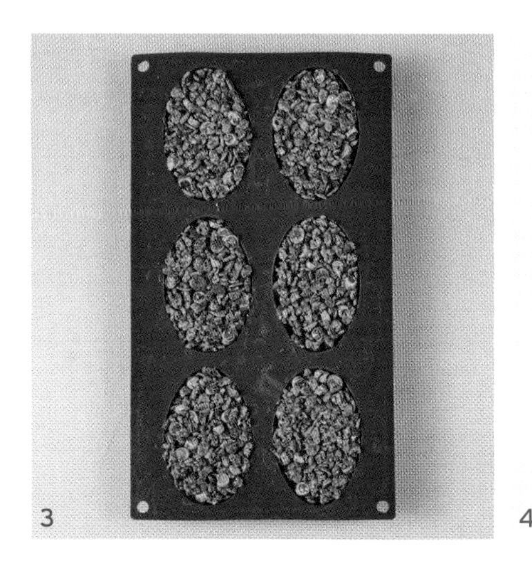
3

4

5

DARK CARAMEL BAVAROIS

다크 캐러멜 바바루아

캐러멜 바바루아

1 냄비에 크렘 앙글레이즈, 젤라틴매스를 넣고 젤라틴매스가 녹을 때까지 가열한다. (약 60℃)

TIP. 바나나가 들어가 무거운 인서트가 들어갔을 때 지탱할 수 있을 정도로 단단하게 완성되어야 한다.

2 다크초콜릿, 캐러멜초콜릿이 담긴 비커에 넣고 바믹서로 블렌딩한다.

3 손으로 가볍게 휘핑한 생크림(60~70%)이 담긴 볼에 넣고 주걱으로 섞는다.

4 바나나 리큐르를 넣고 섞는다.

조립

5 몰드에 85g 채운 후 몰드를 돌려가며 골고루 묻혀준다.

6 얼린 [크런치 + 헤이즐넛 캐러멜 + 캐러멜라이즈 바나나]를 넣고 스패출러로 평평하게 정리하고 급속 냉동(-35 ~ -40℃)한 후 몰드에서 빼내 냉동(-18℃) 보관한다.

Caramel Bavarois

1 Heat crème Anglaise and gelatin mass to 60°C until the gelatin melts.

TIP. It should be firm enough to support a heavy insert with a banana.

2 Blend with dark and caramel chocolate in a beaker using an immersion blender.

3 Pour into the bowl with the heavy cream lightly whipped (60~70%) with a whisk. Combine with a spatula.

4 Mix with banana liqueur.

Assemble

5 Fill 85 grams into the mold and rotate it to coat it evenly.

6 Insert the frozen [crunchy + hazelnut caramel + caramelized banana] and flatten with a spatula. Blast freeze (-35 ~ -40°C), remove from the mold and keep in the freezer (-18°C).

FINISH

마무리

밀크 스프레이, 다크 스프레이를 점사, 분사하여 명암과 음영을 표현하고, 열풍기로 표면에 살짝 열을 가해 질감을 표현한다.

Spray with milk and dark spray mixtures to create contrast and shading. Briefly heat the surface with a heat gun to create texture.

✤ ✤ ✤

1

60℃
2

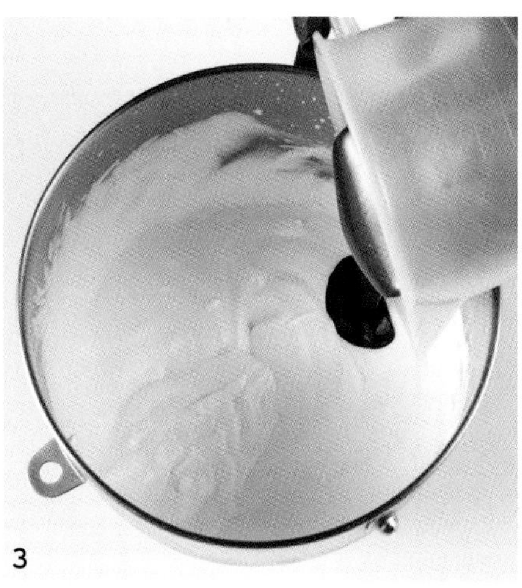
3

4

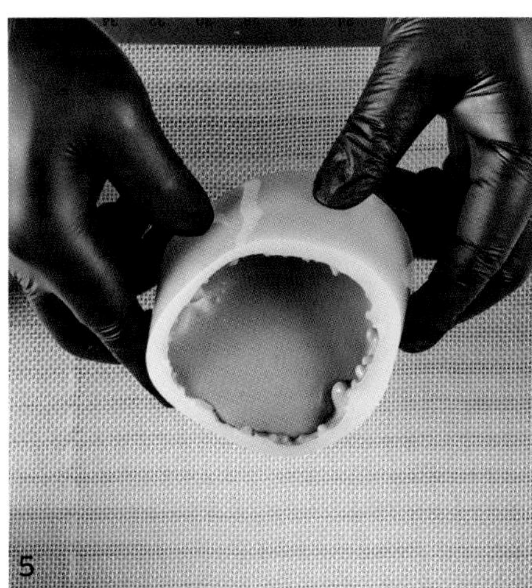
5

6

✤ ✤ ✤ ✤

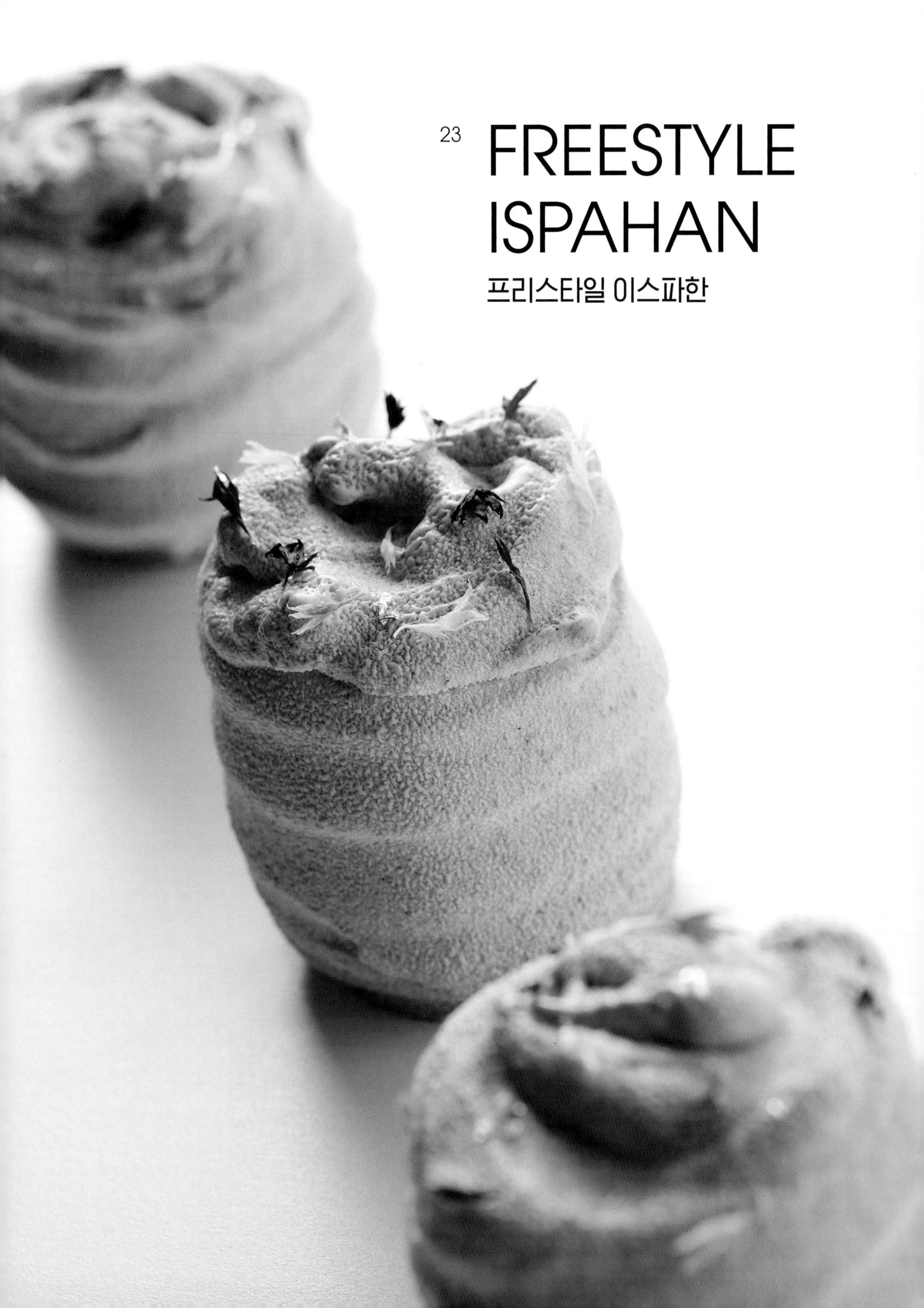

23

FREESTYLE ISPAHAN

프리스타일 이스파한

PAIRING & TEXTURE
페어링 & 텍스처

- 가나슈가 베이스가 되는 일반적인 몽테가 아닌, 초콜릿을 사용하지 않고 과일 퓌레를 넣어 가볍고 신선한 맛과 질감의 몽테를 만들었다.
- 무스 스타일로 만든 제품이지만 일반적인 무스보다는 무겁고 일반적인 몽테보다는 가벼운 중간 식감으로 완성했다. 매일 만들어 바로 사용해야 하는 무스와 다르게 미리 만들어두고 휘핑해 사용할 수 있어 편리하다.

- Rather than the typical ganache-based montée, I used fruit purée instead of chocolate to create a montée that has a light and fresh taste and texture.
- Although it's a mousse-style dessert, it has a medium density that is heavier than a typical mousse and lighter than a typical montée. Unlike a typical mousse, which must be made every day and used right away, this one is convenient because you can make it ahead of time and whip it before use.

TECHNIQUE
테크닉

- 인서트 스펀지를 동일한 사이즈의 틀에 넣어 얼린 후 메인 크림을 무늬 패드에 돌려가며 아이싱하는 독특한 기법을 사용했다.
- 초콜릿을 사용하지 않는 몽테 스타일로 완성했다.
- 여러 가지 과일 퓌레를 이용해 다양한 맛으로 연출할 수 있다.

- I used a unique technique of freezing the insert in the same size mold and then rolling the main cream on the patterned pad to coat.
- It is a montée style dessert, made without chocolate.
- You can use different fruit purée to create different flavors.

DESIGN
디자인

- 초콜릿 색소 배합에 인공적인 색소 대신 동결건조 과일 파우더를 첨가해 자연스러운 컬러를 연출했다.
- 스펀지와 인서트를 얼린 상태에서 수작업으로 자연스러운 회오리 모양을 연출했다.

- I added freeze-dried fruit powder to the chocolate coloring formula instead of artificial colors for natural color.
- The natural swirl shape was made by hand while the sponge cakes and insert were frozen.

✦ HOW TO COMPOSE THIS RECIPE ✦

STEP 1.

메뉴 정하기	프티 가토
Decide on the menu	Petit Gateau

STEP 2.

메인 맛 정하기	라즈베리
Choose the primary flavor	Raspberry

STEP 3.

메인 맛(라즈베리)과의 페어링 선택하기

Select a pairing flavor (Raspberry)

- ☑ 리치 Lychee
- ☑ 장미 Rose
- ☑ 크림치즈 Cream Cheese

- ☑ 사워크림 Sour Cream
- ☑ 바질 Basil

STEP 4.

구성하기

Assemble

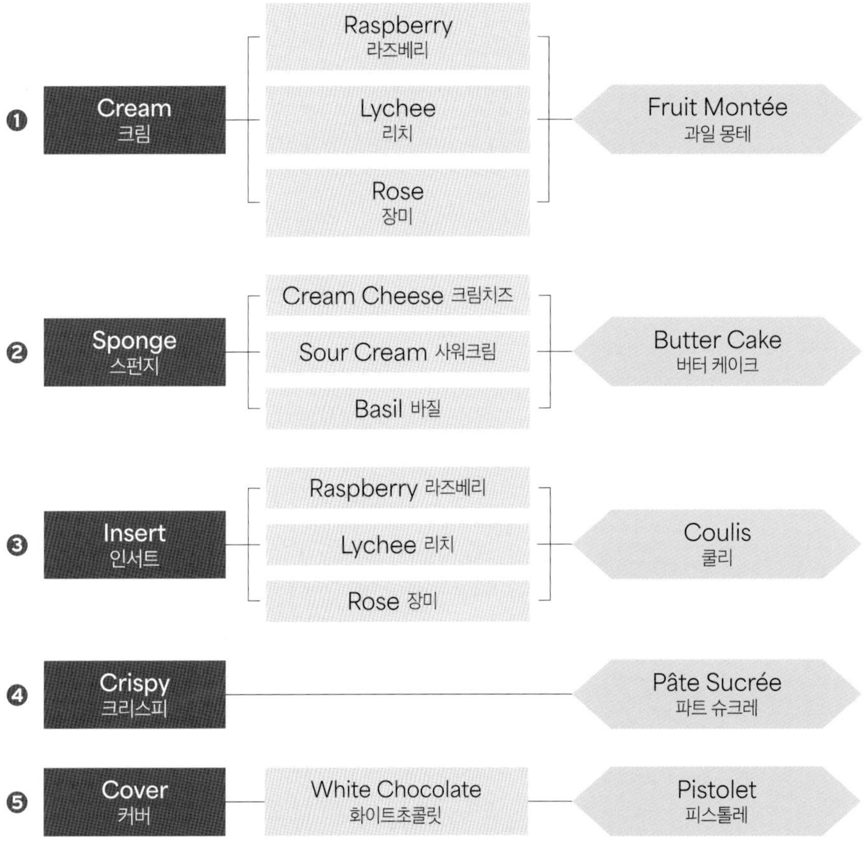

❶ Cream 크림 — Raspberry 라즈베리 / Lychee 리치 / Rose 장미 → Fruit Montée 과일 몽테

❷ Sponge 스펀지 — Cream Cheese 크림치즈 / Sour Cream 사워크림 / Basil 바질 → Butter Cake 버터 케이크

❸ Insert 인서트 — Raspberry 라즈베리 / Lychee 리치 / Rose 장미 → Coulis 쿨리

❹ Crispy 크리스피 → Pâte Sucrée 파트 슈크레

❺ Cover 커버 — White Chocolate 화이트초콜릿 → Pistolet 피스톨레

라즈베리 스프레이
Raspberry Spray
이스파한 몽테
Ispahan Montée
바질 크림치즈 스펀지
Basil Cream Cheese Sponge
리치 라즈베리 쿨리
Lychee Raspberry Coulis
파트 슈크레
Pâte Sucrée

INGREDIENTS 20개 분량/ 20 cakes

바질 크림치즈 스펀지

Basil Cream Cheese Sponge

(철판(60 × 40cm) 1개 분량)
(1 baking tray (60 × 40 cm))

INGREDIENTS		g
버터 (Bridel)	Butter	125g
크림치즈 (kiri)	Cream cheese	75g
사워크림	Sour cream	50g
설탕	Sugar	200g
레몬제스트	Lemon zest	1/2개 분량/ 1/2 lemon
레몬에센스 (Aroma Piu)	Lemon essence	1g
바닐라에센스 (Aroma Piu)	Vanilla essence	7.5g
박력분	Cake flour	175g
아몬드파우더	Almond powder	50g
베이킹파우더	Baking powder	7.5g
달걀	Eggs	175g
바질	Basil	4g
TOTAL		**870g**

*** 만드는 법**

❶ 믹싱볼에 포마드 상태의 버터를 넣고 비터로 풀어준다.
❷ 크림치즈, 사워크림을 넣고 믹싱한다.
❸ 설탕, 레몬제스트, 레몬에센스, 바닐라에센스를 넣고 믹싱한다.
❹ 체 친 [박력분, 아몬드파우더, 베이킹파우더] 절반을 넣고 믹싱한다.
❺ 달걀(30℃)을 두 번 나눠 넣어가며 믹싱한다.
❻ 남은 가루 재료를 넣고 믹싱한다.
❼ 다진 바질을 넣고 믹싱한다.
- 바질은 사용하기 직전에 다져 향을 최대한 살린다.

❽ 테프론시트를 깐 철판(60 × 40cm)에 팬닝한 후 170℃로 예열된 오븐에서 12분간 굽는다.
❾ 냉장고에서 식힌 후 냉동실에서 얼려 지름 4.8cm 원형 커터로 자른다. (높이 1cm)

*** Procedure**

❶ Loosen the softened butter in a mixing bowl using a paddle attachment.
❷ Mix with cream cheese and sour cream.
❸ Add sugar, lemon zest, lemon essence, and vanilla essence.
❹ Mix with half of sifted [cake flour, almond powder, and baking powder].
❺ Mix with eggs (30°C) in two batches.
❻ Mix with the remaining powder ingredients.
❼ Add chopped basil and combine.
- Chop the basil right before use to maximize the flavor.

❽ Spread on a baking tray (60 × 40 cm) lined with a Teflon sheet and bake in an oven preheated to 170°C for 12 minutes.
❾ Cool in the fridge, then freeze. Cut with a 4.8 cm round cutter. (1 cm thick)

파트 슈크레

Pâte Sucrée

• p.348

INGREDIENTS		g
버터 (Bridel)	Butter	165g
슈거파우더	Sugar powder	165g
소금	Salt	2g
박력분	Cake flour	350g
아몬드파우더	Almond powder	65g
레몬제스트	Lemon zest	1/2개 분량/ 1/2 lemon
레몬에센스 (Aroma Piu)	Lemon essence	1g
우유	Milk	65g
TOTAL		**813g**

- 3mm로 밀어 펴 냉장고에서 휴지시킨 후 지름 4.5cm 원형 커터로 재단한다. 170℃로 예열된 오븐에서 15분간 굽는다.
- Roll put to 3 mm and let it rest in a fridge and cut with a 4.5 cm round cutter. Bake for 15 minutes in a oven preheated to 170°C.

리치 라즈베리 쿨리

Lychee Raspberry Coulis

INGREDIENTS		g
냉동 라즈베리	Freeze-dried raspberry	200g
리치 퓌레	Lychee purée	440g
라즈베리 퓌레	Raspberry purée	160g
설탕	Sugar	330g
NH펙틴	NH pectin	23g
로거스트빈검	Locust bean gum	0.5g
라임즙	Lime juice	80g
천연 장미수 (Sosa)	Organic rose water	7g
TOTAL		**1240.5g**

이스파한 몽테

Ispahan Montée

23개 분량/ 23 cakes

INGREDIENTS		g
라즈베리 퓌레	Raspberry purée	150g
리치 퓌레	Lychee purée	100g
설탕	Sugar	50g
전화당	Inverted sugar	120g
젤라틴매스 (×5)	Gelatin mass (×5)	70g
라임즙	Lime juice	20g
산딸기 리큐르 (Dijon Framboises)	Raspberry liqueur	25g
마스카르포네	Mascarpone	100g
생크림	Heavy cream	200g
휘핑크림 (서울우유)	Whipping cream	600g
천연 장미수 (Sosa)	Organic rose water	2g
TOTAL		**1437g**

라즈베리 스프레이

Raspberry Spray

INGREDIENTS		g
화이트초콜릿 (Zephyr white 34%)	White chocolate	600g
카카오버터	Cacao butter	600g
동결건조 라즈베리파우더	Freeze-dried raspberry powder	48g
TOTAL		**1248g**

- 동결건조 라즈베리파우더는 동결건조 딸기 슬라이스 또는 다이스를 곱게 갈아 사용한다.
- 볼에 화이트초콜릿과 카카오버터를 넣고 녹여 바믹서로 블렌딩한 후 동결건조 라즈베리파우더를 섞어 50℃로 맞춰 사용한다.
- Finely grind sliced or diced freeze-dried strawberries to make the freeze-dried raspberry powder.
- Melt white chocolate and cacao butter in a bowl and combine using an immersion blender. Mix with freeze-dried raspberry powder and use at 50°C.

✤

LYCHEE RASPBERRY COULIS

리치 라즈베리 쿨리

1 냄비에 냉동 라즈베리, 리치 퓌레, 라즈베리 퓌레를 넣고 40℃로 가열한다.

2 미리 섞어둔 [설탕, NH펙틴, 로거스트빈검]을 넣고 가열한다.

3 펙틴 반응을 확인하고 라임즙 - 천연 장미수 순서로 넣고 섞은 후 얼음물이 담긴 볼에 받쳐 15℃로 쿨링하고 밀착 랩핑해 냉장 보관한다.

1 Heat frozen raspberries, lychee purée, and raspberry purée to 40°C.

2 Continue to heat with previously mixed [sugar, pectin NH, and locust bean gum].

3 After the pectin activates, add lime juice and then rose water. Cool in an ice bath to 15°C, wrap tightly, and refrigerate.

✤ ✤

ISPAHAN MONTÉE

이스파한 몽테

1 냄비에 라즈베리 퓌레, 리치 퓌레, 설탕, 전화당을 넣고 가열하다가 끓어오르면 불에서 내려 젤라틴매스를 넣고 섞는다

2 얼음물이 담긴 볼에 받쳐 30℃로 쿨링한 후 비커로 옮겨 라임즙, 산딸기 리큐르, 마스카르포네, 생크림, 휘핑크림을 넣고 바믹서로 블렌딩한다.

3 천연 장미수를 넣고 블렌딩한다.

4 바트에 담고 밀착 랩핑한 후 냉장고에서 12시간 숙성시킨다.

5 사용하기 직전 휘퍼로 부드럽게 휘핑한 후 사용한다.

1 Heat raspberry purée, lychee purée, sugar, and inverted sugar. When it comes to a boil, remove from the heat and gelatin mass.

2 Cool in an ice bath to 30°C. Blend with lime juice, raspberry liqueur, mascarpone, heavy cream, and whipping cream in a beaker using an immersion blender.

3 Blend with the rose water.

4 Pour onto a tray, adhere the plastic wrap to the preparation to cover, and refrigerate for 12 hours.

5 Whip to a soft peak right before use.

✤
40℃
1
2
3

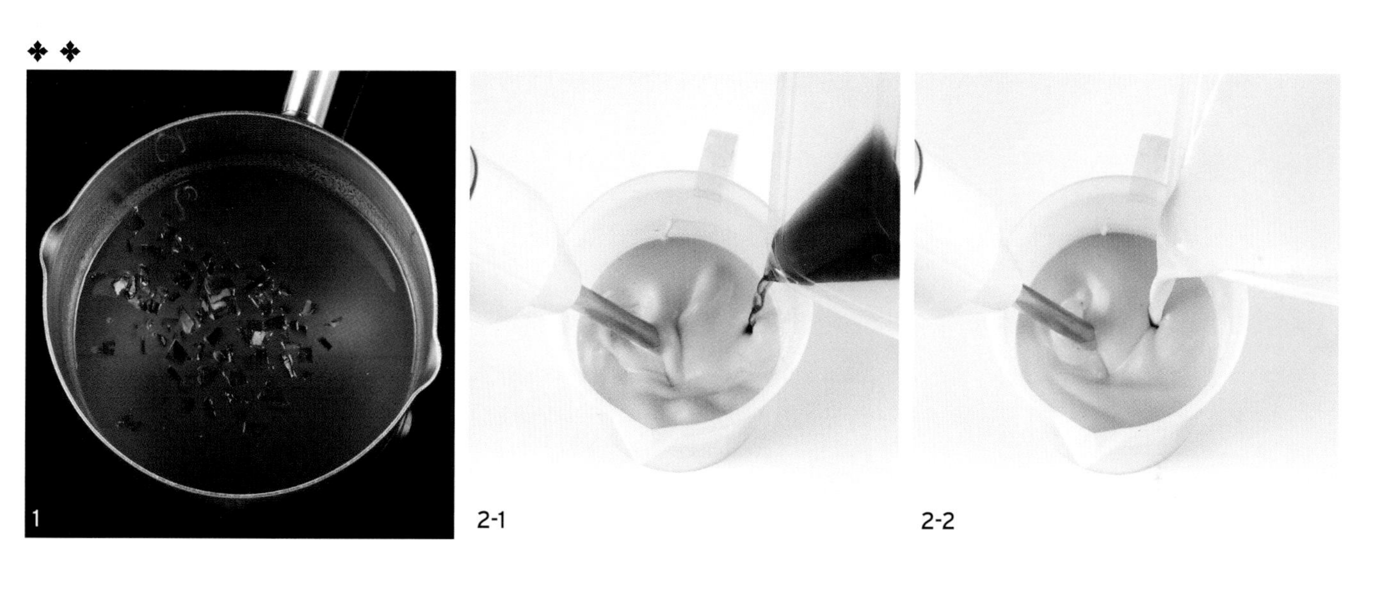
✤ ✤
1
2-1
2-2

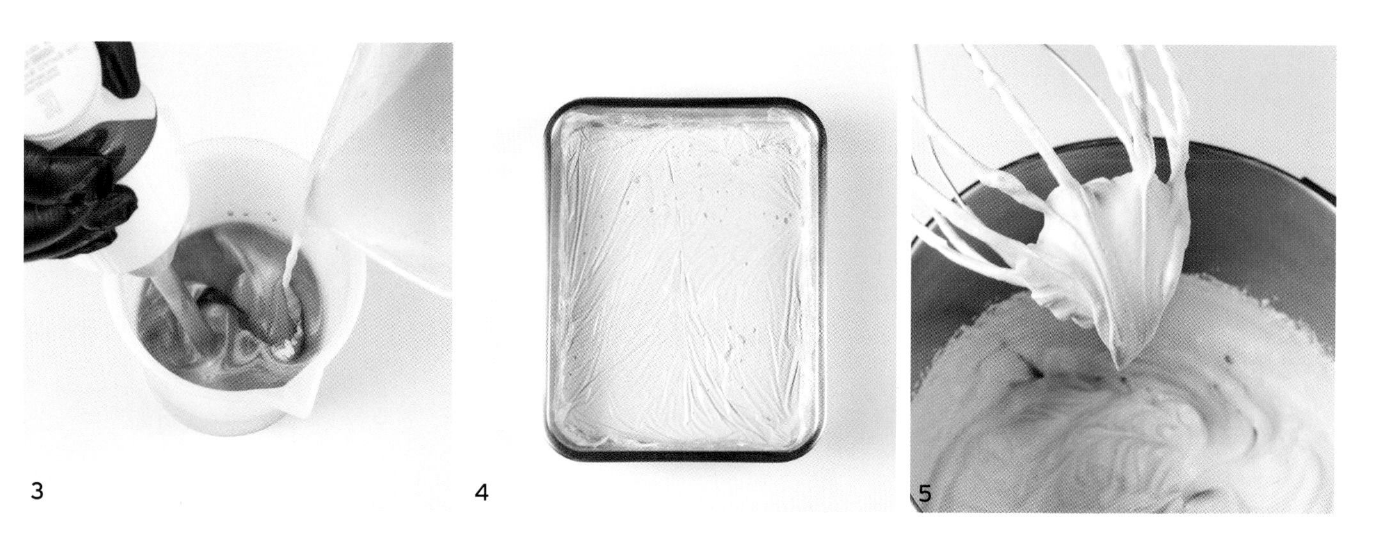
3
4
5

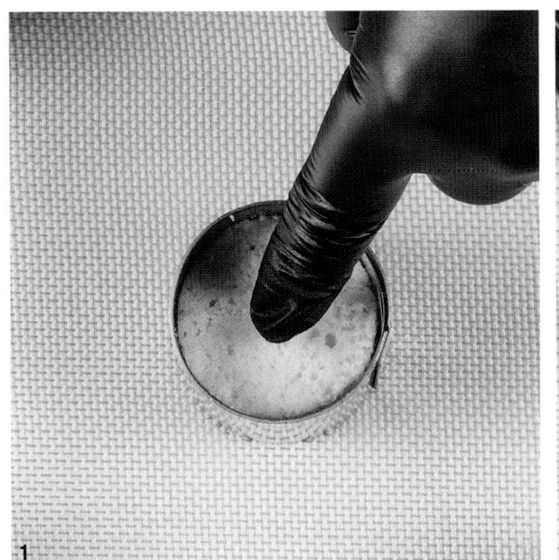
1

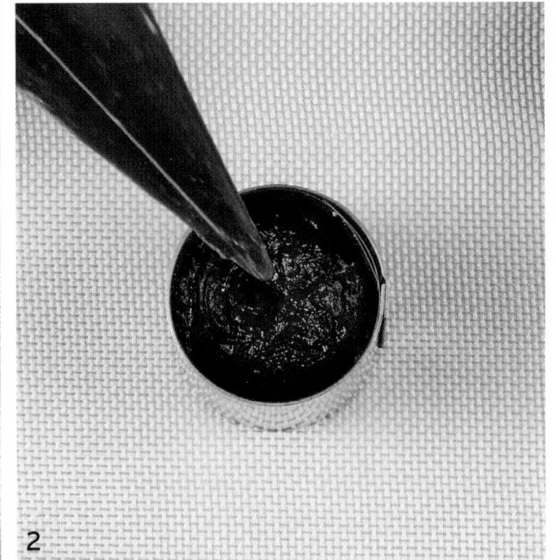
2

3

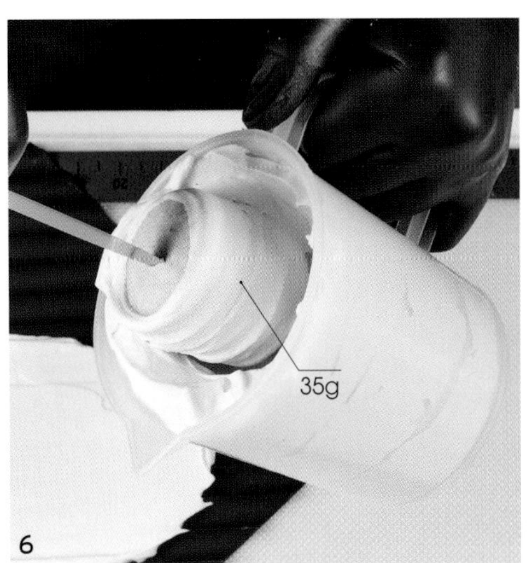

6

7

8-1

✤ ✤ ✤

FINISH 마무리

1 안쪽에 필름을 두른 지름 5cm 무스 링에 잘라둔 바질 크림치즈 스펀지를 넣는다.

2 리치 라즈베리 쿨리를 50g 채운다.

3 바질 크림치즈 스펀지를 하나 더 넣고 냉동한다.

4 무스 링과 필름을 제거한다.

5 무늬 패드에 이스파한 몽테를 도톰하게 펴 바른다.

6 얼린 [바질 크림치즈 스펀지 + 리치 라즈베리 쿨리]를 이스파한 몽테에 담궜다가 뱅글뱅글 돌려가면서 뺀다.

7 이스파한 몽테를 펴 바른 무늬 패드에 무스를 돌려가며 자연스러운 회오리 모양으로 만든다.

8 윗면에 이스파한 몽테 약 22~25g을 파이핑한 후 스패출러로 회오리 모양을 만들고 급속 냉동(-35 ~ -40℃)한 후 냉동(-18℃) 보관한다.

9 라즈베리 스프레이로 두껍게 분사하고 굳힌 후 파트 슈크레에 올리고 식용 꽃잎, 은박을 올려 마무리한다.

TIP. 옆면의 크림이 적은 제품이라 보형성이 떨어지므로 다른 제품보다 분사를 두껍게 하여 유지력과 바삭함을 연출한다.

5

9

1. Insert the cut basil cream cheese sponge in a 5 cm round mousse ring lined with acetate film.
2. Fill 50 grams of lychee raspberry coulis.
3. Insert another sponge and freeze.
4. Remove the ring and the film.
5. Spread a generous amount of Ispahan montée on the patterned pad.
6. Dip the frozen [basil cream cheese sponge + lychee raspberry coulis] into the Ispahan montée and roll it around as you pull it out.
7. Roll the mouse on the pad, spread with Ispahan montée, to make a natural swirling shape.
8. Pipe 22~25 grams of Ispahan montée on the top, stir it with a spatula to make a swirling shape. Blast freeze (-35 ~ -40°C) and keep it frozen (-18°C).
9. Spray the raspberry spray mixture thickly and let it set. Place it on the pâte sucrée and decorate it with edible petals and silver leaves to finish.

   TIP. Because this dessert has less cream on the sides, it has less retention properties. Therefore, it's sprayed thicker than other products to hold itself and have crunchiness.

24

J'ADORE NUTS

자도르 너츠

PAIRING & TEXTURE
페어링 & 텍스처

견과류 디저트의 끝판왕. 견과류가 들어간 시판 초콜릿바를 케이크의 형태와 고급스러운 맛으로 표현한 제품이다. 자칫 너무 달게 완성될 수 있으므로 단맛의 밸런스를 잘 맞추는 것이 관건이다. 여기에서는 설탕 없이 초콜릿의 단맛으로만 맛을 낸 몽테, 당도를 낮춘 라이트 잔두야와 헤이즐넛 캐러멜을 사용해 단맛의 밸런스를 잡았다.

The ultimate nut dessert. This cake represents a commercial chocolate bar with nuts in the form of a cake with an upscale flavor. The trick is to balance the sweetness, as it can easily become too sweet. Here, I used the montée, flavored with only the sweetness of chocolate without added sugar, light gianduja, and hazelnut caramel with reduced sugar to balance the sweetness.

TECHNIQUE
테크닉

- 피스톨레 작업(분사)과 룸 템퍼라처 글레이즈로 질감을 표현했다.
- 견과류라는 재료의 장점을 극대화시켰다.

- I created the texture with the room temperature glaze and spray work.
- I maximized the benefits of nuts as an ingredient.

DESIGN
디자인

- 러블리한 디자인을 생각했을 때 쉽게 볼 수 있는 단순한 하트 모양에서 벗어나고자 하트 문양의 패턴으로 새롭게 디자인해보았다.
- J'adore, 즉, '열렬히 사랑한다'라는 의미를 담아 다양한 재료들을 믹스 매치해보았다.

- I redesigned a heart-shaped pattern to break away from the ordinary heart-shaped dessert that easily comes to mind for a lovely design.
- I mixed and matched various ingredients to go with the name J'adore, meaning 'to love passionately.'

✦ HOW TO COMPOSE THIS RECIPE ✦

STEP 1.	메뉴 정하기	앙트르메
	Decide on the menu	Entremets

STEP 2.	메인 맛 정하기	헤이즐넛
	Choose the primary flavor	Hazelnut

STEP 3. **메인 맛(헤이즐넛)과의 페어링 선택하기**

Select a pairing flavor (Hazelnut)

- ☑ 캐러멜 Caramel
- ☑ 아몬드 Almond
- ☑ 마스카르포네 Mascarpone
- ☑ 캐러멜초콜릿 Caramel Chocolate

STEP 4. **구성하기**

Assemble

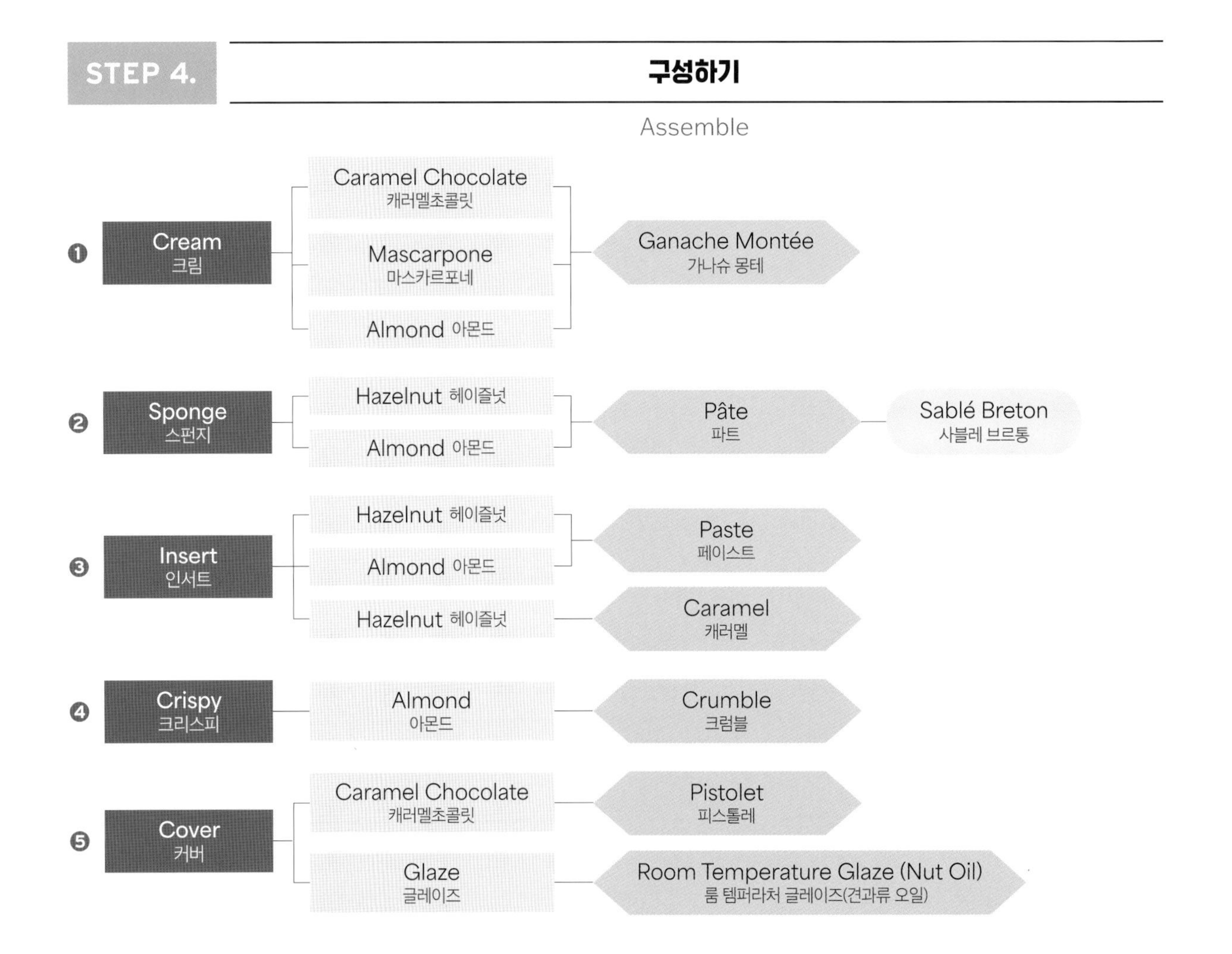

라이트 잔두야 몽테
Light Gianduja Montée
제피르 캐러멜 피스톨레
Zephyr Caramel Pistolet
헤이즐넛 캐러멜
Hazelnut Caramel
드 크럼블
ond Crumble
라이트 잔두야
Light Gianduja
헤이즐넛 아몬드 브르통
Hazelnut Almond Breton
헤이즐넛 아몬드 룸 템퍼라처 글레이즈
Hazelnut Almond Room Temperature Glaze

INGREDIENTS 4개 분량/ 4 cakes

헤이즐넛 아몬드 브르통

Hazelnut Almond Breton

INGREDIENTS		g
캐러멜라이즈 헤이즐넛 (p.494)	Caramelized hazelnut	100g
캐러멜라이즈 아몬드	Caramelized almond	100g
버터 (Bridel)	Butter	250g
황설탕	Brown sugar	210g
소금	Salt	6g
박력분	Cake flour	400g
베이킹파우더	Baking powder	20g
노른자	Egg yolks	130g
바닐라에센스 (Aroma Piu)	Vanilla essence	10g
TOTAL		**1226g**

* 만드는 법

❶ 믹서에 캐러멜라이즈 헤이즐넛, 캐러멜라이즈 아몬드를 넣고 곱게 갈아 준비한다.
❷ 다른 믹서에 차가운 상태의 버터, 황설탕, 소금을 넣고 섞는다.
❸ ❷에 체 친 박력분과 베이킹파우더 1/3을 넣고 섞는다.
❹ 차가운 상태의 노른자, 바닐라에센스를 넣고 섞는다.
❺ 남은 ❸과 ❶을 넣고 섞는다.
❻ 반죽이 뭉쳐지면 랩으로 감싸 냉장고에 보관한다.
❼ 1cm 높이로 밀어 편 후 지름 12cm 원형 무스 링으로 재단한다.
❽ 170℃로 예열된 오븐에서 뎀퍼를 100% 열고 20분간 굽고, 무스 링을 제거한 후 5분 더 굽는다.

* Procedure

❶ Finely grind caramelized hazelnuts and caramelized almonds in a food processor.
❷ Combine cold butter, brown sugar, and salt in a separate bowl.
❸ Add sifted cake flour and 1/3 of baking powder into ❷ and mix.
❹ Mix with cold egg yolks and vanilla essence.
❺ Mix with the remaining ❸ and ❶.
❻ When the dough comes together, wrap it with plastic wrap and refrigerate.
❼ Roll out into 1 cm thickness and cut with a 12 cm round ring.
❽ Bake with the rings in an oven preheated to 170°C and damper open at 100% for 20 minutes; remove the rings and bake for another 5 minutes.

아몬드 크럼블

Almond Crumble

INGREDIENTS		g
버터 (Bridel)	Butter	70g
설탕	Sugar	70g
소금	Salt	2g
아몬드파우더	Almond podwer	70g
박력분	Cake flour	100g
물	Water	10g
캐러멜초콜릿 (Zephyr Caramel 35%)	Caramel chocolate	48g
오렌지제스트	Orange zest	1/2개 분량/ 1/2 orange
TOTAL		**370g**

라이트 잔두야 *

Light Gianduja

INGREDIENTS		g
캐러멜라이즈 아몬드	Caramelized almond	500g
캐러멜라이즈 헤이즐넛	Caramelized hazelnut	500g
밀크초콜릿 (Alunga 41%)	Milk chocolate	50g
캐러멜초콜릿 (Zephyr Caramel 35%)	Caramel chocolate	50g
TOTAL		1100g

라이트 잔두야 몽테

Light Gianduja Montée

INGREDIENTS		g
캐러멜 마스카르포네 몽테 *	Caramel Mascarpone Montée *	900g
라이트 잔두야 *	Light Gianduja *	90g
TOTAL		990g

- 믹싱볼에 모든 재료를 넣고 부드러운 상태로 휘핑해 사용한다.
- Whip all the ingredients to a soft state to use.

캐러멜 마스카르포네 몽테 *

Caramel Mascarpone Montée

INGREDIENTS		g
생크림	Heavy cream	250g
노른자	Egg yolks	160g
젤라틴매스 (×5)	Gelatin mass (×5)	50g
캐러멜초콜릿 (Zephyr Caramel 35%)	Caramel chocolate	120g
연유	Condensed milk	50g
소금	Salt	1.5g
마스카르포네	Mascarpone cheese	225g
골드럼 (PAN RUM)	Gold rum	10g
아몬드 리큐르 (Dijon Almond)	Almond liqueur	10g
휘핑크림	Whipping cream	400g
TOTAL		1276.5g

헤이즐넛 캐러멜

Hazelnut Caramel

INGREDIENTS		g
생크림	Heavy cream	420g
설탕	Sugar	160g
물엿	Corn syrup	260g
헤이즐넛 페이스트	Hazelnut paste	48g
소금	Salt	4g
TOTAL		**892g**

캐러멜 스프레이

Caramel Spray

INGREDIENTS		g
캐러멜초콜릿 (Zephyr Caramel 35%)	Caramel chocolate	300g
카카오버터	Cacao butter	300g
TOTAL		**600g**

- 볼에 모든 재료를 넣고 녹여 바믹서로 블렌딩한 후 50℃로 맞춰 사용한다.
- Melt all the ingredients in a bowl, combine using an immersion blender, and use at 50°C.

헤이즐넛 아몬드 룸 템퍼라처 글레이즈

Hazelnut Almond Room Temperature Glaze

INGREDIENTS		g
캐러멜초콜릿 (Zephyr Caramel 35%)	Caramel chocolate	1000g
밀크초콜릿 (Alunga 41%)	Milk chocolate	1000g
헤이즐넛 오일	Hazelnut oil	400g
아몬드 분태	Chopped almond	적당량/ QS
TOTAL		**2400g**

*** 만드는 법**

❶ 비커에 녹인 캐러멜초콜릿과 밀크초콜릿(45~50℃), 헤이즐넛 오일을 넣고 바믹서로 블렌딩한다.

❷ 아몬드 분태를 넣고 섞어 30℃로 맞춰 사용한다.

*** Procedure**

❶ Blend melted caramel chocolate and milk chocolate (45~50°C) with hazelnut oil in a beaker using an immersion blender.

❷ Mix with chopped almonds and use at 30°C.

NOTE.

✤

ALMOND CRUMBLE

아몬드 크럼블

1 로보쿱에 차가운 상태의 버터, 설탕, 소금, 체 친 아몬드파우더와 박력분 절반을 넣고 갈아준다.

2 보슬보슬한 상태가 되면 물을 넣고 섞은 후 남은 **1**을 넣고 갈아준다.

3 한 덩어리로 뭉쳐지면 테프론시트를 깐 철판 위에서 체(간격 약 0.5cm)에 반죽을 내려 크럼블 상태로 만든다.

4 160℃로 예열된 오븐에서 뎀퍼를 100% 열고 17~18분간 굽는다.

5 볼에 캐러멜초콜릿, 오렌지제스트를 준비한다.

6 오븐에서 나온 뜨거운 상태의 크럼블을 넣고 섞어 코팅한 후 식혀 밀폐용기에 담아 보관한다.

1 Grind cold butter, sugar, salt, sifted almond powder, and half of cake flour in the Robot Coupe.

2 When it becomes crumbly, mix with water, add the remaining (**1**) and grind.

3 When it comes together, pass it through a coarse sieve (about 0.5 cm mesh) over a baking tray lined with a Teflon sheet to make crumbles.

4 Bake in an oven preheated to 160°C and damper open at 100% for 17~18 minutes.

5 Prepare caramel chocolate and orange zest.

6 Add the hot crumbles right out of the oven and toss to coat. Cool it and freeze in an airtight container.

✤ ✤

LIGHT GIANDUJA

라이트 잔두야

1 로보쿱에 캐러멜라이즈 아몬드를 넣고 곱게 갈아준다.

TIP. 아몬드가 갈리면서 살짝 뭉쳐지기 시작했을 때(아직 갈리고 있는 덩어리가 남아 있을 때) 헤이즐넛을 넣는다. 지방이 많은 헤이즐넛은 갈아지면서 바로 기름기를 내뿜기 때문에 처음부터 아몬드와 헤이즐넛을 함께 갈게 되면 아몬드의 형태만 남게 된다.

2 캐러멜라이즈 헤이즐넛을 넣고 고운 입자 상태로 갈아준다.

3 녹인 밀크초콜릿과 캐러멜초콜릿(45~50℃)을 넣고 고르게 섞일 때까지 갈아주고 쿨링한 후 냉장 보관한다.

1 Finely grind caramelized almonds in the Robot Coupe.

TIP. When the almonds start to clump up slightly as they are being ground (still lumps remaining from being ground), add the hazelnuts. Hazelnuts, high in fat, give off oil immediately when ground, so if you grind almonds and hazelnuts together from the start, only the almonds will remain as is.

2 Add caramelized hazelnuts and grind until it's very fine.

3 Add melted milk chocolate and caramel chocolate (45~50°C), grind until combined, and cool to refrigerate.

✤

1

2

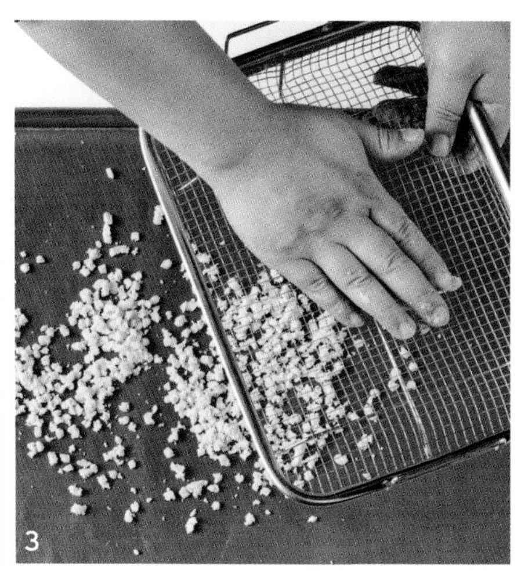
3

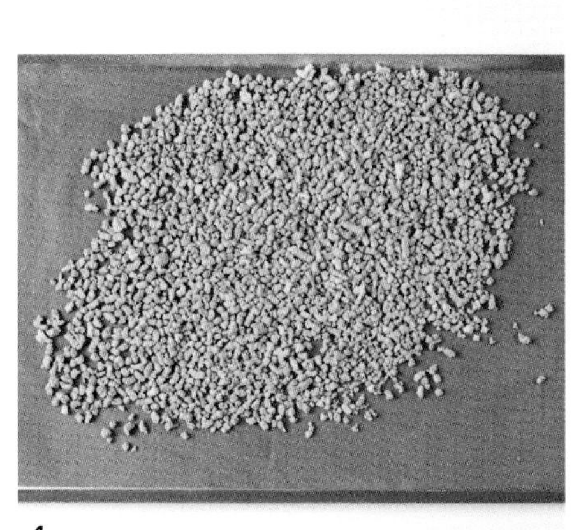
4

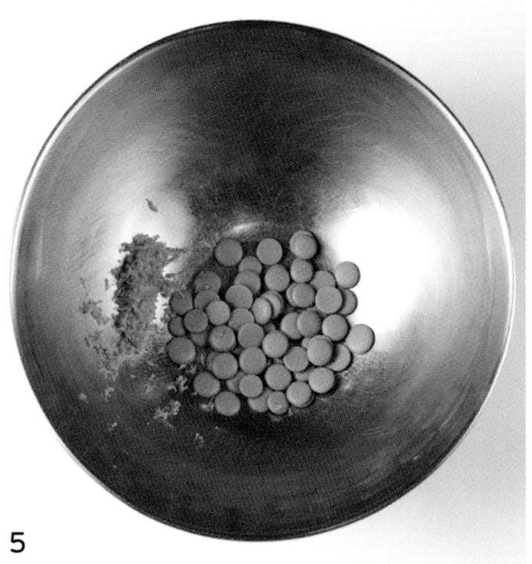
5

6

✤ ✤

1

2

3

✤ ✤ ✤

CARAMEL MASCARPONE MONTÉE

캐러멜 마스카르포네 몽테

1 냄비에 생크림을 넣고 따뜻한 정도로 가열한다. (약 60℃)

2 거품낸 노른자가 담긴 볼에 **1**을 넣고 섞는다.

3 다시 냄비로 옮겨 휘퍼로 저어가며 75℃까지 가열한다.

TIP. 당이 들어가지 않아 75℃에서 노른자가 익는다. 구수한 맛을 위해 노른자를 일부러 익히는 레시피이다.

4 불에서 내린 후 젤라틴매스를 넣고 녹을 때까지 섞는다.

5 캐러멜초콜릿이 담긴 비커에 **4**와 연유, 소금을 넣고 바믹서로 블렌딩한다.

6 마스카르포네, 골드럼, 아몬드리큐르를 넣고 블렌딩한다.

TIP. 휘핑크림 특유의 깔끔하지 않은 끝맛을 없애기 위해 럼과 리큐르를 첨가한다.

7 차가운 상태의 휘핑크림을 넣고 블렌딩한다.

8 바트에 담아 밀착 랩핑하고 12시간 숙성시킨다.

TIP. 사용하기 직전 비터로 부드럽게 풀어 사용한다.

1 Heat heavy cream just until warm (about 60°C).

2 Mix with egg yolks that have been whisked until foamy.

3 Pour back into the saucepan and cook to 75°C.

TIP. The egg yolks cook at 75°C because there is no sugar. This recipe intentionally cooks the egg yolks for an aromatic toastiness.

4 Remove from the heat, add gelatin mass, and stir until the gelatin melts.

5 Blend with caramel chocolate, condensed milk, and salt in a beaker using an immersion blender.

6 Blend with mascarpone, gold rum, and almond liqueur.

TIP. Add rum and liqueur to eliminate the certain aftertaste specific to whipping creams.

7 Add cold whipping cream and continue to blend.

8 Pour onto a tray, adhere the plastic wrap to the preparation to cover, and refrigerate for 12 hours.

TIP. Gently beat with a paddle attachment right before use.

1

2

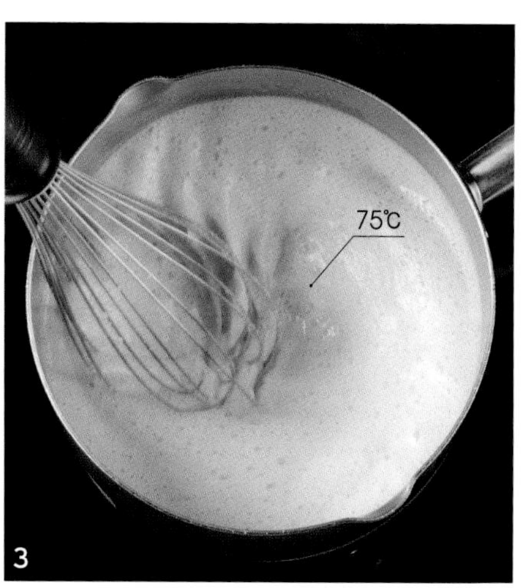

3

4

5

6

7

8

1

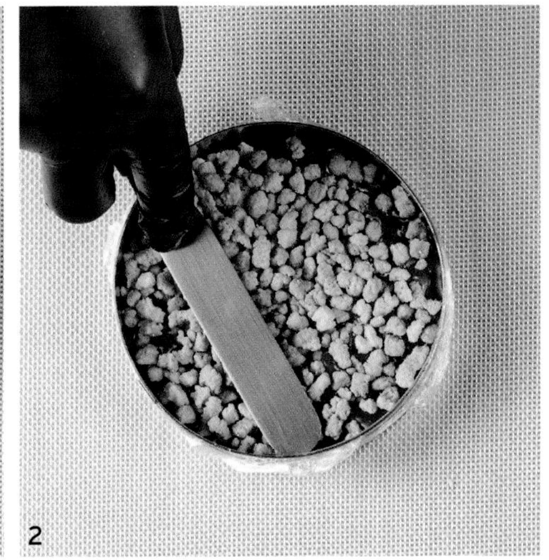
2

3

7

8

9

✤ ✤ ✤ ✤

FINISH 마무리

1 지름 10cm, 높이 2.5cm 원형 무스 링의 아랫면을 타이트하게 랩핑한 후, 헤이즐넛 캐러멜 100g을 채우고 평평하게 펼친다.

2 아몬드 크럼블 25g을 채우고 고르게 펼친다.

3 라이트 잔두야 70g을 채우고 평평하게 펼친다.

4 아몬드 크럼블 20g을 채우고 고르게 펼친다.

5 헤이즐넛 캐러멜 20g을 가운데부터 동그랗게 파이핑하며 채운다.

6 헤이즐넛 아몬드 브르통을 넣고 급속 냉동(-35 ~ -40℃)한 후 히팅건으로 달궈 원형 무스 링을 제거한다.

7 몰드에 라이트 잔두야 몽테를 200g 채운다.

8 얼린 **6**을 넣는다.

9 가장자리를 깔끔하게 정리하고 급속 냉동(-35 ~ -40℃)한 후 몰드에서 빼내 냉동(-18℃) 보관한다.

10 캐러멜 스프레이를 분사한 후 냉동실에서 굳힌다.

11 삼발이를 이용해 높이 2/3 지점까지 헤이즐넛 아몬드 룸 템퍼라처 글레이즈로 코팅한다. 밑면을 깔끔하게 정리한 후 마무리한다.

5

6

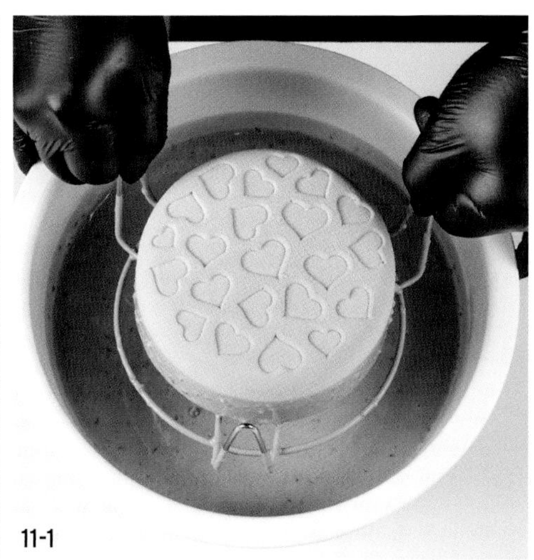

11-1

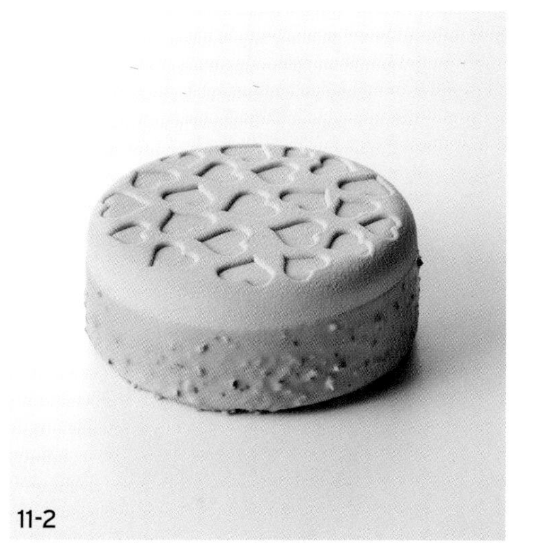

11-2

1 Tightly wrap the bottom of a Ø 10 cm × H 2.5 cm mousse ring. Fill with 100 grams of hazelnut caramel and spread evenly.

2 Fill and evenly spread 25 grams of almond crumble.

3 Fill with 70 grams of light gianduja and spread evenly.

4 Fill and evenly spread 20 grams of almond crumble.

5 Fill 20 grams of hazelnut caramel from the center in a circle.

6 Top with hazelnut almond breton. Blast freeze (-35 ~ -40°C) and heat the ring mold to remove.

7 Fill with 200 grams of light gianduja in the mold.

8 Insert the frozen (**6**).

9 Clean the edges, blast freeze (-35 ~ -40°C), remove from the mold and keep in the freezer (-18°C).

10 Spray with caramel spray mixture and freeze.

11 Use a round wire rack to dip to coat the cake in hazelnut almond room temperature glaze up to 2/3 of the height. Tidy up the bottom to finish.

25 PEACH BASKET

피치 바스켓

PAIRING & TEXTURE
페어링 & 텍스처

복숭아를 메인으로 라즈베리와 사과를 페어링한 메뉴. 수비드 기법을 사용해 사과에 복숭아 향을 입혀 두 가지 과일의 향을 극대화시켰다.

A peach-based menu, paired with raspberry and apple. Using the sous vide technique, we immersed the apples with peach flavor to maximize the flavors of both fruits.

TECHNIQUE
테크닉

- 일반적인 무스 케이크에서 얼린 인서트를 사용하는 반면, 얼리지 않은 인서트를 넣어 과일의 신선함을 그대로 느낄 수 있게 했다.
- 피스톨레 작업(분사)과 룸 템퍼라처 작업으로 두 가지 텍스처를 표현했다.
- 다양한 시판 퓌레들을 페어링해가며 찾아낸 독특한 맛을 사용했다.
- 베이스 몽테 크림에 퓌레를 섞어 다양한 맛으로 연출할 수 있다.

- Unlike typical mousse cakes that use frozen inserts, I composed them with unfrozen inserts for a refreshing, fruity taste.
- Two textures were created with the room temperature glaze and spray work.
- I used a unique flavor discovered by pairing various commercially available purées.
- You can create a variety of flavors by mixing purée with the base montée cream.

DESIGN
디자인

크림의 질감을 최대한 살리면서 바구니 모양이 연상되도록 했고, 그 안에 동그랗게 만든 사과와 젤리를 가득 담았다. (어릴적 알록달록 추억의 사랑방 캔디를 연상하며 만들었다.)

I made it resemble a basket while enhancing the texture of the cream at its best and filled it with apple spheres and jellies. (It was reminiscent of the colorful candy drops from my childhood.)

✦ HOW TO COMPOSE THIS RECIPE ✦

STEP 1.	메뉴 정하기	프티 가토
	Decide on the menu	Petit Gateau

STEP 2.	메인 맛 정하기	복숭아
	Choose the primary flavor	Peach

STEP 3. **메인 맛(복숭아)과의 페어링 선택하기**

Select a pairing flavor (Peach)

- ☑ 라즈베리 Raspberry
- ☑ 사과 Apple
- ☑ 라임 Lime
- ☑ 바질 Basil
- ☑ 마스카르포네 Mascarpone
- ☑ 사워크림 Sour Cream
- ☑ 아몬드 Almond
- ☑ 화이트초콜릿 White Chocolate
- ☑ 루비초콜릿 Ruby Chocolate

STEP 4. **구성하기**

Assemble

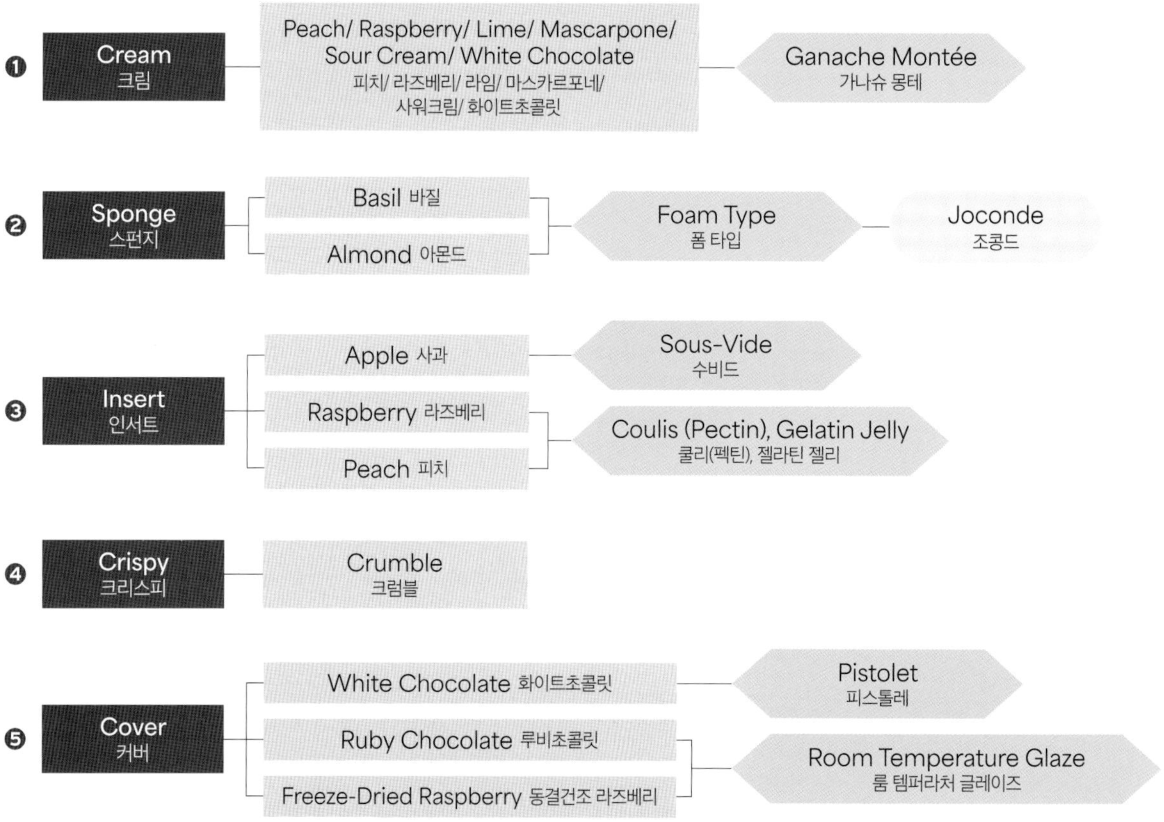

❶ Cream 크림
- Peach/ Raspberry/ Lime/ Mascarpone/ Sour Cream/ White Chocolate
  피치/ 라즈베리/ 라임/ 마스카르포네/ 사워크림/ 화이트초콜릿
  → Ganache Montée 가나슈 몽테

❷ Sponge 스펀지
- Basil 바질 / Almond 아몬드 → Foam Type 폼 타입 → Joconde 조콩드

❸ Insert 인서트
- Apple 사과 → Sous-Vide 수비드
- Raspberry 라즈베리 / Peach 피치 → Coulis (Pectin), Gelatin Jelly 쿨리(펙틴), 젤라틴 젤리

❹ Crispy 크리스피
- Crumble 크럼블

❺ Cover 커버
- White Chocolate 화이트초콜릿 → Pistolet 피스톨레
- Ruby Chocolate 루비초콜릿 / Freeze-Dried Raspberry 동결건조 라즈베리 → Room Temperature Glaze 룸 템퍼라처 글레이즈

화이트 바닐라 스프레이
White Vanilla Spray
라즈베리 룸
템퍼라처 글레이즈
Raspberry Room
emperature Glaze
수비드 사과
Sous-Vide Apple
피치 젤라틴 젤리
Peach Gelatin Jelly
피치 라즈베리
사워크림 몽테
Peach Raspberry
Sour Cream Montée
바질 스펀지
Basil Sponge
피치 쿨리
Peach Coulis
플레인 크럼블
Plain Crumble

INGREDIENTS 18개 분량/ 18 cakes

바질 스펀지

Basil Sponge

(철판(42 × 32cm) 1개 분량)
(1 baking tray (42 × 32 cm))

- 재단한 스펀지 30개 분량
- Makes 30 pieces of cut sponge

INGREDIENTS		g
흰자	Egg whites	85g
설탕	Sugar	85g
달걀	Eggs	40g
노른자	Egg yolks	80g
버터 (Bridel)	Butter	55g
바질	Basil	2g
레몬제스트	Lemon zest	1/3개 분량/ 1/3 lemon
아몬드파우더	Almond powder	62g
박력분	Cake flour	8g
플레인 크럼블 (p.388)	Plain Crumble (p.388)	130g
TOTAL		**547g**

피치 라즈베리 퓌레 *

Peach Raspberry Purée

INGREDIENTS		g
화이트 피치 퓌레	White peach purée	500g
라즈베리 퓌레	Raspberry purée	30g
레몬 퓌레	Lemon purée	22g
피치 아로마 (Sosa)	Peach aroma	5방울/ 5 drops
TOTAL		**552g**

- 비커에 모든 재료를 넣고 바믹서로 블렌딩해 사용한다.
- Blend all the ingredients in a beaker using an immersion blender to use.

피치 쿨리

Peach Coulis

INGREDIENTS		g
피치 라즈베리 퓌레 *	Peach raspberry purée *	310g
설탕	Sugar	120g
NH펙틴	NH pectin	5g
라임즙	Lime juice	10g
피치 리큐르 (Dijon Peach)	Peach liqueur	26g
TOTAL		**471g**

수비드 사과

Sous-Vide Apple

INGREDIENTS		g
피치 라즈베리 퓌레 *	Peach raspberry purée *	100g
물	Water	50g
설탕	Sugar	50g
통후추	Whole black pepper	1g
생강파우더	Ginger powder	0.05g
사과	Apple	5개/ 5 ea
TOTAL		201.05g (사과를 제외한 무게) (without apples)

피치 라즈베리 사워크림 몽테

Peach Raspberry Sour Cream Montée

INGREDIENTS		g
우유	Milk	55g
전화당	Inverted sugar	20g
젤라틴매스 (×5)	Gelatin mass (×5)	50g
화이트초콜릿 (Zephyr white 34%)	White chocolate	150g
마스카르포네	Mascarpone cheese	30g
사워크림	Sour cream	150g
휘핑크림	Whipping cream	450g
레몬 리큐르 (Dijon Lemon)	Dijon lemon liqueur	15g
라임 퓌레	Lime purée	15g
피치 라즈베리 퓌레 *	Peach raspberry purée *	90g
TOTAL		1025g

피치 젤라틴 젤리

Peach Gelatin Jelly

INGREDIENTS		g
피치 라즈베리 퓌레 *	Peach raspberry purée *	60g
물엿	Corn syrup	130g
물	Water	300g
젤라틴매스 (×5)	Gelatin mass (×5)	120g
빨간색 색소 (수용성)	Red color (water soluble)	적당량/ QS
TOTAL		**610g**

* 만드는 법

1. 냄비에 모든 재료를 넣고 가열한다.
2. 젤라틴매스가 녹으면 불을 끄고 색소를 넣고 섞은 후 얼음물이 담긴 볼에 받쳐 15℃로 쿨링한다.
3. 지름 1.5cm 구형 몰드에 채운 후 냉장고에 보관한다.

---

* Procedure

1. Heat all the ingredients in a saucepan.
2. Remove from the heat when the gelatin melts and mix with the food coloring. Cool it in an ice bath to 15°C.
3. Fill the 1.5 cm sphere molds and refrigerate.

화이트 바닐라 스프레이

White Vanilla Spray

INGREDIENTS		g
화이트초콜릿 (Zephyr white 34%)	White chocolate	350g
카카오버터	Cacao butter	350g
하얀색 색소 (이산화티타늄)	Titanium dioxide	3g
바닐라빈	Vanilla bean seeds	1개 분량/ 1 pc
TOTAL		**703g**

- 볼에 모든 재료를 넣고 녹여 바믹서로 블렌딩한 후 50℃로 맞춰 사용한다.
- 국제적인 추세를 반영하여 이산화티타늄은 최소량으로 사용했다.
- Melt all the ingredients in a bowl, combine using an immersion blender, and use at 50°C.
- Titanium dioxide was used in a minimal amount to reflect international trends.

라즈베리 룸 템퍼라처 글레이즈

Raspberry Room Temperature Glaze

INGREDIENTS		g
루비초콜릿 (Callebaut RB2)	Ruby chocolate	100g
화이트초콜릿 (Zephyr white 34%)	White chocolate	250g
식용유	Vegetable oil	100g
동결건조 라즈베리파우더	Freeze-dried raspberry powder	10g
코팅 라즈베리 크리스피 (Sosa)	Wet-proof raspberry crispy	적당량/ QS
TOTAL		**460g**

＊만드는 법

❶ 비커에 녹인 루비초콜릿과 화이트초콜릿(45~50℃), 식용유를 넣고 바믹서로 블렌딩한다.

❷ 동결건조 라즈베리파우더, 코팅 라즈베리 크리스피를 넣고 섞어 30℃로 맞춰 사용한다.

＊Procedure

❶ Blend melted Ruby chocolate, white chocolate (45~50°C), and vegetable oil in a beaker using an immersion blender.

❷ Mix with freeze-dried raspberry powder and wet-proof raspberry crispies, and use at 30°C.

데코젤 시럽

Decorgel Syrup

INGREDIENTS		g
데코젤 뉴트럴	Decorgel neutral	100g
30° 시럽	30°B syrup	30g
TOTAL		**130g**

● 모든 재료를 섞어 사용한다.

● Mix all the ingredients to use.

✤

BASIL SPONGE

바질 스펀지

1 믹싱볼에 흰자(30℃)를 넣고 설탕을 나눠 넣어가며 기공이 조밀하고 단단한 상태로 휘핑한다.

2 볼에 달걀과 노른자(30℃)를 넣고 가볍게 섞은 후 녹인 버터(45℃)를 넣고 섞는다.

3 다진 바질을 넣고 가볍게 섞는다.

TIP. 바질은 사용하기 직전 다져 향을 최대한 살린다.

4 레몬제스트를 넣고 가볍게 섞는다.

TIP. 거품이 살짝 올라오는 정도에서 마무리한다.

5 **4**에 **1**의 절반을 넣고 섞는다.

6 체 친 아몬드파우더, 박력분을 넣고 섞는다.

7 남은 **1**과 함께 섞는다.

8 테프론시트를 깐 철판(42 × 32cm) 에 고르게 팬닝한다.

9 플레인 크럼블 130g을 골고루 뿌리고 160℃로 예열된 오븐에서 23분간 굽고 식힌 후, 지름 5.5cm 원형 틀로 잘라 냉동한다.

1 Whip egg whites (30°C) in a mixing bowl while adding sugar in batches until dense and stiff peaks form.

2 In a separate bowl, whisk eggs, egg yolks (30°C), and mix with melted butter (45°C).

3 Gently mix with chopped basil.

TIP. Chop the basil right before use to maximize the flavor.

4 Gently mix with lemon zest.

TIP. Stop when it becomes frothy.

5 Mix half of (**1**) into (**4**).

6 Mix with sifted almond powder and cake flour.

7 Mix with the remaining (**1**).

8 Spread evenly on a baking tray (42 × 32 cm) lined with a Teflon sheet.

9 Sprinkle 130 grams of plain crumble and bake in an oven preheated to 160°C for 23 minutes. Cool and cut with a 5.5 cm round cutter and freeze.

1

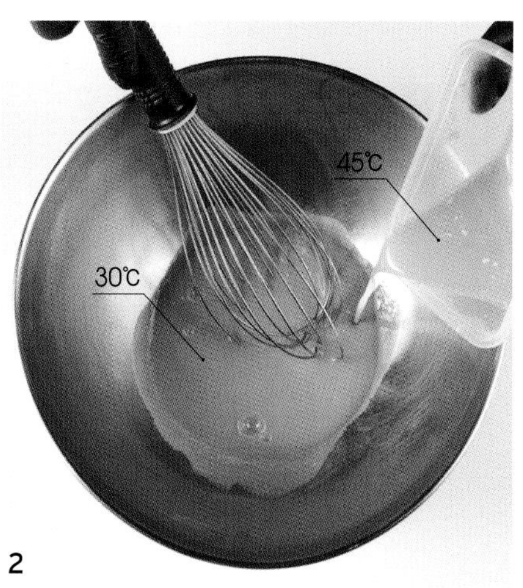

2

3

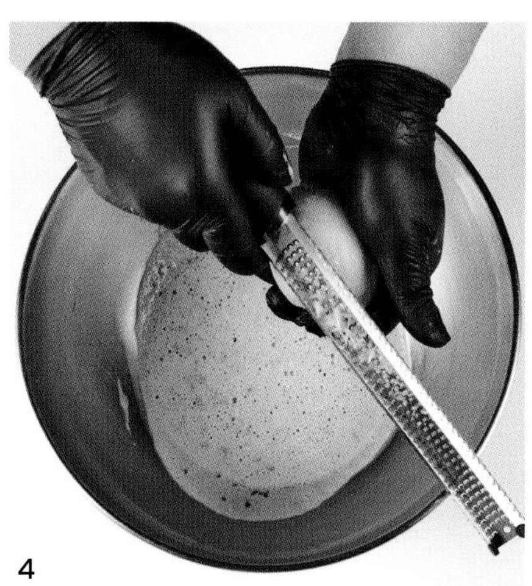
4

5

6

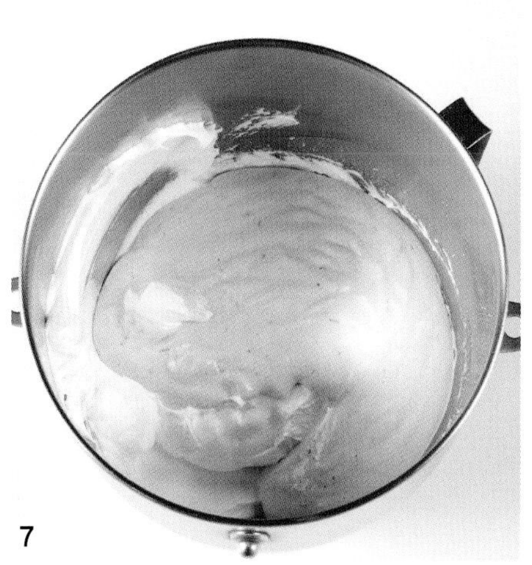
7

8

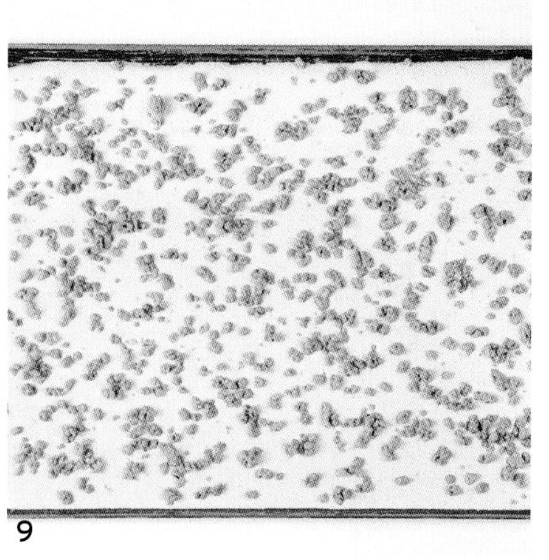
9

✤ ✤

PEACH COULIS

피치 쿨리

1 냄비에 피치 라즈베리 퓌레, 미리 섞어둔 설탕과 NH펙틴을 넣고 주걱으로 저어가며 가열한다.

2 펙틴 반응을 확인한 후 라임즙을 넣고 가열하다가 한번 더 반응이 일어나면 불에서 내린다.

3 피치 리큐르를 넣고 얼음물이 담긴 볼에 받쳐 쿨링한다.

1 Heat peach raspberry purée and previously mixed sugar and pectin NH while stirring.

2 After the pectin activates, add lime juice. Remove from the heat when the pectin activates again.

3 Mix with peach liqueur and lime juice. Cool in an ice bath.

✤ ✤ ✤

SOUS-VIDE APPLE

수비드 사과

1 냄비에 사과를 제외한 모든 재료를 넣고 가열한 후 끓어오르면 쿨링한다.

2 화채 스쿱을 이용해 사과를 동그랗게 파낸다. (사과 약 5개 분량)

3 **1**과 함께 섞는다.

4 사과를 건져내 진공백에 담아 진공한 후 수비드한다.

TIP. 85℃에서 50분간 수비드한 후 곧바로 얼음물에 담궈 빠르게 식혀 냉장 보관한다.

1 Heat all the ingredients except apples. Once it boils, remove from the heat and cool.

2 Scoop out apples using a melon baller. (about 5 apples)

3 Mix with (**1**).

4 Put the apples in a vacuum bag, vacuum, and sous-vide the bag.

TIP. Sous-vide at 85°C for 50 minutes, then immediately immerse in ice water to cool rapidly.

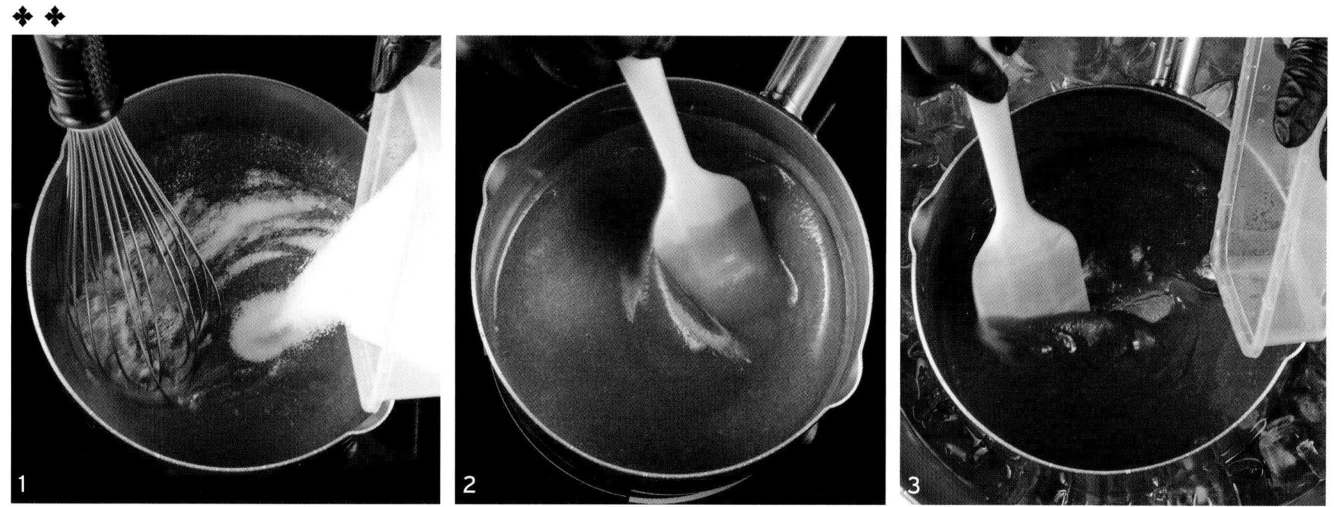
1
2
3

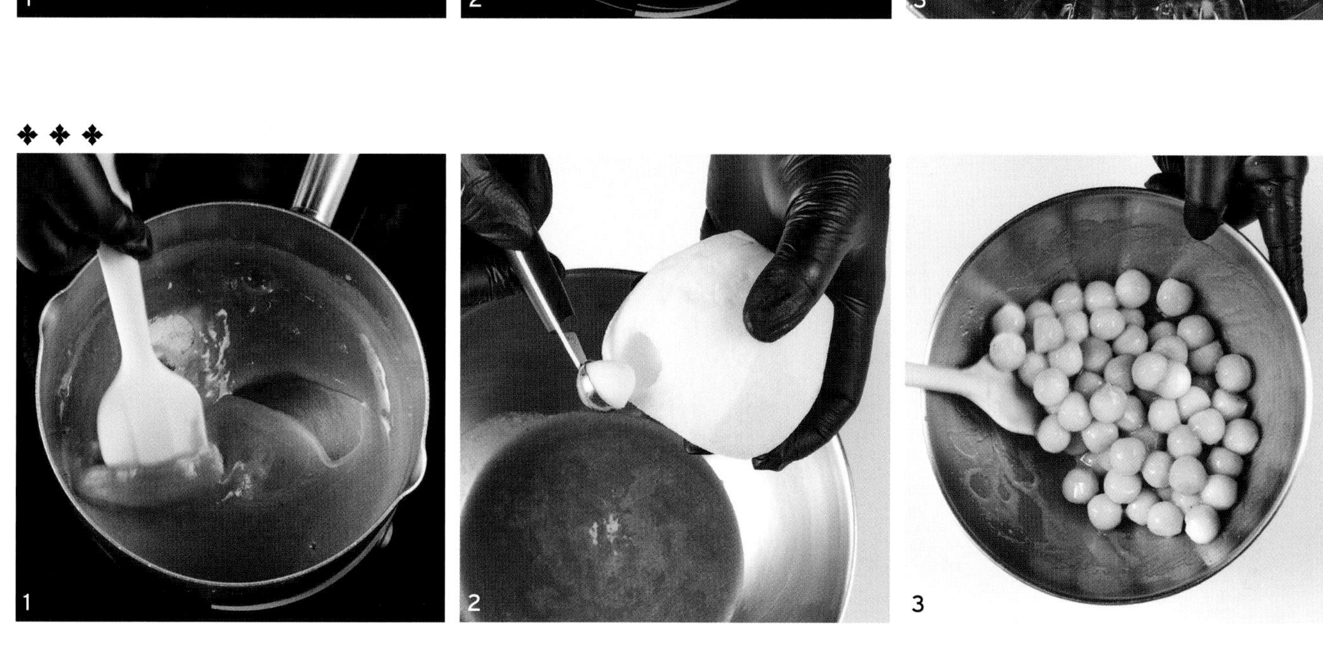
1
2
3

4-1
4-2

✤ ✤ ✤ ✤

PEACH RASPBERRY SOUR CREAM MONTÉE

피치 라즈베리 사워크림 몽테

1 냄비에 우유, 전화당, 젤라틴매스를 넣고 가열한다. (약 65℃)

2 화이트초콜릿이 담긴 비커에 붓고 바믹서로 블렌딩한다.

3 마스카르포네, 사워크림을 넣고 블렌딩한다.

4 휘핑크림을 넣고 블렌딩한다.

5 레몬 리큐르, 라임 퓌레를 넣고 블렌딩한다.

6 바트에 담고 밀착 랩핑해 냉장고에서 12시간 숙성시킨다.

7 숙성시킨 **6**의 900g과 피치 라즈베리 퓌레 90g을 휘핑해 사용한다

1 Heat milk, inverted sugar, and gelatin mass to about 65°C.

2 Blend with white chocolate in a beaker.

3 Blend with mascarpone and sour cream.

4 Add whipping cream and blend.

5 Add lemon liqueur and lime purée and continue to blend.

6 Pour onto a stainless-steel tray, adhere the plastic wrap to the preparation to cover, and refrigerate for 12 hours.

7 Whip 900 grams of the refrigerated (**6**) with 90 grams of peach raspberry purée to use.

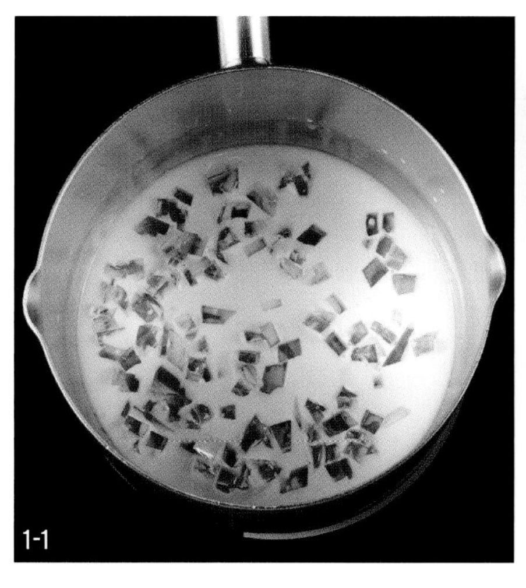
1-1

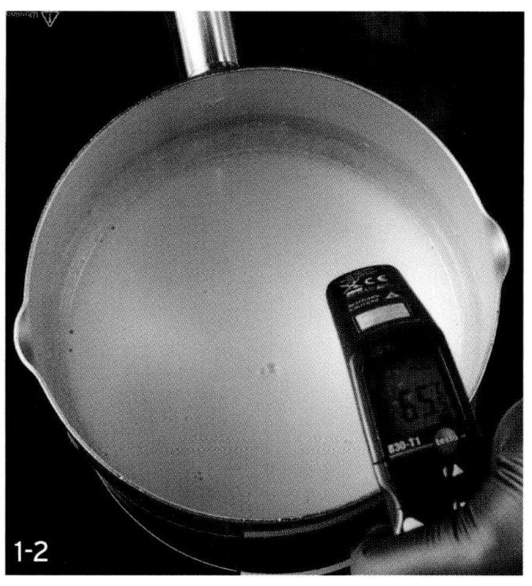
1-2

2

3

4

5

6

7

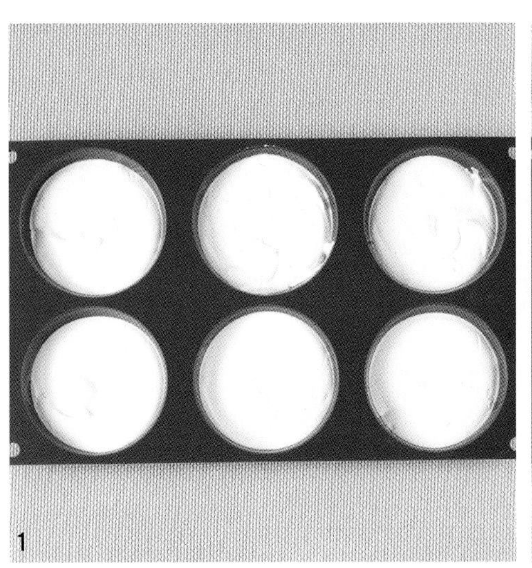

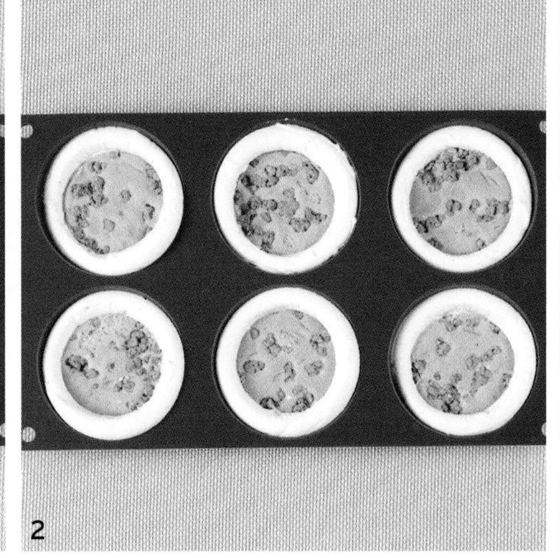

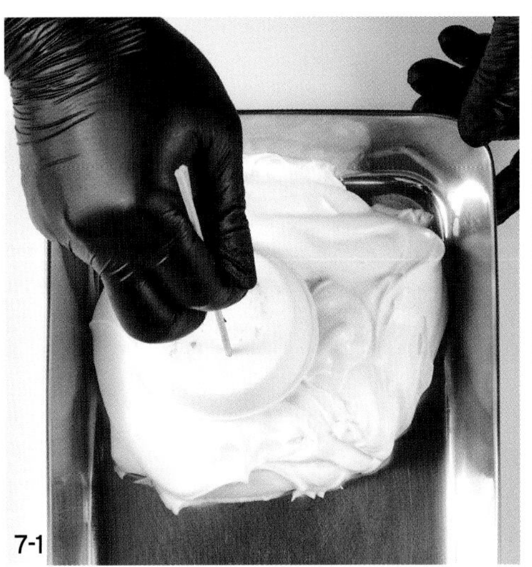

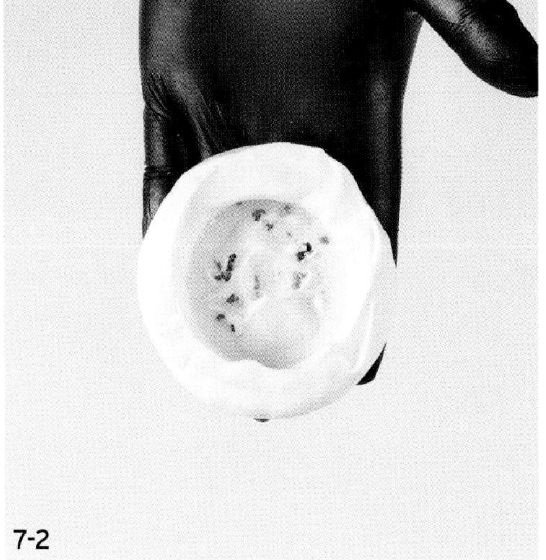

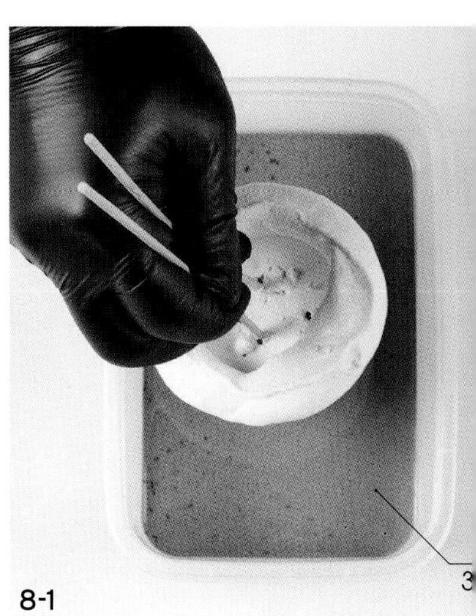

✤ ✤ ✤ ✤ ✤

FINISH 마무리

1 몰드(실리코마트 SF047)에 피치 라즈베리 사워크림 몽테를 24g씩 채운 후 스패출러로 평평하게 정리한다.

2 얼린 바질 스펀지를 넣고 꾹 눌러준다.

3 피치 라즈베리 사워크림 몽테를 18g 올린 후 스패출러로 평평하게 정리한다.

4 지름 5.5cm, 높이 1cm 원형 실리콘 몰드를 중앙에 올린다.

5 밀대로 꾹 눌러 바스켓 모양을 만든 후 급속 냉동(-35 ~ -40℃)한다.

6 몰드에서 뺀 후 꼬치를 꽂는다.

7 피치 라즈베리 사워크림 몽테에 약 15g을 묻혀 형태를 만들어 냉동한다. 화이트 바닐라 스프레이를 분사한 후 냉동실에서 굳힌다.

8 라즈베리 룸 템퍼라처 글레이즈(30℃)로 밑부분을 코팅한 후 냉장 보관한다.

9 장식해 마무리한다.

❶ 피치 쿨리 26g을 파이핑한 후 수비드 사과 6개를 올린다.
❷ 피치 젤라틴 젤리를 군데군데 올린다.
❸ 피치 쿨리를 조금씩 파이핑한다.
❹ 붓으로 데코젤 시럽을 바르고 허브를 올려 마무리한다.

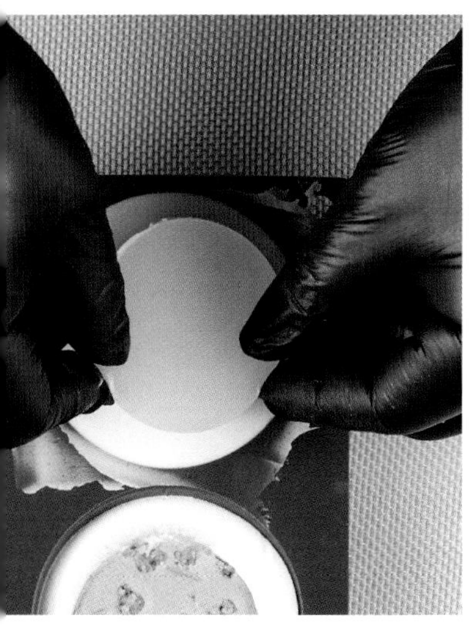

5

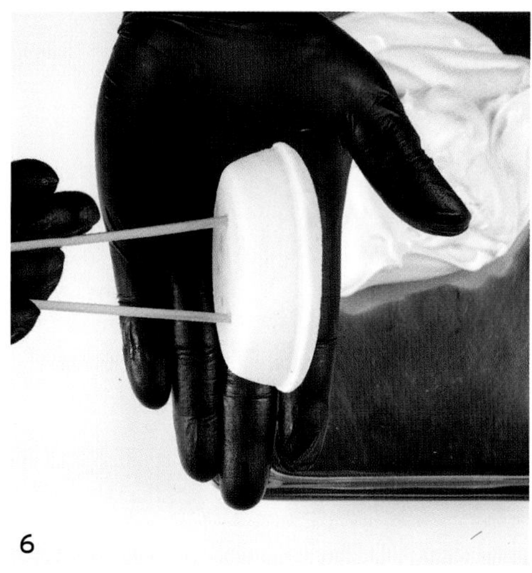

6

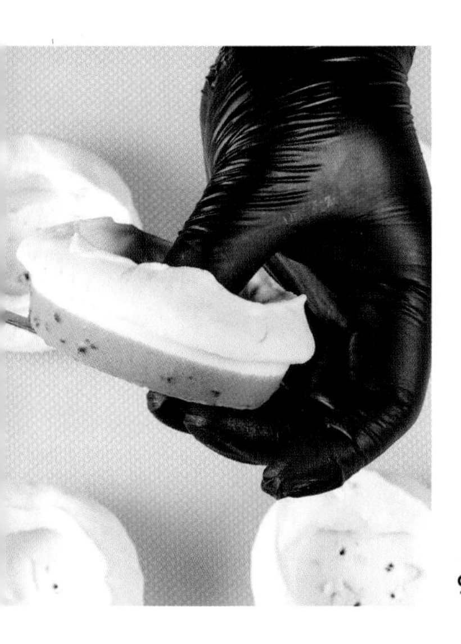

9-1

9-2

1 Fill 24 grams of peach raspberry sour cream montée in the mold (Silikomart SF047) and flatten with a spatula.

2 Insert the frozen basil sponge and press them firmly.

3 Top with 18 grams of peach raspberry sour cream montée and flatten with a spatula.

4 Place a 5.5 cm diameter and 1 cm high silicon mold in the center.

5 Press firmly with a rolling pin to form a basket shape, then blast freeze (-35 ~ -40°C) them.

6 Remove from the mold and insert a skewer.

7 Dip it in the peach raspberry sour cream montée, shape it, and freeze. Spray with white vanilla spray mixture and keep frozen.

8 Coat the bottom with raspberry room temperature glaze (30°C) and refrigerate.

9 Decorate to finish.

❶ Pipe 26 grams of peach coulis and top with 6 sous-vide apples.

❷ Randomly arrange the peach gelatin jellies.

❸ Pipe small amounts of peach coulis.

❹ Brush decorgel syrup and top with herbs to finish.

26

TINGLING STRAWBERRY CAKE

팅글링 딸기 케이크

PAIRING & TEXTURE
페어링 & 텍스처

산초는 디저트에서 조금은 생소하게 느껴지는 재료라고 생각할 수도 있지만, 마라탕의 열풍으로 한국에서도 이제는 더 이상 낯선 재료가 아니라고 생각해 딸기, 바닐라, 바질, 초콜릿과 함께 페어링해 기존 디저트에서 느끼지 못했던 독창적인 맛을 연출했다.

Sichuan pepper may be a somewhat unusual ingredient for dessert, but with the craze of malatang*, I thought it was no longer a bizarre ingredient in Korea. Hence, I paired it with strawberry, vanilla, basil, and chocolate to create a unique taste unlike anything you've ever experienced in dessert.

* Malatang: A common type of Chinese street food known as hot pot.

TECHNIQUE
테크닉

제과에서 허브를 사용할 때 흔히 인퓨징 기법으로 맛을 내는 것이 일반적이지만, 여기에서는 100% 착즙을 통해 퓌레 형태로 만들고 소스처럼 케이크 위에 얹어 페어링에 포인트를 주었다.

When using herbs in confectionery, it's common to flavor them with infusion techniques. However, the herbs are 100% extracted into purée and used on the cake like a sauce to accentuate the pairing.

DESIGN
디자인

- 개성 있으면서도 유행을 타지 않는 디자인을 고민하다가 탄생한 메뉴이다.
- 곡선이 주는 익숙함을 표현하기 위해 흔히 사용하는 지름 1cm 원형 깍지로 파이핑한 듯한 느낌이 드는 몰드를 사용했다. 특히 몰드 가운데에 깊게 들어가는 홈을 만들어 프레시한 과일을 듬뿍 올릴 수 있도록 디자인했다.

- This menu was created while searching for an uncommon yet timeless design.
- To depict the ease of the curves, I used a mold that looks like it was piped with the usual 1 cm round tip. In particular, it was designed with a deep indentation in the center to hold plenty of fresh fruit.

✦ HOW TO COMPOSE THIS RECIPE ✦

STEP 1.	메뉴 정하기 Decide on the menu	앙트르메 Entremets
STEP 2.	메인 맛 정하기 Choose the primary flavor	딸기 Strawberry

STEP 3. 메인 맛(딸기)과의 페어링 선택하기

Select a pairing flavor (Strawberry)

☑ 라즈베리	Raspberry
☑ 바질	Basil
☑ 코코넛	Coconut
☑ 바닐라	Vanilla
☑ 마스카르포네	Mascarpone
☑ 화이트초콜릿	White Chocolate
☑ 캐러멜초콜릿	Caramel Chocolate
☑ 산초	Sichuan Pepper

STEP 4. 구성하기

Assemble

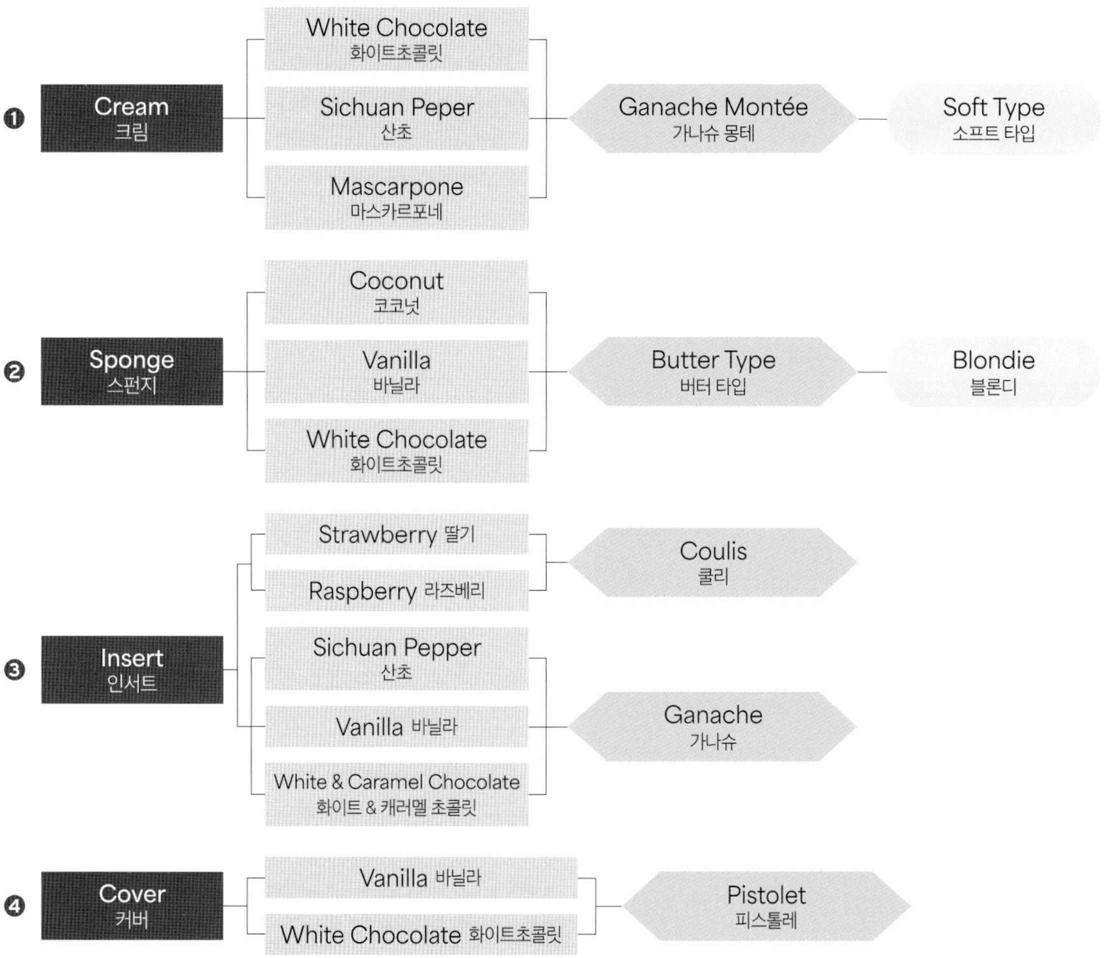

❶ Cream 크림
- White Chocolate 화이트초콜릿
- Sichuan Peper 산초
- Mascarpone 마스카르포네

→ Ganache Montée 가나슈 몽테 — Soft Type 소프트 타입

❷ Sponge 스펀지
- Coconut 코코넛
- Vanilla 바닐라
- White Chocolate 화이트초콜릿

→ Butter Type 버터 타입 — Blondie 블론디

❸ Insert 인서트
- Strawberry 딸기
- Raspberry 라즈베리

→ Coulis 쿨리

- Sichuan Pepper 산초
- Vanilla 바닐라
- White & Caramel Chocolate 화이트 & 캐러멜 초콜릿

→ Ganache 가나슈

❹ Cover 커버
- Vanilla 바닐라
- White Chocolate 화이트초콜릿

→ Pistolet 피스톨레

바질 겔
Basil Gel
초 마스카르포네 몽테
Sichuan Pepper
Mascarpone
Montée
딸기 쿨리 B
Strawberry Coulis B
바닐라 화이트 스프레이
Vanilla White
Spray
코코넛 블론디
Coconut Blondie
산초 바닐라 가나슈
Sichuan Pepper
Vanilla Ganache
딸기 쿨리 A
Strawberry Coulis A

INGREDIENTS

10개 분량/ 10 cakes

코코넛 블론디

Coconut Blondie

(철판(60 × 40cm) 1개 분량)
(1 baking tray (60 × 40 cm))

INGREDIENTS		g
코코넛롱	Shredded dried coconut	300g
달걀	Eggs	360g
설탕	Sugar	400g
전화당	Inverted sugar	60g
바닐라빈 씨	Vanilla bean seeds	1개 분량/ 1 pc
화이트초콜릿 (Zephyr white 34%)	White chocolate	200g
캐러멜초콜릿 (Zephyr Caramel 35%)	Caramel chocolate	100g
버터 (Bridel)	Butter	410g
바닐라에센스 (Aroma Piu)	Vanilla essence	10g
박력분	Cake flour	225g
아몬드파우더	Almond powder	90g
베이킹파우더	Baking powder	4g
TOTAL		**2159g**

산초 마스카르포네 몽테

Sichuan Pepper Mascarpone Montée

INGREDIENTS		g
생크림A	Heavy cream A	300g
산초	Sichuan pepper	6g
노른자	Egg yolks	192g
연유	Condensed milk	60g
젤라틴매스 (×5)	Gelatin mass (×5)	55g
화이트초콜릿 (Zephyr white 34%)	White chocolate	150g
마스카르포네	Mascarpone	300g
휘핑크림	Whipping cream	240g
생크림B	Heavy cream B	240g
바닐라에센스 (Aroma Piu)	Vanilla essence	24g
TOTAL		**1567g**

딸기 쿨리

Strawberry Coulis

INGREDIENTS		g
딸기 퓌레	Strawberry purée	360g
냉동 라즈베리	Freeze-dried raspberry	260g
설탕	Sugar	210g
NH펙틴	NH pectin	13g
레몬즙	Lemon juice	45g
TOTAL		**888g**

바질 퓌레 *

Basil Purée

INGREDIENTS		g
바질	Basil	200g
물기를 짠 바질	Blanched and squeezed basil	152g
물	Water	228g
설탕	Sugar	50g
TOTAL		**630g**

*** 만드는 법**

❶ 바질을 잎과 줄기로 분리하고 끓는 물에 10초간 데친 후, 얼음물에 담가 빠르게 식혀 손으로 물기를 꼭 짜준다.

❷ 냄비에 물기를 짠 바질 152g, 물, 설탕을 담고 가열한다.

❸ 끓어오르면 불에서 내려 얼음물이 담긴 볼에 받쳐 쿨링한다.

❹ 착즙기로 착즙한 후 필요한 만큼 계량해 사용한다.

● 대량으로 작업하는 경우 냉동 보관하며 사용한다.

*** Procedure**

❶ Separate the basil into leaves and stems, blanch for 10 seconds, cool it rapidly in ice water, and squeeze out the water with your hands.

❷ Heat 152 grams of the squeezed basil, water, and sugar in a pot.

❸ When it boils, remove from the heat and cool in an ice bath.

❹ Extract it with a juicer and use it as needed.

● Store it frozen when working in large quantities.

바질 겔

Basil Gel

INGREDIENTS		g
바질 퓌레 *	Basil purée *	430g
물엿	Corn syrup	50g
아가아가	Agar agar	5g
젤라틴매스 (×5)	Gelatin mass (×5)	18g
TOTAL		**503g**

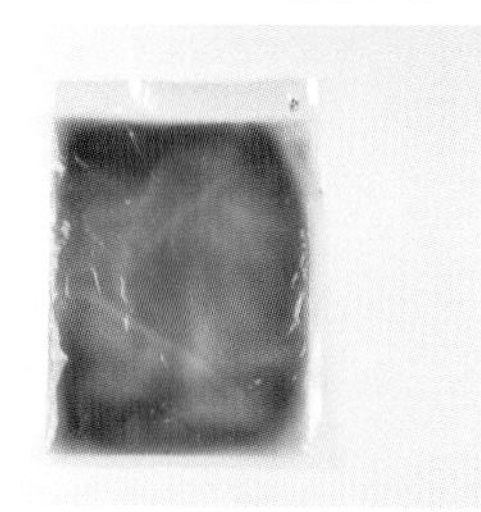

*** 만드는 법**

❶ 냄비에 모든 재료를 넣고 휘퍼로 저어가며 가열한다.

❷ 끓어오르면 불에서 내려 쿨링한다.

● 바질 겔은 쉽게 단단해지므로 끓어오르자마자 불에서 내린다.

❸ 바믹서로 블렌딩한 후 진공 압축해 냉장고에서 보관한다.

❹ 사용하기 전 바믹서로 한 번 더 부드럽게 풀어준 후 짤주머니에 담아 사용한다.

*** Procedure**

❶ Heat all the ingredients in a pot while stirring with a whisk.

❷ When it boils, remove from the heat and cool.

● Basil gel sets easily, so remove from the heat as soon as it reaches to a boil.

❸ Blend using an immersion blender, vacuum seal, and refrigerate.

❹ To use, blend again with an immersion blender and use in a piping bag.

산초 바닐라 가나슈

Sichuan Pepper
Vanilla Ganache

INGREDIENTS		g
산초	Sichuan pepper	8g
생크림	Heavy cream	280g
소르비톨	Sorbitol	96g
바닐라빈 껍질	Vanilla bean pod	2개 분량/ 2 pcs
화이트초콜릿 (Zephyr white 34%)	White chocolate	160g
캐러멜초콜릿 (Zephyr Caramel 35%)	Caramel chocolate	120g
TOTAL		**664g**

바닐라 화이트 스프레이

Vanilla White Spray

INGREDIENTS		g
화이트초콜릿 (Zephyr white 34%)	White chocolate	350g
카카오버터	Cacao butter	350g
하얀색 색소 (이산화티타늄)	Titanium dioxide	3g
바닐라빈	Vanilla bean	1개 분량/ 1 pc
TOTAL		**703g**

- 볼에 모든 재료를 넣고 녹여 바믹서로 블렌딩한 후 50℃로 맞춰 사용한다.
- 국제적인 추세를 반영하여 이산화티타늄은 최소량으로 사용했다.

- Melt all the ingredients in a bowl, combine using an immersion blender, and use at 50°C.
- Titanium dioxide was used in a minimal amount to reflect international trends.

NOTE.

✤

COCONUT BLONDIE

코코넛 블론디

1 로보쿱에 코코넛롱을 넣고 곱게 갈아 준비한다.

TIP. 고운 입자의 코코넛 가루를 사용하는 것보다 코코넛롱을 사용하는 것이 코코넛 고유의 향을 더 살릴 수 있다.
코코넛롱은 오븐에서 살짝 로스팅해 사용한다.

2 믹싱볼에 달걀, 설탕, 전화당, 바닐라빈 씨를 넣고 뜨거운 물이 담긴 볼에 받쳐 휘퍼로 저어가며 45℃까지 온도를 올린다.

TIP. 거품이 너무 많이 올라오지 않도록 주의하며 천천히 저어가며 온도를 올려준다.

3 중탕볼에서 내린 후 45℃로 온도를 맞춘 [화이트초콜릿, 캐러멜초콜릿, 버터], 바닐라에센스를 넣고 바믹서로 블렌딩한다.

4 **1**과 체 친 박력분, 아몬드파우더, 베이킹파우더를 넣고 휘퍼로 섞는다.

5 테프론시트를 깐 철판(60 × 40cm)에 고르게 팬닝한 후 댐퍼를 100%로 열고 160℃로 예열된 오븐에서 22분간 굽는다.

TIP. 구워져 나온 후 유산지를 올린 식힘망 위에 뒤집어 빼 냉장고에서 식혀 랩핑해 냉동 보관한다.

6 지름 9.7cm 원형 커터로 자른 후 냉동한다.

1 Finely grind shredded dried coconut in the Robot Coupe.

TIP. Using the shredded type of coconut rather than the finer powder type enhances the coconut flavor.
Lightly roast the shredded coconut before use.

2 Add eggs, sugar, inverted sugar, and vanilla bean seeds in a mixing bowl. Place it over a double boiler and stir with a whisk until it reaches 45°C.

TIP. Bring to temperature while stirring slowly, being careful not to make too many bubbles.

3 Remove from the double boiler. Blend with [white chocolate, caramel chocolate, and butter] melted to 45°C, with vanilla essence, using an immersion blender.

4 Mix with (**1**) and sifted cake flour, almond powder, and baking powder using a whisk.

5 Spread evenly on a baking tray (60 × 40 cm) lined with a Teflon sheet. Bake in an oven preheated to 160°C, damper opened at 100%, for 22 minutes.

TIP. When it's baked, invert it onto a cooling rack lined with parchment paper and remove the baking tray. Cool it in a refrigerator, wrap it in plastic wrap, and freeze.

6 Cut with a 9.7 cm round cutter and freeze.

1-1
1-2
2
CALLEBAUT

45℃
3-1
3-2
4

5
6

✤ ✤

SICHUAN PEPPER MASCARPONE MONTÉE

산초 마스카르포네 몽테

1 냄비에 생크림A와 산초를 담고 가열한 후 끓어오르면 불에서 내려 냄비 입구를 랩핑한 후 약 5분간 두어 인퓨징한다.

2 노른자(30℃)를 볼에 담아 가볍게 거품 내고 1을 체에 걸러가며 휘퍼로 섞는다.

3 다시 냄비로 옮겨 약불에서 가열한다. (크렘 앙글레이즈)

4 75℃가 되면 불에서 내려 연유, 젤라틴매스를 넣고 섞는다.

5 화이트초콜릿이 담긴 볼에 넣고 바믹서로 블렌딩한다.

6 마스카르포네를 두 번 나눠 넣어가며 블렌딩한다.

7 휘핑크림과 생크림B - 바닐라에센스 순서로 넣고 충분히 블렌딩한다.

8 바트에 담고 밀착 랩핑해 냉장고에서 12시간 숙성시킨다.

TIP. 사용하기 직전 비터로 부드럽게 풀어 사용한다.

1 Heat heavy cream A and Sichuan pepper. Once it boils, remove from the heat, cover the opening of the pot, and infuse for 5 minutes.

2 Strain (**1**) into a bowl with egg yolks (30°C) that have been whisked until foamy.

3 Pour it back into the pot and cook over a low heat. (Crème Anglais)

4 Remove from the heat when it reaches 75°C and mix with condensed milk and gelatin mass.

5 Blend with white chocolate in a beaker.

6 Blend with mascarpone in two batches.

7 Blend thoroughly in the order of whipping cream with heavy cream B and vanilla essence.

8 Pour onto a stainless-steel tray, adhere the plastic wrap to the preparation to cover, and refrigerate for 12 hours.

TIP. Gently beat using a paddle attachment right before use.

1

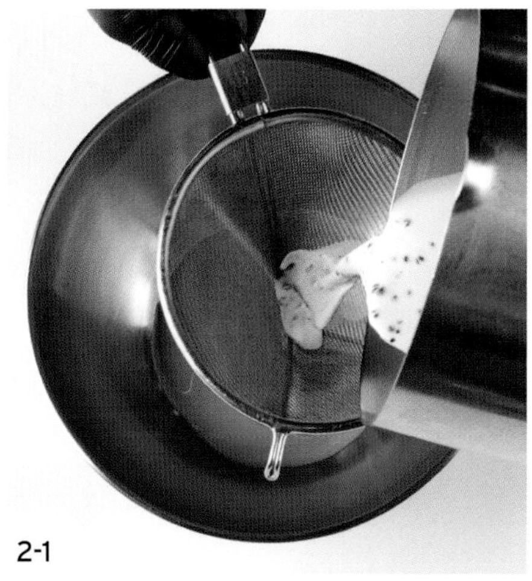
2-1

2-2

3

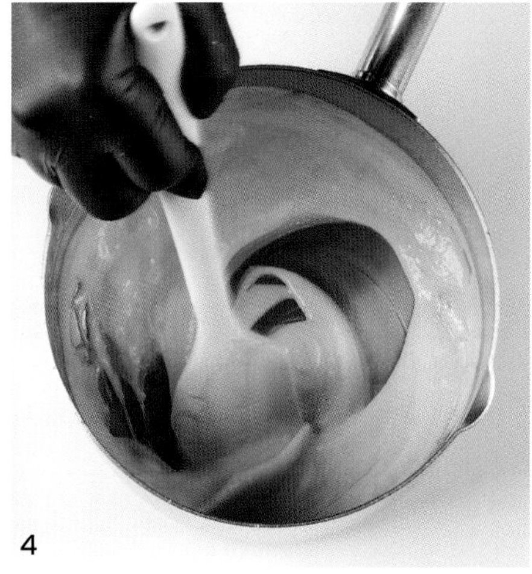
4

5

6

7

8

✤ ✤ ✤

STRAWBERRY COULIS

딸기 쿨리

1 냄비에 딸기 퓌레, 냉동 라즈베리를 넣고 가열한다.

2 미리 섞어둔 설탕과 NH펙틴을 넣고 휘퍼로 저어가며 가열한다.

3 가열하면서 펙틴 반응을 확인한 후 불에서 내려 레몬즙을 넣고 섞고 쿨링한다.

TIP. 쿨링한 쿨리는 바트에 담고 밀착 랩핑해 냉장 보관한다.

1 Heat strawberry purée and frozen raspberries.

2 Continue to heat while stirring in previously mixed sugar and pectin NH with a whisk.

3 After the pectin activates, remove from the heat, mix with lemon juice, and cool.

TIP. Pour the cooled coulis onto a stainless-steel tray, adhere the plastic wrap to the preparation to cover, and refrigerate.

✤ ✤ ✤ ✤

SICHUAN PEPPER VANILLA GANACHE

산초 바닐라 가나슈

1 짤주머니에 산초를 담고 밀대로 으깨 준비한다.

2 냄비에 생크림, 소르비톨, 바닐라빈 껍질, 으깬 산초를 넣고 가열한다.

3 끓어오르면 불에서 내려 냄비 입구를 랩핑한 후 5분간 두어 인퓨징한다.

4 화이트초콜릿과 캐러멜초콜릿이 담긴 비커에 60℃의 **3**을 붓고 바믹서로 블렌딩한다. (최종 온도 28~33℃)

TIP. 체에 내려 산초와 바닐라빈을 걸러낸다.

5 바트에 담고 밀착 랩핑해 냉장고에 보관한다.

1 Put Sichuan peppers in a piping bag and crush them with a rolling pin.

2 Heat heavy cream, sorbitol, vanilla bean pod (without the seeds), and crushed Sichuan pepper.

3 When it boils, remove from the heat, cover the opening of the pot, and infuse for 5 minutes.

4 Pour (**3**) set to 60°C into a beaker and blend with white and caramel chocolates using an immersion blender. (Final temperature: 28~33°C)

TIP. Stain to sieve the Sichuan pepper and vanilla bean pod.

5 Pour onto a stainless-steel tray, adhere the plastic wrap to the preparation to cover, and refrigerate.

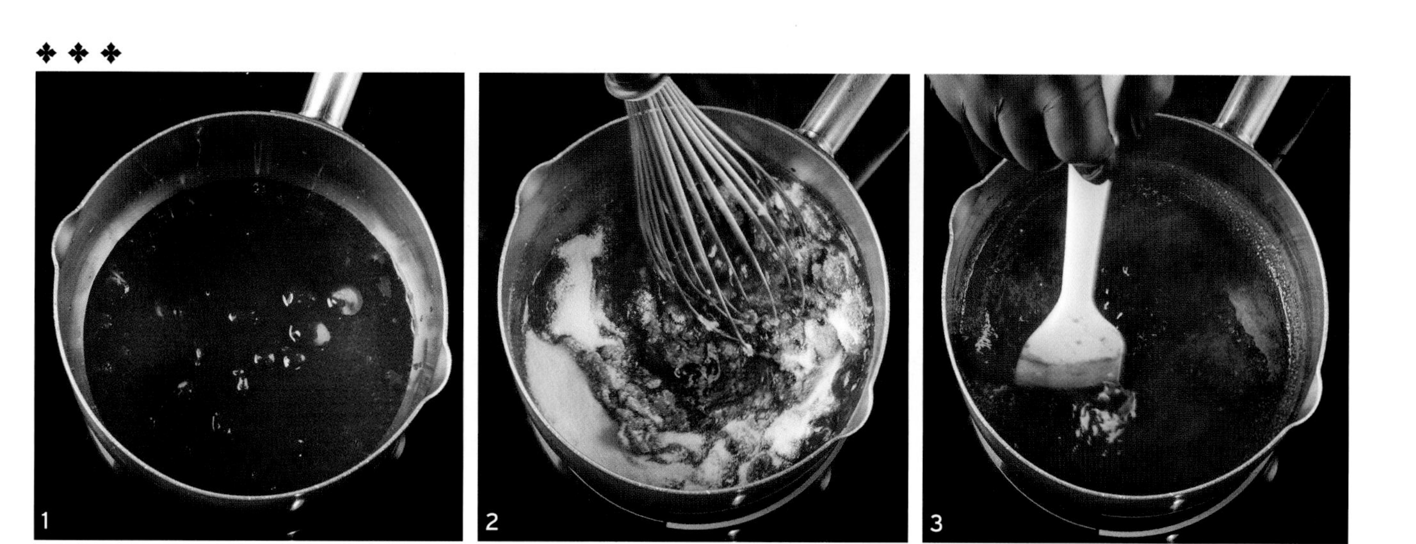
✤ ✤ ✤
1
2
3

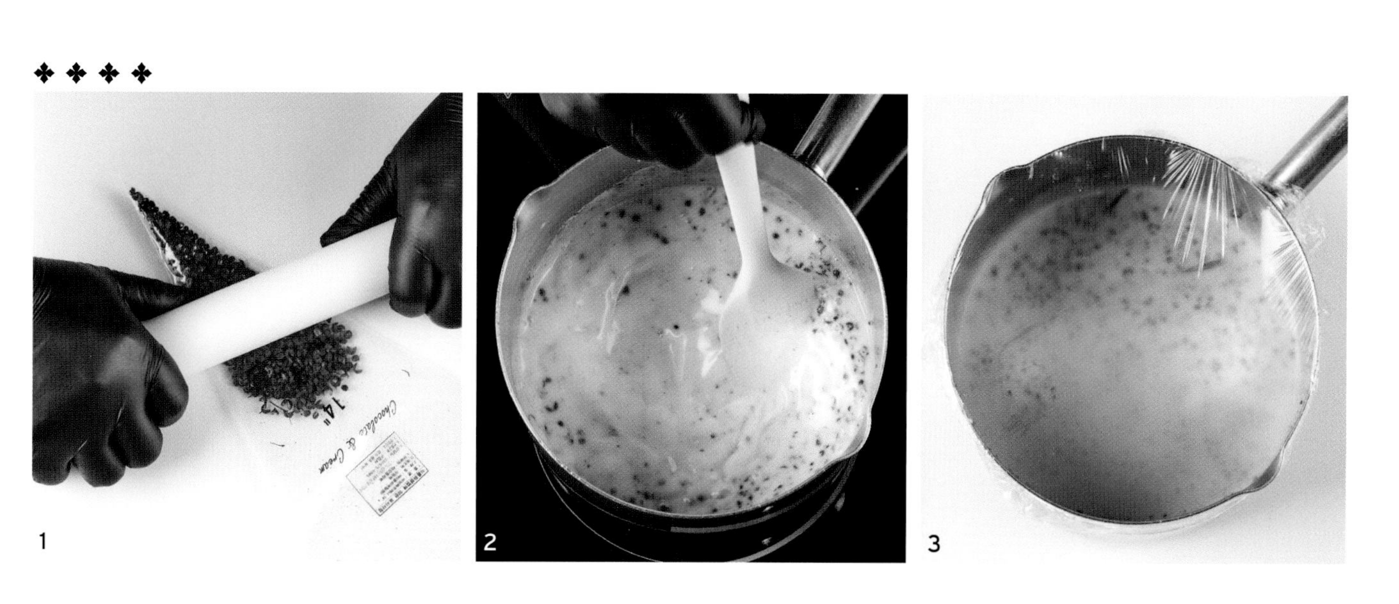
✤ ✤ ✤ ✤
1
2
3

4-1
4-2
28~33℃
5

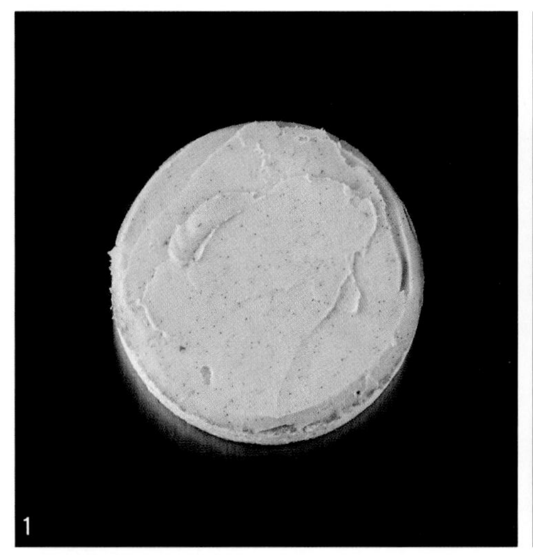

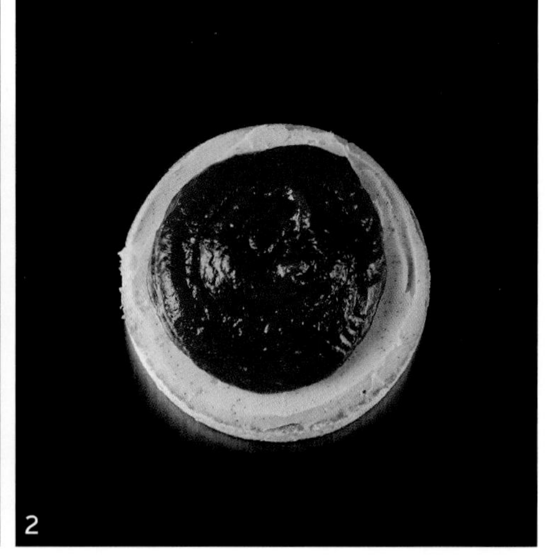

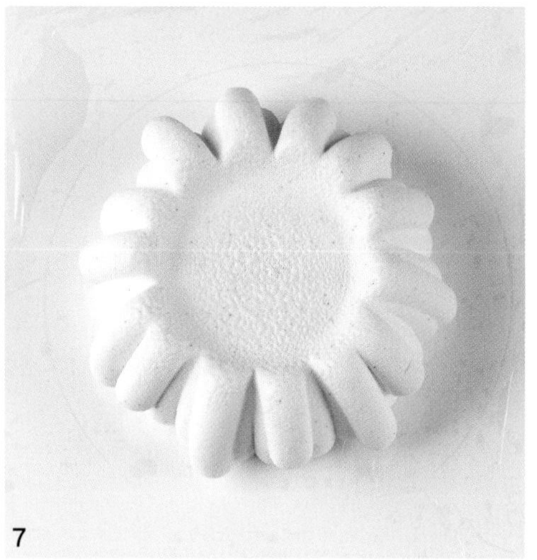

✤ ✤ ✤ ✤ ✤

FINISH 마무리

1　코코넛 블론디에 산초 마스카르포네 바닐라 가나슈 35g을 파이핑한 후 평평하게 정리한다.

2　딸기 쿨리 50g을 파이핑한 후 평평하게 정리해 냉동한다.

3　산초 마스카르포네 몽테를 부드럽게 휘핑한다.

　TIP. 휘핑을 멈춘 후에도 비터로 만들어진 크림의 선이 남아 있을 때까지만 휘핑한다.
　너무 많이 휘핑하면 단단해지고 크림도 느끼해지므로 주의한다.

4　몰드에 산초 마스카르포네 몽테를 140g 채운 후 바닥에 쳐 기포를 빼준다.

5　얼린 **2**를 넣는다.

6　윗면을 깔끔하게 정리하고 급속 냉동(-35 ~ -40℃)한 후 몰드에서 빼내 냉동(-18℃) 보관한다.

7　바닐라 화이트 스프레이를 분사한 후 냉동실에서 굳힌다.

8　딸기 쿨리 15g을 파이핑한 후 평평하게 펼친다.

9　바질 겔 4g을 파이핑한다.

10　딸기를 가득 올리고 데코젤 뉴트럴을 군데군데 파이핑해 이슬을 표현한 후 금박을 올려 마무리한다.

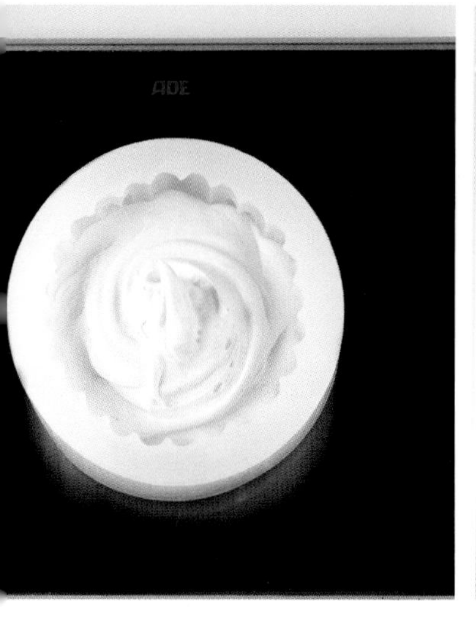

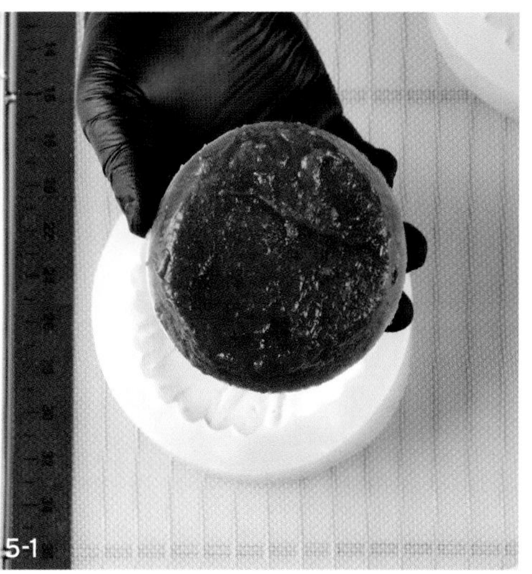

5-1

5-2

Make sure to place the insert slightly higher than the mold so that when you lift the finished cake with the spatula to transfer, the spatula won't touch the montée cream and damage it.

10

1 Evenly spread 35 grams of Sichuan mascarpone vanilla ganache on the coconut blondie.

2 Pipe and spread 50 grams of strawberry coulis and freeze.

3 Gently whip Sichuan mascarpone montée using a paddle attachment.

TIP. Whip only until a line of cream the paddle makes remains when stopped. Be careful not to whip it too much, as it will become stiff and taste greasy.

4 Fill 140 grams of Sichuan mascarpone montée in the mold and tap it on the table to remove the bubbles.

5 Insert the frozen (**2**).

6 Neatly organize the top, blast freeze (-35 ~ -40°C), remove from the mold, and keep frozen (-18°C).

7 Spray with vanilla white spray mixture and freeze.

8 Pipe and spread 15 grams of strawberry coulis.

9 Pipe 4 grams of basil gel.

10 Overload with strawberries, randomly pipe Decorgel Neutral to depict dew, and finish with gold leaves.

Chef Eric's WORLD CHOCOLATE MASTERS

(2022)

WORLD CHOCOLATE MASTERS

월드 초콜릿 마스터즈(World Chocolate Masters, 이하 WCM)는 2005년부터 시작된 세계적인 초콜릿 대회로, 약 20여 개 국가가 참가하며 파티세리, 제과 분야에 종사하는 21세 이상의 성인을 대상으로 진행된다.

3년마다 한 번씩 열리는 WCM은 세계를 목표로 삼는 셰프들에게 매우 중요한 대회다. 이 대회는 카카오바리가 주관, WCM이 주최하며 각지의 초콜릿 예술가들에게 무한한 가능성과 아티스트로서의 명성, 전 세계적으로 자신의 초콜릿 예술 작품을 알릴 수 있는 기회의 무대이기도 하다.

The World Chocolate Masters (WCM) is a global chocolate competition that began in 2005. It is open to adults over the age of 21 who work in the field of patisserie or confectionery and represents nearly 20 countries.

Held every three years, the WCM is a crucial competition for aspiring chefs. Organized by Cacao Barry and hosted by the WCM, it offers chocolate artists from all over the world endless possibilities, fame as artists, and the opportunity to showcase their chocolate art on a global scale.

이 대회에서 셰프들은 초콜릿을 다루는 기술과 창의성을 겨루는데, 각 나라에서 지원한 선수들은 예선전을 거쳐 대표 선수로 선출된 후 프랑스 파리에서 열리는 본선에 출전한다. 본선은 초콜릿 및 제과 전문가로 구성된 국제 심사위원단이 있는 대회장에서 총 3일간 진행하며, 1, 2일차 합산 점수로 총 10명의 선수만이 3일차 슈퍼 파이널 결승 무대에 진출한다.

초콜릿을 활용한 제품의 맛, 디자인, 혁신성, 테크닉 등 다양한 요소들을 평가하여 3일차 슈퍼 파이널 무대에서 우승자를 결정하게 되는데, 2022년 WCM 한국 대표로 출전한 김동석 셰프는 세계 6위, 아시아 최초 1위라는 눈부신 성과를 이뤄냈다.

TOMORROW(내일, 미래)을 주제로 한 2022 WCM. 그중 TRANSFORM 부문에서 심사위원단의 호평을 받은 김동석 셰프의 파블로바 3종. 건강 추구, 쾌락주의, 탐험가라는 콘셉트를 주제로 독특하게 풀어냈다. 특히 파블로바는 습기에 취약하다는 인식을 완전히 뒤집은, 수 일간 실온에 두어도 바삭함을 유지하는 제품을 선보여 높은 점수를 받았다.

The competition challenges chefs against each other in a test of skill and creativity with chocolate, with competitors from each country competing in preliminary rounds before being selected to represent their country at the main event in Paris, France. The competition takes place over three days in front of an international jury of chocolate and confectionery experts, with only 10 competitors advancing to the Super Final on Day 3 based on their combined scores from Days 1 and 2.

The winners were judged on various factors, including flavor, design, innovation, and technic, to determine the winner at the Super Final stage on Day 3. Chef Kim Dong-seok (Eric), who represented South Korea at the 2022 WCM, achieved a remarkable ranking of sixth in the world and first in Asia.

The theme of the 2022 WCM was TOMORROW (future), and Chef Kim Dong-seok (Eric)'s three pavlovas won the jury's favor in the TRANSFORM category. The concept of health-seeker, hedonist, and explorer was uniquely interpreted in the theme. Thus, the pavlova received high marks for its ability to remain crispy even after being left at room temperature for several days, completely reversing the perception that pavlova is vulnerable to moisture.

이 책에서는 김동석 셰프가 2022 WCM TRANSFORM 부문에서 선보인 '파블로바 3종'과, TASTE 부문에서 선보인 '무한의 씨앗'을 소개한다. '파블로바 3종'의 경우 기존의 파블로바와 달리 습기에 강하게 만든 레시피이므로 잘 활용한다면 업장에서 판매하기에도 좋을 것이다. '무한의 씨앗'의 경우 한국 식재료인 쌍화차와 화요(증류주)를 다크초콜릿과 페어링한 제품으로 자칫 호불호가 나뉠 수 있는 독특한 맛을 전 세계인 모두가 낯설지 않게 즐길 수 있도록 맛을 잡은 레시피이므로 디저트에서 재료를 사용하는 데 있어 한계가 없음을 확인할 수 있는 좋은 예시일 것이다.

In this book, Chef Dong-seok (Eric) Kim introduces the 'Three Pavlovas' presented at WCM TRANSFORM 2022 and 'Infinite Seeds' presented at TASTE category. Unlike traditional pavlova, the recipe for '3 kinds of pavlova' is made to resist moisture, so It would be a perfect recipe to sell in a restaurant if utilized well. In the case of 'Infinite Seeds,' the Korean ingredients Ssanghwacha and Hwayo (distilled liquor) are paired with dark chocolate. The unique flavors that may divide people into likes and dislikes are flavored in such a way that everyone around the world can enjoy, so it is an exceptional example that there are no limits to utilizing ingredients in desserts.

1

INFINITY SEED

인피니티 씨드

DESCRIPTION

특징

- '무한의 씨앗'이라는 주제로 연출해보았다. 이는 지구상에 존재하는 모든 식물들의 분자 구조를 결합하여 만든 슈퍼 씨드를 의미하며, 멸종한 식물의 분자 구조를 뽑아서 사용할 수 있는 무한한 씨앗이라는 의미도 함께 담았다.
- 각 나라의 로컬 식자재를 사용해야 하는 규정에 따라 화요(시판 증류주, 25%)와 쌍화차를 선택했다.
- 국가대표 선수들이 본인의 초콜릿을 만드는 과제가 있어 'IAN'이라는 71% 다크 초콜릿을 개발해 사용했다.
- 폴리페놀파우더를 사용해 콜레스테롤 수치를 낮추는 효과가 있는 메뉴로 완성했다.
- 인서트를 얼리지 않는 형태로 만들기 위해 머랭 무스를 얼릴 때 인서트가 들어가는 공간을 만들어 주었다.

- I directed it under the theme 'Infinite Seed.' It refers to a super seed created by combining the molecular structures of all the plants on Earth, and it's an infinite seed that can be used by extracting the molecular structures of extinct plants.
- In accordance with the regulations requiring the use of local ingredients from each country, we chose Hwayo (commercially available spirits, 25%) and Ssanghwacha.
- The national team chefs were to create their own chocolate, so we developed a 71% dark chocolate called 'IAN.'
- We used polyphenol powder to create a cholesterol-lowering menu.
- To make a non-frozen insert, I made a cavity for the insert to fit into when freezing the meringue mousse.

✦ HOW TO COMPOSE THIS RECIPE ✦

STEP 1.	메뉴 정하기 Decide on the menu	프티 가토 Petit Gateau
STEP 2.	메인 맛 정하기 Choose the primary flavor	다크초콜릿 Dark Chocolate

STEP 3. 메인 맛(다크초콜릿)과의 페어링 선택하기

Select a pairing flavor (Dark Chocolate)

- ☑ 패션푸르트 Passion Fruit
- ☑ 아몬드 Almond
- ☑ 헤이즐넛 Hazelnut
- ☑ 화요(한국 증류주) Hwayo (Korean distilled spirits)
- ☑ 홍차 Black Tea
- ☑ 요거트 Yogurt
- ☑ 쌍화차 Ssanghwacha
- ☑ 딜 Dill
- ☑ 오렌지 Orange

STEP 4. 구성하기

Assemble

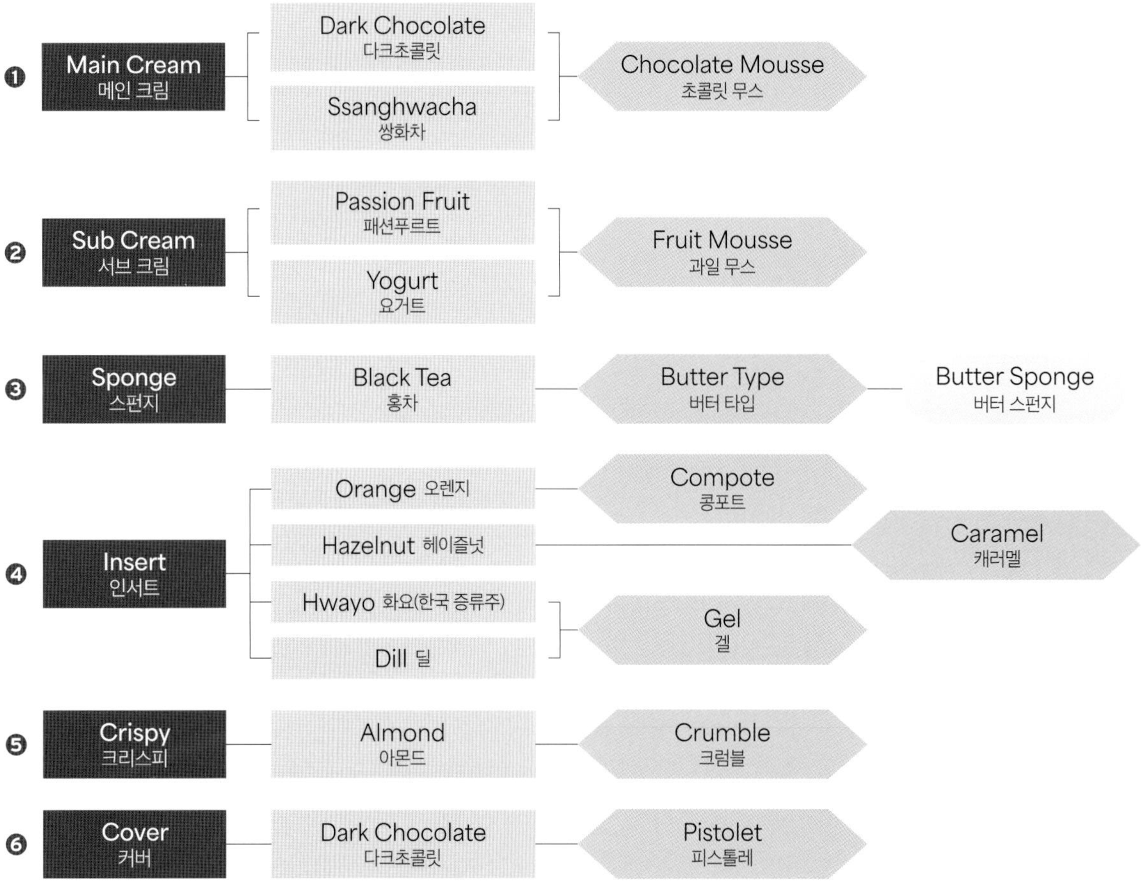

	Component	Flavor	Type	
❶	Main Cream 메인 크림	Dark Chocolate 다크초콜릿 Ssanghwacha 쌍화차	Chocolate Mousse 초콜릿 무스	
❷	Sub Cream 서브 크림	Passion Fruit 패션푸르트 Yogurt 요거트	Fruit Mousse 과일 무스	
❸	Sponge 스펀지	Black Tea 홍차	Butter Type 버터 타입	Butter Sponge 버터 스펀지
❹	Insert 인서트	Orange 오렌지	Compote 콩포트	
		Hazelnut 헤이즐넛		Caramel 캐러멜
		Hwayo 화요(한국 증류주) Dill 딜	Gel 겔	
❺	Crispy 크리스피	Almond 아몬드	Crumble 크럼블	
❻	Cover 커버	Dark Chocolate 다크초콜릿	Pistolet 피스톨레	

화요 딜 겔
Hwayo Dill Gel
패션 요거트 무스
Passion Yogurt Mousse
오렌지 콩포트
Orange Compote
아몬드 크럼블
Almond Crumble
이안 초콜릿 무스
IAN Chocolate Mousse
홍차 스펀지
Black Tea Sponge
헤이즐넛 캐러멜
Hazelnut Caramel
초콜릿 스프레이
Chocolate Spray

INGREDIENTS 40개 분량/ 40 cakes

홍차 스펀지

Black Tea Sponge

(철판(60 × 40cm) 1개 분량)
(1 baking tray (60 × 40 cm))

INGREDIENTS		g
설탕	Sugar	320g
홍차 잎 (TWG)	Black tea leaves	12g
달걀	Eggs	290g
전화당	Inverted sugar	48g
밀크초콜릿 (Ghana 40%)	Milk chocolate	200g
버터 (Bridel)	Butter	330g
아몬드 페이스트	Almond paste	70g
오렌지제스트	Orange zest	2개 분량/ 2 oranges
박력분	Cake flour	150g
베이킹파우더	Baking powder	4g
TOTAL		**1424g**

아몬드 크럼블

Almond Crumble

• p.388

INGREDIENTS		g
버터 (Bridel)	Butter	70g
소금 (Fleur de Sel)	Salt	2g
설탕	Sugar	70g
아몬드파우더	Almond powder	70g
박력분	Cake flour	100g
물	Water	10g
캐러멜초콜릿 (Zephyr Caramel 35%)	Caramel chocolate	48g
오렌지제스트	Orange zest	1/2개 분량/ 1/2 orange
TOTAL		**370g**

패션 요거트 무스

Passion Yogurt Mousse

INGREDIENTS		g
흰자	Egg whites	55g
물	Water	22g
설탕A	Sugar A	44g
꿀	Honey	121g
패션푸르트 퓌레	Passion fruit purée	50g
설탕B	Sugar B	80g
레몬즙	Lemon juice	20g
젤라틴매스 (×5)	Gelatin mass (×5)	56g
마스카르포네	Mascarpone cheese	100g
플레인 요거트	Plain yogurt	150g
생크림	Heavy cream	200g
TOTAL		**898g**

오렌지 콩포트

Orange Compote

INGREDIENTS		g
오렌지즙	Orange juice	156g
패션푸르트 퓌레	Passion fruit purée	90g
설탕	Sugar	84g
NH펙틴	Pectin NH	13.5g
꿀	Honey	51g
트레할로스	Trehalose	60g
생강즙	Ginger juice	4.5g
오렌지에센스 (Aroma Piu)	Orange essence	3방울/ 3 drops
세그먼트한 오렌지 (오렌지 약 9개 분량)	Fresh orange segments (about 9 oranges)	270g
TOTAL		**729g**

헤이즐넛 캐러멜

Hazelnut Caramel

• p.446

INGREDIENTS		g
설탕	Sugar	160g
물엿	Corn syrup	260g
생크림	Heavy cream	420g
헤이즐넛 페이스트	Hazelnut paste	48g
소금	Salt	4g
TOTAL		**892g**

화요 딜 겔

Hwayo Dill Gel

INGREDIENTS		g
물	Water	220g
화이트와인 (Moscato)	White wine	200g
화요	Hwayo	120g
아가아가	Agar-agar	4.5g
설탕	Sugar	200g
NH펙틴	Pectin NH	7g
로거스트빈검	Locust bean gum	1g
레몬즙	Lemon juice	50g
딜	Dill	7g
TOTAL		**809.5g**

이안 초콜릿 무스

IAN Chocolate Mousse

INGREDIENTS		g
휘핑크림	Whipping cream	250g
전화당	Inverted sugar	100g
쌍화탕 (광동) 졸인 것	Reduced Ssanghwacha (Gwangdong brand)	100g
젤라틴매스 (×5)	Gelatine mass (×5)	60g
다크초콜릿 (Ian 71%)	Dark chocolate	150g
밀크초콜릿 (Ghana 40%)	Milk chocolate	150g
폴리페놀파우더	Polyphenol powder	1.5g
헤이즐넛 페이스트	Hazelnut paste	50g
생크림	Heavy cream	450g
TOTAL		**1311.5g**

- 쌍화탕은 100g 기준 46g 이하로 졸여 사용한다. 여기에서는 쌍화탕 220g을 끓이면서 약 100g으로 졸여 사용했다.
- Boil Ssanghwatang to reduce to less than 46 grams per 100 grams. Here, 220 grams of Ssanhwatang was reduced to about 100 grams.

초콜릿 스프레이

Chocolate Spray

INGREDIENTS		g
다크초콜릿 (Guayaquil 64%)	Dark chocolate	150g
밀크초콜릿 (Lactee Superieure 38%)	Milk chocolate	150g
카카오버터	Cacao butter	300g
TOTAL		**600g**

- 볼에 모든 재료를 넣고 녹여 바믹서로 블렌딩한 후 50℃로 맞춰 사용한다.
- Melt all the ingredients in a bowl, combine with an immersion blender, and use at 50°C.

NOTE.

✤

BLACK TEA SPONGE

홍차 스펀지

1 써머믹서에 설탕, 홍차 잎을 넣고 곱게 간다.

2 달걀(30℃), 전화당을 넣고 간다.

3 녹인 밀크초콜릿(45~50℃)과 녹인 버터(45℃), 아몬드 페이스트, 오렌지제스트, 체 친 박력분, 베이킹파우더를 넣고 곱게 간다.

4 테프론시트를 깐 철판(60 × 40cm)에 반죽을 붓고 L자 스패출러로 평평하게 정리한다.

5 160℃로 예열된 오븐에서 뎀퍼를 100% 열고 30분간 굽는다.

6 3 × 4.5cm 크기의 물방울 모양 커터로 자른 후 냉동한다.

TIP. 시트가 너무 차가우면 재단할 때 부서질 수 있으니 너무 차갑지 않은 상태로 사용한다.

1 Finely grind sugar and black tea leaves in Thermomix.

2 Blend with eggs (30°C) and inverted sugar.

3 Add melted milk chocolate (45~50°C) and melted butter (45°C), almond paste, orange zest, sifted cake flour, and baking powder; grind finely.

4 Pour the batter onto a baking sheet (60 × 40 cm) lined with a Teflon sheet and even out with an offset spatula.

5 Bake in an oven preheated to 160°C and damper open at 100% for 30 minutes.

6 Cut with a 3 × 4.5 cm tear-drop-shaped cake ring.

TIP. Make sure the sponge is not too cold, as they may crumble when you cut.

30℃
1
2-1
2-2

3-1
3-2
4

5

6

✤ ✤

PASSION YOGURT MOUSSE

패션 요거트 무스

1. 믹싱볼에 흰자를 넣고 거품이 올라오는 상태까지 휘핑한 다음 118℃로 끓인 시럽(물+설탕A+꿀)을 서서히 흘려가며 이탈리안 머랭을 만든다. (1권 p.258 '이탈리안 머랭' 참고)

   TIP. 시럽 가열 시 110℃ 정도가 되었을 때 흰자를 휘핑하기 시작하면 작업 속도가 비슷하게 맞춰진다.

2. 냄비에 패션프루트 퓌레, 설탕B, 레몬즙, 젤라틴매스를 넣고 젤라틴매스가 녹을 때까지 주걱으로 저어가며 50~60℃로 가열한다.

   TIP. 1과 2를 동시에 작업한다.

3. 비커에 마스카르포네, 플레인 요거트와 **2**를 넣고 바믹서로 블렌딩한다.
4. **1**에 **3**을 두세 번 나눠 넣어가며 휘퍼로 섞는다. (최종 온도 17~18℃)
5. 60~70% 정도로 휘핑한 차가운 상태의 생크림을 준비한다.
6. **5**에 **4**를 넣고 섞는다.
7. 완성된 무스를 몰드에 소량(약 5g) 파이핑한다.
8. 몰드 무늬 사이사이에 무스가 잘 들어갈 수 있도록 미니 주걱을 이용해 골고루 펼쳐준다.

1. In a mixing bowl, whip egg whites until frothy peaks form, then gradually whisk in the syrup (water, sugar A, and honey) boiled to 118°C to make an Italian meringue. (See "Italian Meringue" on p.258 in Book 1)

   TIP. If you start whipping the whites when the syrup reaches about 110°C, the work pace will match up.

2. In a saucepan, add passion fruit purée, sugar B, lemon juice, and gelatin mass. Stir with a spatula until the gelatin mass melts while heating to 50~60°C.

   TIP. Make (1) and (2) at the same time.

3. In a beaker, blend mascarpone, plain yogurt, and (**2**) with an immersion blender.
4. Add (**3**) into (**1**) in two or three batches and combine with a whisk. (Final temperature: 17~18°C)
5. Whip the cold whipping cream to 60~70% (soft~medium peaks).
6. Mix (**4**) into (**5**).
7. Pipe a small amount of the mousse (about 5 g) into the molds.
8. Use a mini spatula to spread the mousse to fill the patterns.

1

2

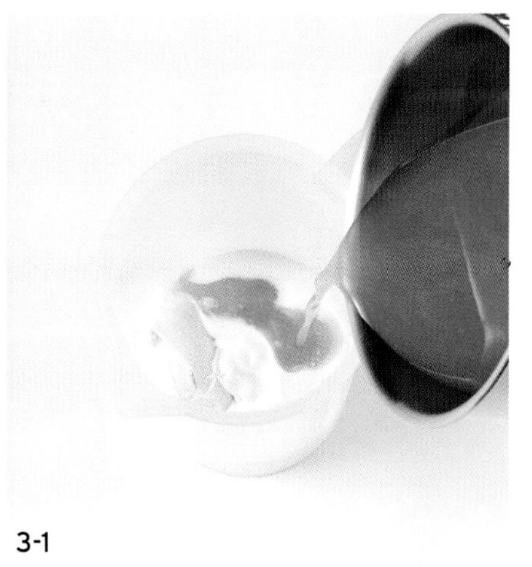
3-1

3-2

4

5

6

7

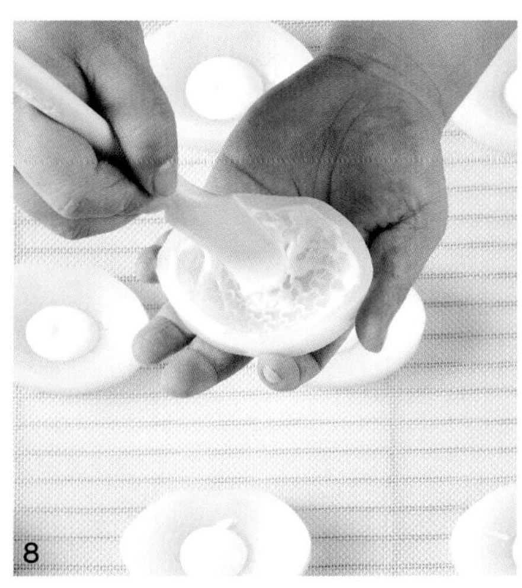
8

9 몰드에 패션 요거트 무스를 15g씩 파이핑한다.

10 인서트 몰드를 넣는다. (인서트 몰드 무게 15g)

11 미니 L자 스패출러로 윗면을 정리한다.

12 급속냉동고(-30 ~ -40℃)에 넣고 30분 뒤에 뒤집은 후 다시 완전히 얼린다.

13 인서트 몰드를 먼저 제거하고 바깥쪽 몰드를 제거한 후, 냉동고(-18℃)에 보관한다.

9 Pipe in 15 grams of mousse.

10 Embed the insert mold. (Weight of the insert mold: 15 grams)

11 Flatten the top with a mini offset spatula.

12 Put it in a blast freezer (-30 ~ -40°C) for 30 minutes, flip it upside down, and freeze completely.

13 Remove the insert mold first, then the outer mold. Store in the freezer (-18°C).

✤ ✤ ✤

ORANGE COMPOTE

오렌지 콩포트

1 오렌지는 세그먼트한 후 키친타월에 받쳐 물기를 제거해 준비한다.

TIP. 대회에서 시판 과일 퓌레나 주스를 사용하지 못해 직접 만들어야 하므로 과일(원재료)의 상태가 매우 중요하다. 세그먼트하고 남은 오렌지는 착즙기로 즙을 내어 중량을 맞춰 준비한다. 대회에서는 음식물 쓰레기가 최소한으로 나오는 것이 중요하므로 패션 퓌레도 대회에서는 과일을 직접 착즙해 사용했다.

2 비커에 오렌지즙, 패션푸르트 퓌레, 미리 섞어둔 [설탕, NH펙틴, 트레할로스], 꿀, 생강즙, 오렌지에센스를 넣고 바믹서로 블렌딩한 후 냄비로 옮겨 펙틴 반응을 확인하며 가열한다.

3 펙틴 반응을 확인한 후 불에서 내려 블렌딩한 다음 바트에 담아 밀착 랩핑해 냉장고에 보관한다.

TIP. 사용하기 직전 푸드프로세서로 블렌딩한 다음 세그먼트한 오렌지 270g을 섞는다.

1 Segment the oranges and set them on a paper towel to drain.

TIP. Because you cannot use commercial fruit purée or juice in the competition, the condition of the fruit (raw ingredient) is critical. After segmenting the oranges, juice the leftovers with a juicer. It's essential to minimize food waste in competitions, so we also juiced the passion fruit purée ourselves.

2 In a beaker, blend orange juice, passion fruit purée, pre-mixed [sugar, pectin NH and trehalose], honey, ginger juice, and orange essence with an immersion blender. Transfer into a saucepan and cook while checking the reaction of pectin.

3 After checking on the pectin, remove from the heat. Blend and transfer into a tray and cover it with plastic wrap, make sure the wrap is in contact with the compote; refrigerate.

TIP. Blend with an immersion blender just before use and mix with 270 grams of orange segments.

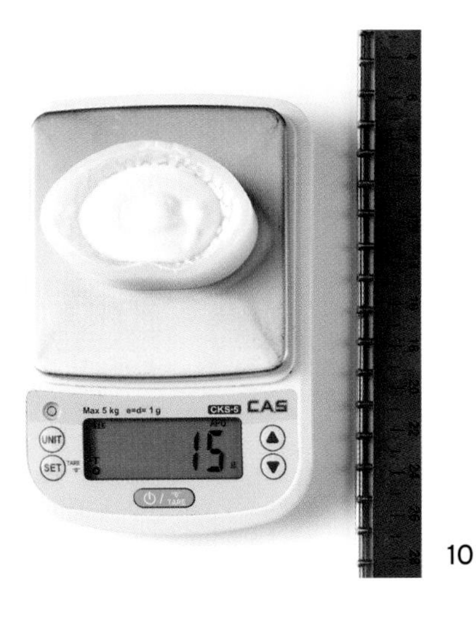

9

10

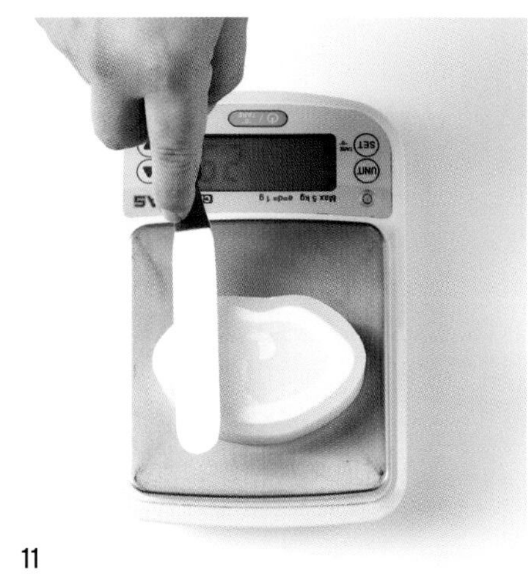
11

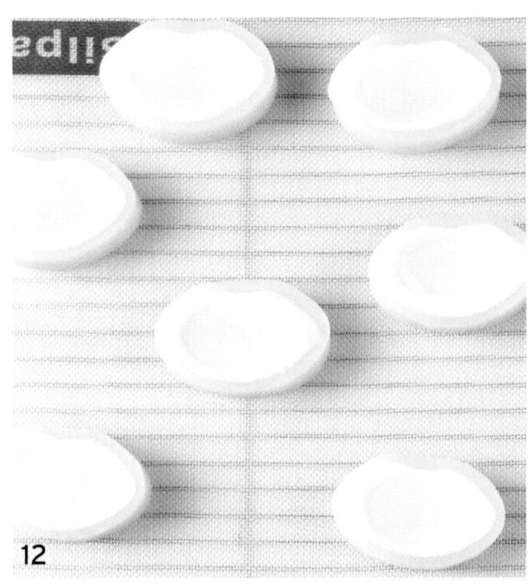
12

13-1

13-2

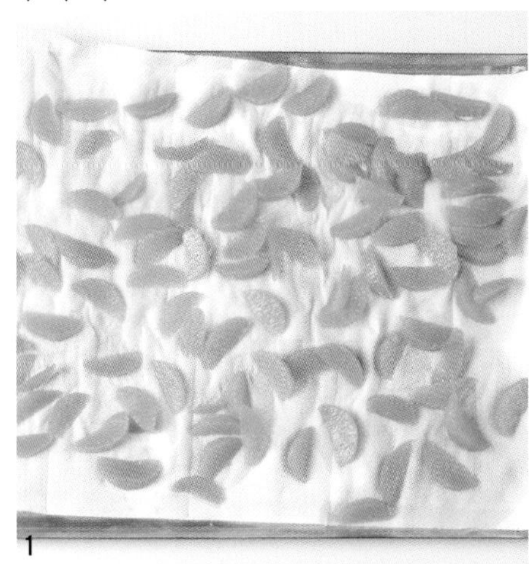
1

2

3

✤ ✤ ✤ ✤

HAZELNUT CARAMEL

헤이즐넛 캐러멜

1 예열한 냄비에 설탕, 물엿을 넣고 주걱으로 저어가며 가열한다.

TIP. 설탕은 세 번에 나눠 넣어가며 작업한다.

2 진한 갈색으로 캐러멜라이징되면 뜨겁게 데운 생크림을 두세 번 나눠 넣고 가열한다.

3 106℃가 되면 불에서 내린다.

4 헤이즐넛 페이스트, 소금을 넣고 주걱으로 섞는다.

5 얼음물이 담긴 볼에 받쳐 쿨링한 후 바믹서로 블렌딩한다.

6 바트에 담고 밀착 랩핑한 후 냉장고에 보관한다.

1 Heat sugar and corn syrup in a preheated saucepan while stirring with a spatula.

TIP. Add the sugar one-third at a time.

2 When it caramelizes to a deep brown color, add the hot heavy cream in two or three batches.

3 Remove from the heat when it reaches 106°C.

4 Stir in hazelnut paste and salt with a spatula.

5 Reduce the temperature over an ice bath and mix with an immersion blender.

6 Pour it into a tray and cover it with plastic wrap, make sure the wrap is in contact with the caramel; refrigerate.

1

2

3

4

5

6

✤ ✤ ✤ ✤ ✤

HWAYO DILL GEL

화요 딜 겔

1. 냄비에 레몬즙과 딜을 제외한 모든 재료를 넣은 후 알코올이 날아가고 겔화가 될 때까지 휘퍼로 저어가며 충분히 가열한다.

   TIP. 가열하기 전 딜과 레몬즙을 제외한 모든 재료를 비커에 담고 블렌딩해 사용한다. 브랜드별로 로거스트빈검 입자가 달라 가열하는 시간도 달라지므로, 항상 일정한 상태로 완성하기 위한 과정이다.
   가루 재료는 미리 통에 담아 섞어 준비한다.

2. 겔화 반응을 확인한 후 불에서 내린다.
3. 바트에 부어 밀착 랩핑한 후 냉장고에 보관한다.
4. 사용하기 직전 로보쿱에 레몬즙과 다진 딜을 넣고 갈아 사용한다.

1. In a saucepan, combine all the ingredients except lemon juice and dill. Heat sufficiently while stirring until the alcohol evaporates and the mixture jellifies.

   TIP. Blend all the ingredients in a beaker except dill and lemon juice before cooking. Different brands of locust bean gum have different particles and require different heating time. Therefore, it's important to be consistent.
   Prepare the powdered ingredients in advance by mixing them in a container.

2. Check for the gelling reaction and remove from the heat.
3. Pour it into a tray, cover tightly with plastic wrap, and refrigerate.
4. Just before using, blend with lemon juice and dill in the Robot Coupe.

✤ ✤ ✤ ✤ ✤ ✤

IAN CHOCOLATE MOUSSE

이안 초콜릿 무스

1. 냄비에 휘핑크림, 전화당, 졸인 쌍화탕(광동), 젤라틴매스를 넣고 젤라틴매스가 녹을 때까지 주걱으로 저어가며 50~60℃로 가열한다.
2. 다크초콜릿, 밀크초콜릿, 폴리페놀파우더가 담긴 비커에 **1**을 넣고 바믹서로 블렌딩한다.
3. 헤이즐넛 페이스트를 넣고 블렌딩한다.
4. 손으로 가볍게 휘핑한 생크림(60~70%)이 담긴 볼에 **3**을 두세 번 나눠 넣어가며 섞는다.

1. Heat whipping cream, inverted sugar, reduced Ssanghwacha (Gwangdong), and gelatin mass in a saucepan. Stir with a spatula until the gelatin mass melts while heating to 50~60°C.
2. Pour (**1**) into a beaker with dark chocolate, milk chocolate, and polyphenol powder; mix with an immersion blender.
3. Continue blending with hazelnut paste.
4. In a bowl with lightly hand-whipped whipping cream (60~70%), fold in (**3**) in two or three batches.

✤ ✤ ✤ ✤ ✤

1

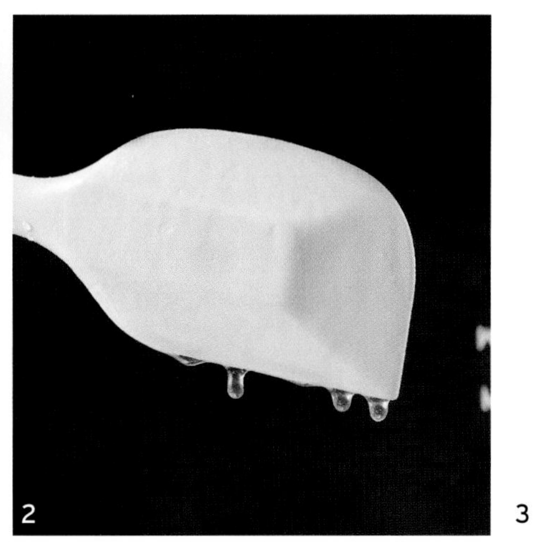
2

3

✤ ✤ ✤ ✤ ✤ ✤

1

2

3

4-1

4-2

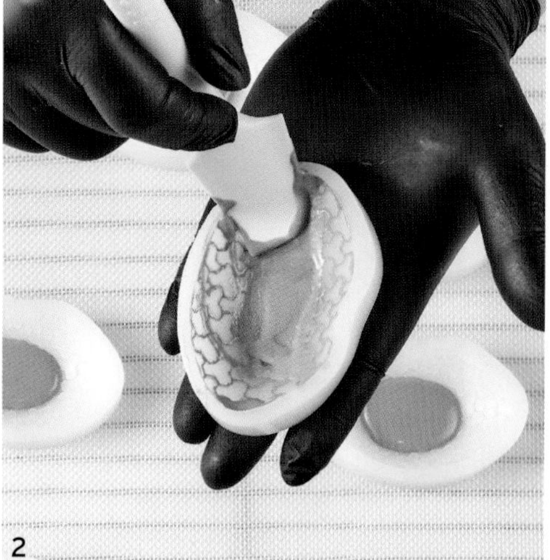

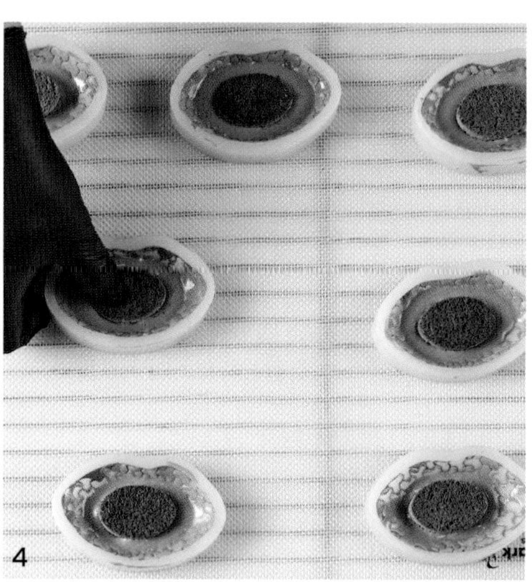

✤ ✤ ✤ ✤ ✤ ✤ ✤

FINISH 마무리

1 완성된 이안 초콜릿 무스를 몰드에 소량 파이핑한다.

2 몰드 무늬 사이사이에 무스가 잘 들어갈 수 있도록 미니 주걱으로 펴준다.

3 이안 초콜릿 무스를 약 5g 파이핑한다.

TIP. 무스를 넣고 스펀지를 넣어야 스펀지가 뜨지 않는다.

4 얼린 홍차 스펀지를 꾹 눌러 넣는다.

5 이안 초콜릿 무스를 파이핑한다.

6 윗면을 깔끔하게 정리한다. 급속 냉동(-35 ~ -40℃)한 후 몰드에서 빼내 냉동(-18℃) 보관한 다음 초콜릿 스프레이를 분사한다.

7 얼린 무스 위에 헤이즐넛 캐러멜 2g, 아몬드 크럼블 2g, 오렌지 콩포트 5g, 화요 딜 겔 15g을 올린다.
얼린 패션 요거트 무스를 덮고 화요 딜 겔을 파이핑한 후 허브와 금박을 올려 마무리한다.

7-1

7-2

1 Pipe a small amount of IAN chocolate mousse in the mold.

2 Spread the mousse with a mini spatula to fill in between the creases of the pattern.

3 Pipe about 5 grams of IAN chocolate mousse.

TIP. Insert the sponge into the mousse to prevent the sponge from floating.

4 Place the frozen black tea sponge and press into the mousse.

5 Pipe IAN chocolate mousse.

6 Flatten the top with a small offset spatula and blast freeze them (-30 ~ -40°C). Remove from the molds and store in the freezer (-18°C). Spray the completely frozen mousse with the chocolate spray mixture.

7 Top the frozen mousse with 2 grams of hazelnut caramel, 2 grams of almond crumble, 5 grams of orange compote, and 15 grams of Hwayo dill gel. Cover with the frozen passion yogurt mousse, pipe a small drop of dill gel, and finish with gold leaves.

2

INFINITE PAVLOVA (HEDONIST)

인피니트 파블로바 (쾌락주의자)

DESCRIPTION

특징

- 'Transform'이라는 대회 주제에 따라 한 가지 제품을 다양하게 표현하라는 과제를 받고 만든 메뉴. 나는 파블로바를 선택했고 쾌락주의자, 건강 추구자, 탐험가 세 가지 콘셉트로 제품을 구상했다.
- 쾌락주의를 생각했을 때 자극적이면서 만족감 높은 것이라 생각하여 'Summer Dessert'의 느낌으로 호불호 없는 재료들을 선택하여 디저트를 만들었다.
- 파블로바의 장점을 더 부각시키고 단점을 보완한 새로운 형태의 파블로바이다.
- 작은 알갱이 형태로 만들었기 때문에 머랭의 굽는 시간을 단축시킬 수 있다.
- 머랭을 초콜릿으로 코팅하여 보존 기간을 혁신적으로 늘렸다.
- 한국의 전통 간식인 '쌀 강정'에서 영감을 받아 머랭을 작은 알갱이로 만들어 초콜릿과 버무려 강정 형태로 만들었다.
- 사용한 틀은 다양하게 연출이 가능하다.
- 사용한 재료들을 더욱 돋보이게 하기 위해 가운데서 튀어나오는 듯한 느낌으로 연출했다.

- This menu was created in accordance with the competition's theme, the Transform. I selected pavlova and designed the product with three concepts: hedonist, health seeker, and explorer.
- When I think of hedonism, I think of something both stimulating and satisfying. With that in mind, I created a dessert with a 'Summer Dessert' concept, choosing ingredients with no likes or dislikes.
- It's a new type of pavlova that highlights its strengths and complements its weaknesses.
- The baking time of the meringue can be reduced thanks to its small size.
- The shelf life of meringue has been innovatively extended by coating it with chocolate.
- Inspired by a traditional Korean snack called 'rice gangjeong,' the meringue was made into small pellets and mixed with chocolate to form gangjeong.
- You can use the mold in various ways.
- To make the ingredients stand out even more, I created them as if they were protruding from the center.

✦ HOW TO COMPOSE THIS RECIPE ✦

STEP 1.	메뉴 정하기 Decide on the menu	파블로바 Pavlova
STEP 2.	메인 맛 정하기 Choose the primary flavor	코코넛 Coconut

STEP 3. 메인 맛(코코넛)과의 페어링 선택하기

Select a pairing flavor (Coconut)

- ☑ 코코넛 Coconut
- ☑ 오렌지 Orange
- ☑ 파인애플 Pineapple
- ☑ 패션푸르트 Passion Fruit
- ☑ 헤이즐넛 Hazelnut
- ☑ 밀크초콜릿 Milk Chocolate
- ☑ 캐러멜초콜릿 Caramel Chocolate

STEP 4. 구성하기

Assemble

❶ Cream 크림 — Coconut 코코넛 — Chocolate Mousse 초콜릿 무스

❷ Insert 인서트
- Pineapple, Passion Fruit 파인애플, 패션푸르트 — Compote 콩포트
- Hazelnut 헤이즐넛 — Caramel 캐러멜
- Milk & Caramel Chocolate 밀크 & 캐러멜 초콜릿 — Ganache 가나슈

❸ Crispy 크리스피
- Orange 오렌지
- Caramel Chocolate 캐러멜초콜릿

— Meringue 머랭 — Pavlova 파블로바

파인 패션 콩포트
Pine Passion Compote
인서트 가나슈
Insert Ganache
코코넛 무스
Coconut Mousse
제피르 캐러멜 머랭
Zephyr Caramel Meringue
헤이즐넛 캐러멜
Hazelnut Caramel

INGREDIENTS 20개 분량/ 20 cakes

머랭 베이스 *
Meringue Base

INGREDIENTS		g
흰자	Egg whites	380g
이소말트파우더	Isomalt powder	190g
설탕	Sugar	190g
폴리페놀파우더	Polyphenol powder	2g
알부민파우더	Albumin powder	13g
슈거파우더	Sugar powder	250g
탈지분유	Skim milk powder	40g
옥수수전분	Cornstarch	40g
TOTAL		**1105g**

제피르 캐러멜 머랭
Zephyr Caramel Meringue

INGREDIENTS		g
머랭 베이스 *	Meringue base *	200g
캐러멜초콜릿 (Zephyr Caramel 35%)	Caramel chocolate	105g
카카오버터	Cacao butter	66g
오렌지제스트	Orange zest	1개 분량/ 1 orange
TOTAL		**371g**

코코넛 무스
Coconut Mousse

INGREDIENTS		g
코코넛밀크	Coconut milk	190g
코코넛 퓌레	Coconut purée	65g
코코넛롱	Shredded dried coconut	60g
젤라틴매스 (×5)	Gelatin mass (×5)	45g
화이트초콜릿 (Zephyr white 34%)	White chocolate	120g
바닐라빈 씨	Vanilla bean seeds	1/4개 분량/ 1/4 pc
코코넛 럼 (Malibu)	Coconut rum	15g
생크림	Heavy cream	300g
TOTAL		**795g**

인서트 가나슈

Insert Ganache

INGREDIENTS		g
버터 (Bridel)	Butter	15g
생크림	Heavy cream	120g
밀크초콜릿 (Alunga 41%)	Milk chocolate	75g
캐러멜초콜릿 (Zephyr Caramel 35%)	Caramel chocolate	45g
TOTAL		**255g**

- 냄비에 버터와 생크림을 넣고 60℃로 가열한 후 초콜릿이 담긴 볼에 담아 바믹서로 블렌딩한다.
- Heat butter and heavy cream in a saucepan to 60°C. Pour it over the chocolates in a beaker and blend using an immersion blender.

헤이즐넛 캐러멜

Hazelnut Caramel

• p.446

INGREDIENTS		g
생크림	Heavy cream	420g
설탕	Sugar	160g
물엿	Corn syrup	260g
헤이즐넛 페이스트	Hazelnut paste	48g
소금	Salt	4g
TOTAL		**892g**

파인 패션 콩포트

Pine Passion Compote

INGREDIENTS		g
착즙한 파인애플 퓌레	Pineapple purée, juiced	170g
패션푸르트 퓌레	Passionfruit purée	10g
바닐라빈 껍질	Vanilla bean pod	1/5개 분량/ 1/5 pc
설탕	Sugar	70g
NH펙틴	NH pectin	7g
파인애플 (Dole)	Pineapple	300g
TOTAL		**557g**

- 파인애플 퓌레의 고형분 함량이 높아 착즙해 사용한다.
- Pineapple was juiced to use due to its high solids content.

✤

MERINGUE BASE

머랭 베이스

1 믹싱볼에 흰자, 이소말트파우더를 넣고 충분히 섞는다.

2 뜨거운 물이 담긴 볼 위에 **1**을 올리고 휘퍼로 저어가며 45℃까지 온도를 올린다.

3 45℃가 되면 설탕, 폴리페놀파우더, 알부민파우더를 두세 번 나눠 넣어가며 휘핑한다.

TIP. 사용할 설탕 양의 절반을 이소말트파우더로 대체하면서 당은 낮아지고, 식감은 더 바삭하며, 습도에도 강한 머랭으로 만들었다.

4 35℃가 되면 저속으로 휘핑해 기공을 정리한다.

5 체 친 [슈거파우더, 탈지분유, 옥수수전분]을 넣고 주걱으로 섞는다.

6 스패출러를 이용해 타공 매트에 고르게 펼친다.

TIP. 너무 많이 왔다갔다 하면 머랭이 사그라들 수 있으니 주의한다.

7 타공 매트를 제거한다.

8 120℃로 예열된 오븐에서 뎀퍼를 100%로 열고 30분간 구운 후 식혀 밀폐용기에 보관한다.

1 Thoroughly mix egg whites and isomalt powder in a mixing bowl.

2 Place (1) over hot water and whisk until it reaches 45°C.

3 When it reaches 45°C, whip it with sugar, polyphenol powder, and albumin powder in two or three batches.

TIP. By replacing half of the sugar with isomalt powder, I made the meringue with lower sugar content, crispier texture, and resistance to humidity.

4 At 35°C, whip at low speed to regulate the bubbles.

5 Mix with sifted [sugar powder, skim milk powder, and cornstarch] with a spatula.

6 Spread evenly on the perforated mat using an offset spatula.

TIP. Be careful not to go back and forth too much because the meringue may deflate.

7 Remove the perforated mat.

8 Bake in a deck oven preheated to 120°C with the damper opened at 100% for 30 minutes. Cool completely and keep them in an airtight container.

1

2

3

4

5

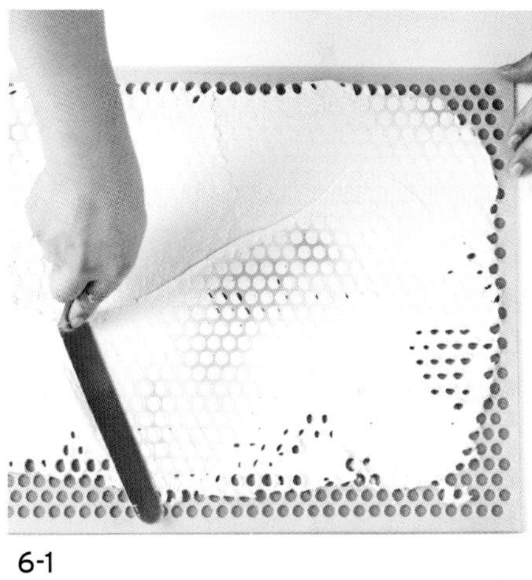
6-1

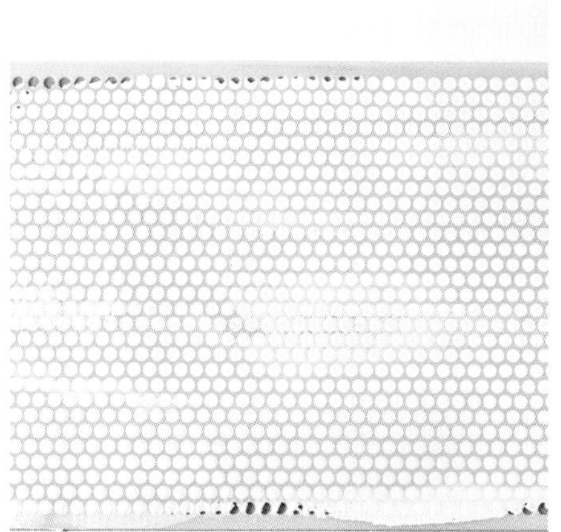
6-2

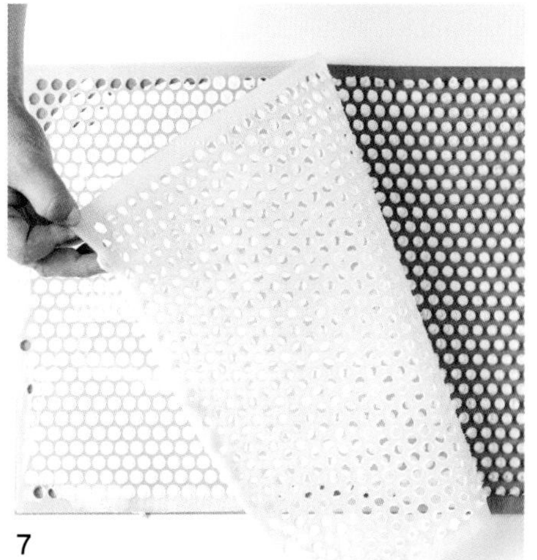
7

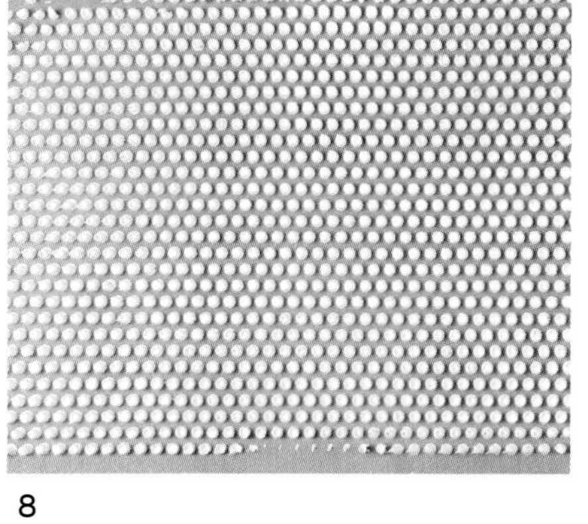
8

✤ ✤

ZEPHYR CARAMEL MERINGUE

제피르 캐러멜 머랭

1 머랭 베이스를 전자레인지에서 돌려 따뜻한 상태(35℃)로 준비한다.

2 볼에 녹인 캐러멜초콜릿(45~50℃)과 50℃ 이하로 녹인 카카오버터를 넣고 섞는다.

3 오렌지제스트를 넣고 섞는다.

4 **1**에 **3**을 넣고 섞은 후 온도를 확인한다. 만약 35℃ 보다 낮다면 전자레인지에서 온도를 35℃로 맞춰 사용한다.

5 35℃로 온도를 유지하며 몰드에 15g씩 채운다.

6 스패출러로 윗면을 평평하게 정리한다.

7 약 10분간 냉동(-18℃)한 후 몰드를 제거한다.

8 밀폐용기에 담아 습기가 없는 곳에서 보관한다.

TIP. 하루 종일 쇼케이스에 들어가 있어도 눅눅해지지 않는 강력한 머랭이다. 한 알 한 알은 단단하지만 작은 머랭 여러 개가 뭉쳐져 전체적인 식감은 바스러지는 느낌이다.

1 Microwave the meringue base warm to 35°C.

2 In a separate bowl, combine melted caramel chocolate(45~50°C) and cacao butter melted to below 50°C.

3 Mix with orange zest.

4 Add (**3**) into (**1**) and check the temperature. If it's below 35°C, microwave to adjust the temperature.

5 Maintain the temperature at 35°C and fill 15 grams in the molds.

6 Flatten the top with an offset spatula.

7 Freeze (-18°C) them for about 10 minutes and remove the molds.

8 Put them in an airtight container and store in a dry place.

TIP. It's a sturdy meringue that does not become soggy even if it's kept in the showcase all day. Each piece is solid, but overall, it gives a crumbly texture because many small meringues are clustered together.

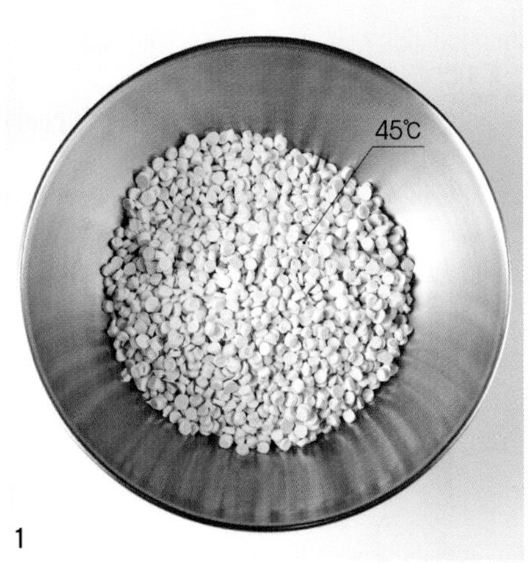

1

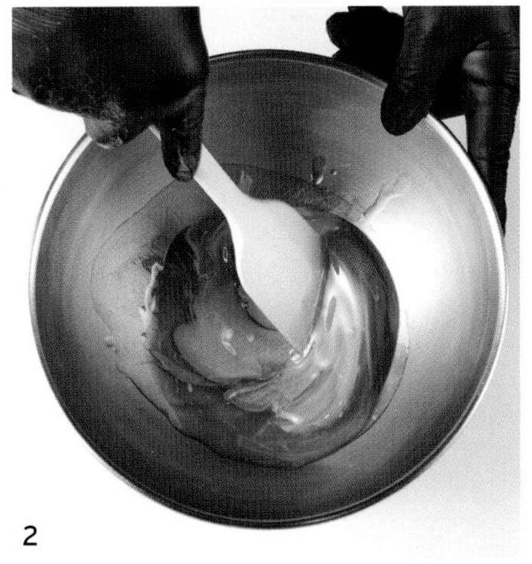

2

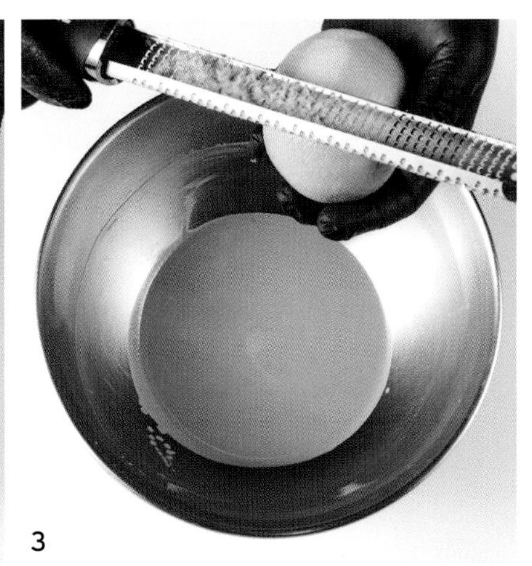

3

4-1

35℃ 유지
Maintain 35°C

4-2

5

6

7

8

✤ ✤ ✤

COCONUT MOUSSE

코코넛 무스

1 냄비에 코코넛밀크, 코코넛 퓌레, 코코넛롱을 넣고 가열한다.

TIP. 코코넛롱은 오븐에서 살짝 로스팅해 사용한다.

2 끓어오르면 불에서 내려 10분 동안 인퓨징한 후 200g으로 계량해 준비한다.

TIP. 만약 200g이 안되면 코코넛밀크를 추가해 200g으로 맞춰 사용한다.

3 다시 불에 올려 젤라틴매스를 넣고 녹을 때까지 가열한다. (약 60℃)

4 화이트초콜릿이 담긴 비커에 **3**과 바닐라빈 씨를 넣은 후 바믹서로 블렌딩한다.

5 코코넛 럼을 넣고 블렌딩한다.

6 손으로 가볍게 휘핑한 생크림(60~70%)을 준비한다.

TIP. 차가운 상태로 사용한다.

7 **6**에 **5**를 두세 번 나눠 넣어가며 섞는다.

8 몰드에 25g씩 채우고 미니 L자 스패출러로 윗면을 평평하게 정리한다.

9 급속 냉동(-35 ~ -40℃)한 후 몰드에서 빼내 냉동(-18℃) 보관한다.

1 Heat coconut milk, coconut purée, and shredded coconut.

TIP. Lightly roast the shredded coconut before use.

2 When it comes to a boil, remove from the heat and infuse for 10 minutes. Measure 200 grams and set aside.

TIP. If it's less than 200 grams, add coconut milk to adjust to 200 grams.

3 Heat it again with gelatin mass until it melts to about 60°C.

4 Blend (**3**) with white chocolate and vanilla bean seeds in a beaker using an immersion blender.

5 Blend with coconut rum.

6 In a separate bowl, lightly whip the cream (60~70%) with a whisk.

TIP. Use it cold.

7 Mix (**5**) into (**6**) in two or three batches.

8 Fill 25 grams in the mold and flatten the top using a small offset spatula.

9 Blast freeze (-35 ~ -40°C), remove from the mold and keep frozen (-18°C).

1
2
UNIT
SET
200
CAS
3
60℃
4
5
6
7
8
9

✤ ✤ ✤ ✤

PINE PASSION COMPOTE

파인 패션 콩포트

1 냄비에 착즙한 파인애플 퓌레, 패션푸르트 퓌레, 바닐라빈 껍질을 넣고 가열한다.

2 45℃가 되기 전에 미리 섞어둔 설탕과 NH펙틴을 넣고 주걱으로 저어가며 가열한다.

TIP. 1의 액체 재료, 설탕과 NH펙틴은 미리 블렌딩해 사용할 수 있다.

3 펙틴 반응을 확인한 후 불에서 내려 쿨리를 완성한다.

4 완성된 쿨리 절반은 밀착 랩핑해 냉장 보관한다.

5 남은 쿨리의 절반은 파인애플과 함께 냄비에서 주걱으로 저어가며 가열한다.

6 파인애플이 투명하게 졸아들면 바트에 담아 밀착 랩핑해 냉장 보관한다.

7 냉장고에 보관한 파인애플을 사방 1cm로 자른다.

8 남은 쿨리와 함께 버무린다.

1 Heat fresh squeezed pineapple purée, passion fruit purée, and vanilla bean pod (without the seeds).

2 Add previously mixed sugar and pectin NH before it reaches 45°C and stir to cook.

TIP. You can blend the liquid ingredients from (1), sugar, and pectin NH in advance.

3 Remove from the heat when the pectin activates.

4 Pour half the finished coulis onto a stainless-steel tray, adhere the plastic wrap to the preparation to cover, and refrigerate.

5 Heat half of the remaining coulis with pineapple while stirring with a spatula.

6 When it's reduced and the pineapple becomes transparent, pour it onto a stainless-steel tray, adhere the plastic wrap to the preparation to cover, and refrigerate.

7 Cut the refrigerated pineapple mixture into 1 cm cubes.

8 Toss with the remaining coulis.

1

2

3

4

5

6

7

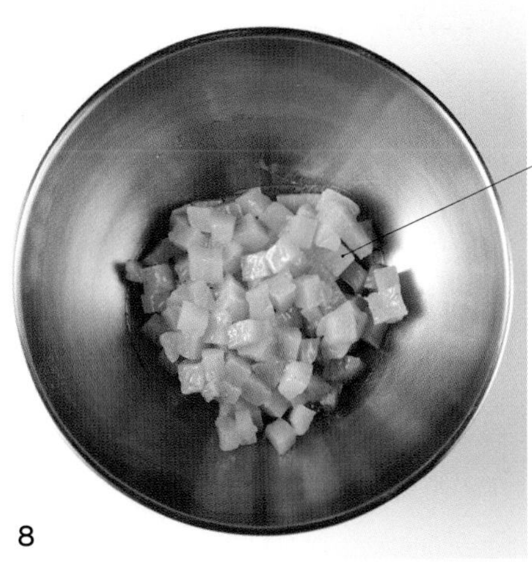

8

✤ ✤ ✤ ✤ ✤

FINISH 마무리

제피르 캐러멜 머랭 위에 얼린 코코넛 무스를 얹고, 그 안에 인서트 가나슈 12g - 헤이즐넛 캐러멜 2g - 파인 패션 쿨리 5g - 파인 패션 콩포트 18g을 순서대로 채우고 라임제스트를 뿌린 후 금박과 스테비아로 장식해 마무리한다.

Place the frozen coconut mousse on the Zephyr caramel meringue. Fill in the order of 12 grams insert ganache – 2 grams hazelnut caramel – 5 grams pine passion coulis – 18 grams pine passion compote. Sprinkle lime zest, decorate with gold flakes and stevia leaves to finish.

NOTE.

3

INFINITE PAVLOVA (HEALTHSEEKER)

인피니트 파블로바 (건강 추구자)

DESCRIPTION

특징

- 건강하게 만들면 그만큼 맛은 떨어질 것이라는 생각을 뒤집었다. 폴리페놀 파우더를 가지고 콜레스테롤 수치를 낮추는 방법을 선택하여 메뉴를 개발하였다.
- 설탕의 일부를 대체당으로 바꿔 단맛을 줄이고, 크리스피한 식감은 살렸다.
- 파블로바의 장점을 더 부각시키고 단점을 보완한 새로운 형태의 파블로바이다.
- 작은 알갱이 형태로 만들었기 때문에 머랭의 굽는 시간을 단축시킬 수 있다.
- 머랭을 초콜릿으로 코팅하여 보존 기간을 혁신적으로 늘렸다.
- 한국의 전통 간식인 '쌀 강정'에서 영감을 받아 머랭을 작은 알갱이로 만들어 초콜릿과 버무려 강정 형태로 만들었다.
- 사용한 틀은 다양하게 연출이 가능하다.
- 사용한 재료들을 더욱 돋보이게 하기 위해 가운데서 튀어나오는 듯한 느낌으로 연출했다.

- This menu overturns the idea that healthy food doesn't taste good. It was developed by selecting the method to lower cholesterol levels using polyphenol powder.
- Some of the sugar was replaced with an alternative sugar to reduce sweetness and preserve the crispiness.
- It's a new type of pavlova that highlights its strengths and complements its weaknesses.
- The baking time of the meringue can be reduced thanks to its small size.
- The shelf life of meringue has been innovatively extended by coating it with chocolate.
- Inspired by a traditional Korean snack called 'rice gangjeong,' the meringue was made into small pellets and mixed with chocolate to form gangjeong.
- You can use the mold in various ways.
- To make the ingredients stand out even more, I created them as if they were protruding from the center.

✦ HOW TO COMPOSE THIS RECIPE ✦

STEP 1.	메뉴 정하기 Decide on the menu	파블로바 Pavlova
STEP 2.	메인 맛 정하기 Choose the primary flavor	메밀 Buckwheat

STEP 3. 메인 맛(메밀)과의 페어링 선택하기

Select a pairing flavor (Buckwheat)

- ☑ 유자 Yuja
- ☑ 아몬드 Almond
- ☑ 헤이즐넛 Hazelnut
- ☑ 검정깨 Black Sesame
- ☑ 다크초콜릿 Dark Chocolate
- ☑ 밀크초콜릿 Milk Chocolate

STEP 4. 구성하기

Assemble

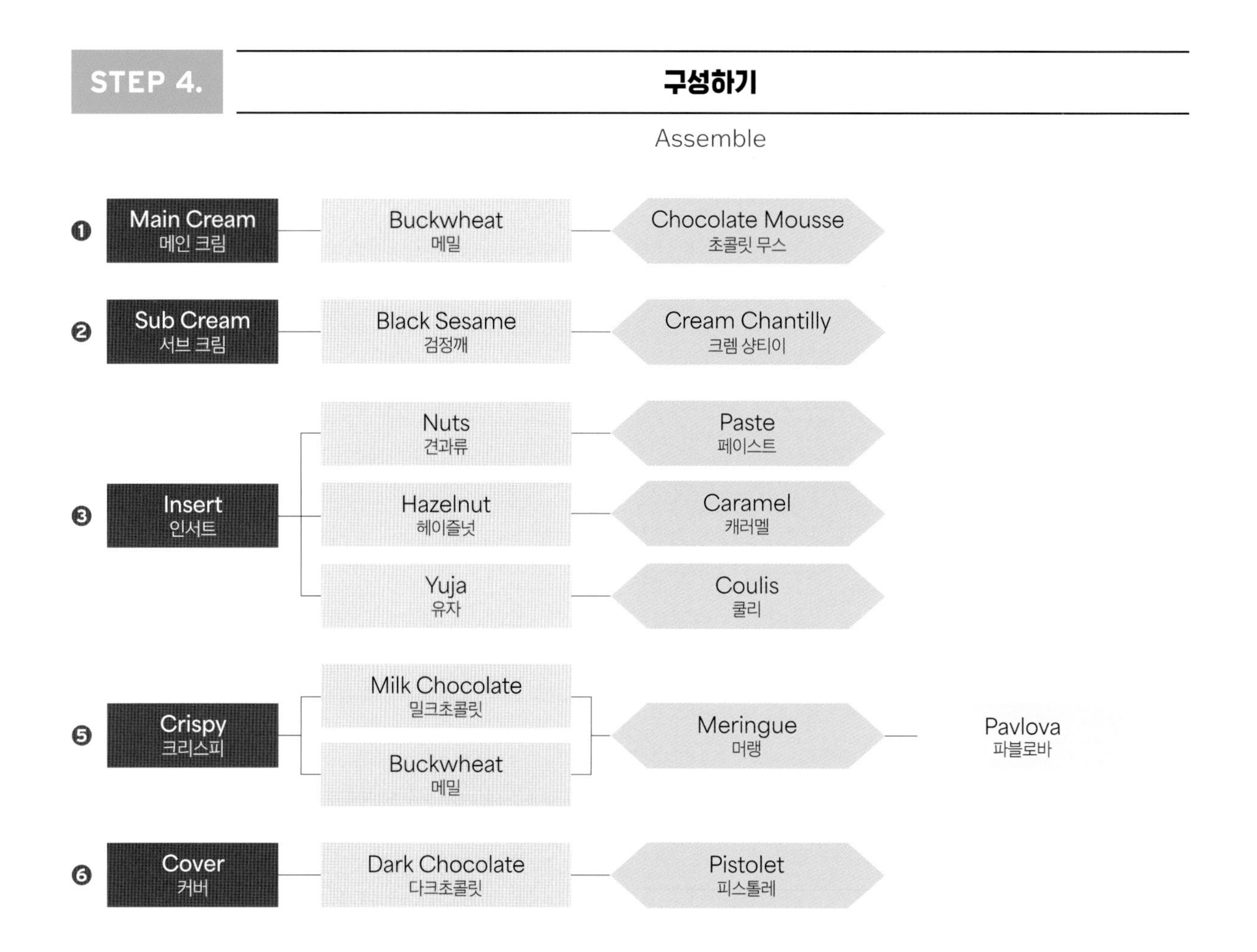

검정깨 샹티이 크림
Black Sesame Chantilly Cream
유자 쿨리
Yuja Coulis
메밀 무스
Buckwheat Mousse
검정깨 너츠 페이스트
Black Sesame Nuts Paste
가나 머랭
Ghana Meringue
헤이즐넛 캐러멜
Hazelnut Caramel

INGREDIENTS
20개 분량/ 20 cakes

머랭 베이스 *
Meringue Base

INGREDIENTS		g
흰자	Egg whites	380g
이소말트파우더	Isomalt Powder	190g
설탕	Sugar	190g
알부민파우더	Albumin powder	13g
슈거파우더	Sugar powder	250g
탈지분유	Skim milk powder	40g
옥수수전분	Cornstarch	40g
폴리페놀파우더	Polyphenol powder	2g
TOTAL		**1105g**

가나 머랭
Ghana Meringue

INGREDIENTS		g
머랭 베이스 *	Meringue base *	160g
밀크초콜릿 (Ghana 40%)	Milk chocolate	60g
카카오버터	Cacao butter	30g
카카오매스	Cacao mass	15g
소바차파우더	Powdered buckwheat tea	1g
소바차 (홀)	Whole buckwheat tea	10g
TOTAL		**276g**

- 소바차파우더는 로스팅한 소바차를 곱게 갈아 사용한다.
- p.458~461를 참고한다.
- Finely grind roasted buckwheat tea to make the buckwheat tea powder.
- Refer to p.458~461.

메밀 무스

Buckwheat Mousse

INGREDIENTS		g
생크림A	Heavy cream A	150g
소바차파우더	Powdered buckwheat tea	15g
우유	Milk	60g
전화당	Inverted sugar	20g
설탕	Sugar	45g
젤라틴매스 (×5)	Gelatin mass (×5)	24g
밀크초콜릿 (Ghana 40%)	Milk chocolate	30g
다크초콜릿 (Ian 71%)	Dark chocolate	70g
생크림B	Heavy cream B	240g
폴리페놀파우더	Polyphenol powder	0.5g
TOTAL		**654.5g**

* 만드는 법

❶ 냄비에 생크림A, 소바차우더, 우유, 전화당, 설탕을 넣고 가열한다.
❷ 끓어오르면 불에서 내려 젤라틴매스를 넣고 섞는다.
❸ 체에 걸러 밀크초콜릿, 다크초콜릿, 폴리페놀파우더가 담긴 비커에 넣고 바믹서로 블렌딩한다.
❹ 차가운 상태의 믹싱볼에 생크림B를 담고 손으로 가볍게 휘핑한다.
❺ ❹에 ❸을 붓고 휘퍼로 섞는다.
❻ 소스건에 담아 몰드에 25g씩 깔끔하게 몰딩한다.
❼ 급속냉동고(-35 ~ -40℃)에 넣고 얼린다.
(30분 얼린 후 뒤집어 다시 얼린다.)
❽ 차가운 상태의 바트에 옮긴 후 냉동(-18℃) 보관한다.

* Procedure

❶ Heat heavy cream A, buckwheat tea powder, milk, inverted sugar, and sugar.
❷ When it comes to a boil, remove from the heat and mix with gelatin mass.
❸ Strain into a beaker with milk chocolate, dark chocolate, and polyphenol powder; blend using an immersion blender.
❹ Lightly whip the heavy cream in a cold mixing bowl with a whisk.
❺ Pour ❸ into ❹ and combine with a whisk.
❻ Use a sauce gun to fill 25 grams in the mold.
❼ Blast freeze (-35 ~ -40°C). (Freeze for 30 minutes, turn them over and freeze again.)
❽ Transfer onto a cold stainless-steel tray and freeze (-18°C).

헤이즐넛 캐러멜

Hazelnut Caramel

• p.446

INGREDIENTS		g
설탕	Sugar	160g
물엿	Corn syrup	260g
생크림	Heavy cream	420g
헤이즐넛 페이스트	Hazelnut paste	48g
소금	Salt	4g
TOTAL		**892g**

유자 쿨리

Yuja Coulis

INGREDIENTS		g
유자 퓌레	Yuja purée	60g
TPT 시럽	TPT syrup	150g
설탕	Sugar	70g
NH펙틴	NH pectin	6g
젤라틴매스 (×5)	Gelatin mass (×5)	20g
TOTAL		**306g**

- TPT 시럽 = 물과 설탕을 1:1 비율로 끓여 만든 것으로, 식혀 사용한다.
- TPT syrup: A syrup made by boiling water and sugar in a 1:1 ratio. Use after cooling.

＊만드는 법

❶ 비커에 유자 퓌레, TPT시럽, 미리 섞어둔 설탕과 NH펙틴을 넣고 블렌딩한다.
❷ 냄비에 옮겨 가열한다.
❸ 펙틴 반응을 확인한 후 불에서 내려 젤라틴매스를 넣고 섞는다.
❹ 얼음물이 담긴 볼에 받쳐 쿨링한 후 바믹서로 블렌딩한다.
❺ 바트에 담아 밀착 랩핑하고 냉장 보관한 후 짤주머니에 담아 사용한다.

＊Procedure

❶ Blend yuja purée, TPT syrup, and previously mixed sugar and pectin NH in a beaker using an immersion blender.
❷ Pour into a pot and heat.
❸ After the pectin activates, remove from the heat and mix with gelatin mass.
❹ Cool in an ice bath and blend using an immersion blender.
❺ Pour it onto a stainless-steel tray, adhere the plastic wrap to the preparation to cover, and refrigerate. Use in a piping bag.

검정깨 너츠 페이스트

Black Sesame Nuts Paste

INGREDIENTS		g
검정깨 페이스트	Black sesame paste	100g
헤이즐넛 페이스트	Hazelnut paste	50g
아몬드 페이스트	Almond paste	50g
TOTAL		**200g**

- 믹서에 모든 재료를 넣고 블렌딩한 후 짤주머니에 담아 사용한다.
- Blend all the ingredients in a mixer and use in a piping bag.

검정깨 샹티이 크림

Black Sesame Chantilly Cream

INGREDIENTS		g
생크림	Heavy cream	300g
연유	Condensed milk	18g
설탕	Sugar	18g
검정깨 프랄린	Black sesame praline	50g
TOTAL		**386g**

*** 검정깨 프랄린 만들기**

❶ 설탕을 냄비에 넣고 캐러멜화시킨 후 테프론시트 위에 펼쳐 굳힌다.

❷ 푸드프로세서에 ❶과 검정깨를 넣고 곱게 갈아 사용한다.

- 설탕과 검정깨는 1:1 비율로 사용한다.

*** Black Sesame Praline**

❶ Caramelize sugar in a pot and spread over a Teflon sheet to set.

❷ Finely grind it with black sesame seeds to use.

- Use sugar and black sesame seeds in a 1:1 ratio.

다크 스프레이

Dark Spray

INGREDIENTS		g
카카오버터	Cacao butter	100g
다크초콜릿 (Mi-Amère 58%)	Dark chocolate	100g
TOTAL		**200g**

- 볼에 모든 재료를 넣고 녹여 바믹서로 블렌딩한 후 50℃로 맞춰 사용한다.
- Melt all the ingredients in a bowl, combine with an immersion blender, and use at 50°C.

4

INFINITE PAVLOVA (EXPLORER)

인피니트 파블로바 (탐험가)

DESCRIPTION

특징

- 도전을 즐기는 사람들을 탐험가라고 이해했다. 재미있는 도전을 생각해봤는데 한국의 매운맛을 가지고 해외 사람들이 챌린지하는 것을 보고 아이디어를 얻어 매운맛 제품을 만들었고 매운맛을 내기 위해 스모크 파프리카, 카이엔페퍼를 사용했다.
- 파블로바의 장점을 더 부각시키고 단점을 보완한 새로운 형태의 파블로바이다.
- 클래식한 스위스 머랭을 사용해 다른 두 가지 제품(건강추구, 쾌락주의)과 머랭의 식감에 차별을 두었다.
- 라즈베리 스프레이를 분사하여 맛, 색감, 코팅을 통해 습기에 강하게 만들었다.
- 사용한 재료들을 더욱 돋보이게 하기 위해 가운데서 튀어 나오는 듯한 느낌으로 연출하였다.
- 매운맛이 폭발하는 듯한 느낌을 받을 수 있도록 디자인하였다.

- I understand that those who enjoy challenges are explorers. I was thinking of a fun challenge, and after seeing people around the world challenging themselves with Korean spicy flavors, I got the idea to create a spicy dessert. I used smoked paprika and cayenne pepper to make it spicy.
- It's a new type of pavlova that highlights its strengths and complements its weaknesses.
- A classic Swiss meringue was used to differentiate the textures from the HeathSeeker and the Hedonist.
- The shelf life of meringue has been innovatively extended by coating it with raspberry spray.
- To make the ingredients stand out even more, I created them as if they were protruding from the center.
- I designed it to make it feel like an explosion of spicy flavor.

✦ HOW TO COMPOSE THIS RECIPE ✦

STEP 1.	메뉴 정하기 Decide on the menu	파블로바 Pavlova
STEP 2.	메인 맛 정하기 Choose the primary flavor	라즈베리 Raspberry

STEP 3. 메인 맛(라즈베리)과의 페어링 선택하기

Select a pairing flavor (Raspberry)

☑ 라즈베리	Raspberry	☑ 마스카르포네	Mascarpone	☑ 스모크 파프리카	Smoked Paprika
☑ 딸기	Strawberry	☑ 사워크림	Sour Cream	☑ 카이엔페퍼	Cayenne Pepper
☑ 바질	Basil	☑ 화이트초콜릿	White Chocolate	☑ 발사믹식초	Balsamic Vinegar
		☑ 밀크초콜릿	Milk Chocolate	☑ 올리브오일	Olive Oil

STEP 4. 구성하기

Assemble

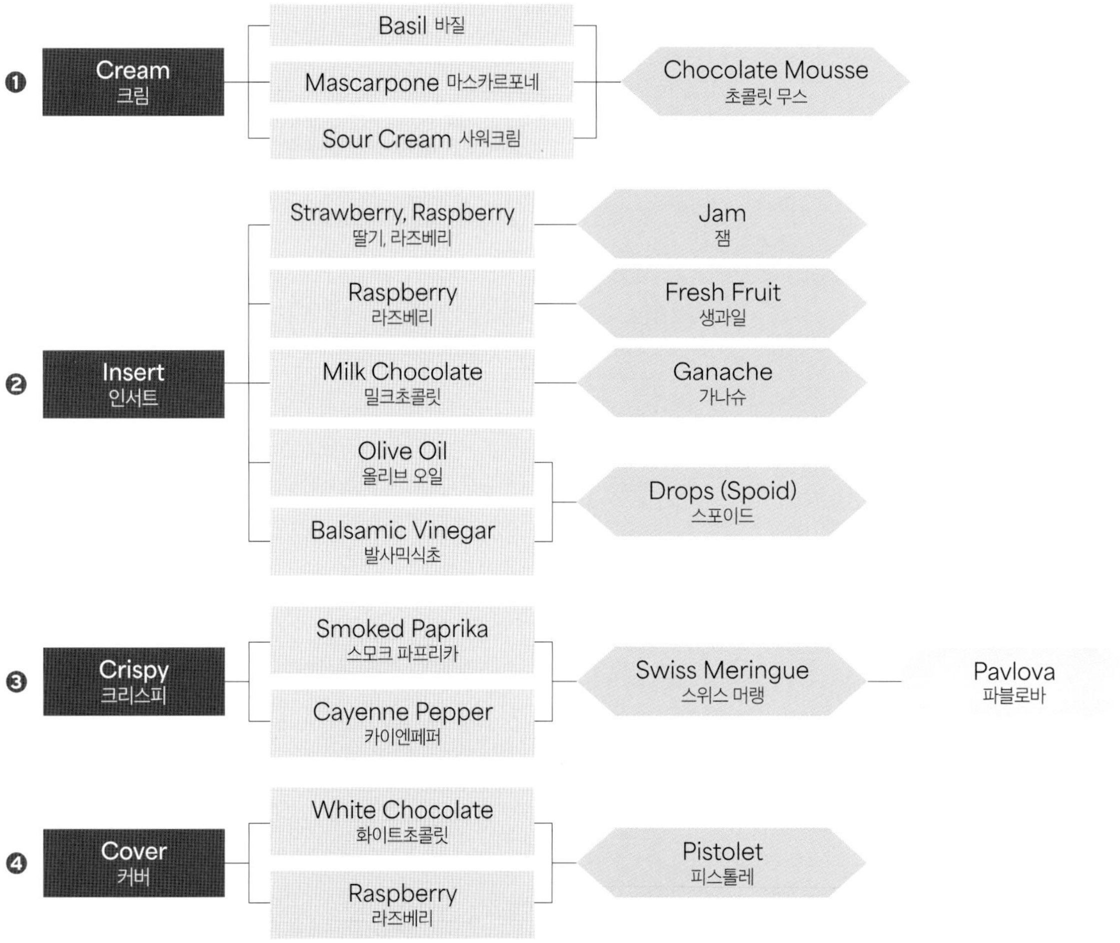

❶ Cream 크림
- Basil 바질 / Mascarpone 마스카르포네 / Sour Cream 사워크림 → Chocolate Mousse 초콜릿 무스

❷ Insert 인서트
- Strawberry, Raspberry 딸기, 라즈베리 → Jam 잼
- Raspberry 라즈베리 → Fresh Fruit 생과일
- Milk Chocolate 밀크초콜릿 → Ganache 가나슈
- Olive Oil 올리브 오일 / Balsamic Vinegar 발사믹식초 → Drops (Spoid) 스포이드

❸ Crispy 크리스피
- Smoked Paprika 스모크 파프리카 / Cayenne Pepper 카이엔페퍼 → Swiss Meringue 스위스 머랭 → Pavlova 파블로바

❹ Cover 커버
- White Chocolate 화이트초콜릿 / Raspberry 라즈베리 → Pistolet 피스톨레

발사믹식초
Balsamic Vinegar
라즈베리 & 라즈베리 시럽
Raspberry & Raspberry Syrup
올리브오일
Olive Oil
바질 사워크림 무스
Basil Sour Cream
Mousse
라즈베리 스프레이
aspberry Spray
스파이시 스위스 머랭
Spicy Swiss Meringue
인서트 가나슈
Insert Ganache
딸기잼
Strawberry Jam

INGREDIENTS 20개 분량/ 20 cakes

스파이시 스위스 머랭

Spicy Swiss Meringue

INGREDIENTS		g
흰자	Egg whites	150g
이소말트파우더	Isomalt powder	150g
슈거파우더	Sugar powder	150g
스모크 파프리카파우더	Smoked paprika powder	0.6g
카이엔페퍼파우더	Cayenne pepper powder	0.6g
TOTAL		**451.2g**

라즈베리 스프레이

Raspberry Spray

INGREDIENTS		g
화이트초콜릿 (Zephyr white 34%)	White chocolate	200g
카카오버터	Cacao butter	200g
동결건조 라즈베리파우더	Freeze-dried raspberry powder	5g
TOTAL		**405g**

- 동결건조 라즈베리파우더는 동결건조 라즈베리를 곱게 갈아 사용했다.
- 볼에 화이트초콜릿과 카카오버터를 넣고 녹여 바믹서로 블렌딩한 후 동결건조 라즈베리파우더를 섞어 50℃로 맞춰 사용한다.
- Finely grind freeze-dried raspberries to make the powder.
- Melt white chocolate and cacao butter in a bowl and combine using an immersion blender. Mix with freeze-dried raspberries and use at 50°C.

바질 사워크림 무스

Basil Sour Cream Mousse

INGREDIENTS		g
우유	Milk	60g
생크림A	Heavy cream A	120g
바질	Basil	3g
젤라틴매스 (×5)	Gelatin mass (×5)	60g
화이트초콜릿 (Zephyr white 34%)	White chocolate	160g
마스카르포네	Mascarpone cheese	100g
사워크림	Sour cream	30g
생크림B	Heavy cream B	260g
TOTAL		**793g**

딸기잼
Strawberry Jam

INGREDIENTS		g
딸기 퓌레	Strawberry purée	300g
라즈베리 퓌레	Raspberry purée	30g
설탕	Sugar	170g
NH펙틴	NH pectin	8g
젤라틴매스 (×5)	Gelatin mass (×5)	24g
라임즙	Lime juice	15g
카이엔페퍼파우더	Cayenne pepper	0.5g
TOTAL		**547.5g**

인서트 가나슈
Insert Ganache

INGREDIENTS		g
버터 (Bridel)	Butter	15g
생크림	Heavy cream	135g
밀크초콜릿 (Alunga 41%)	Milk chocolate	75g
다크초콜릿 (Mi-Amère 58%)	Dark chocolate	15g
캐러멜초콜릿 (Zephyr Caramel 35%)	Caramel chocolate	45g
TOTAL		**285g**

- 냄비에 버터와 생크림을 넣고 60℃로 가열한 후 초콜릿이 담긴 볼에 담아 바믹서로 블렌딩한다.
- Heat butter and cream to 60°C in a saucepan. Pour it into a beaker with the chocolates and blend using an immersion blender.

라즈베리 시럽
Raspberry Syrup

INGREDIENTS		g
라즈베리 퓌레	Raspberry purée	20g
데코젤 뉴트럴	Decorgel neutral	200g
30° 시럽	30˚B syrup	40g
TOTAL		**260g**

- 모든 재료를 함께 섞어 사용한다.
- Combine all the ingredients to use.

✤

SPICY SWISS MERINGUE

스파이시 스위스 머랭

1 뜨거운 물이 담긴 냄비에 받친 믹싱볼에 흰자, 이소말트파우더를 넣고 50℃가 될 때까지 휘퍼로 저어가며 가열한다.

2 믹싱볼에 옮겨 30℃까지 휘핑한다.

3 체 친 슈거파우더와 스모크 파프리카파우더를 넣고 섞는다.

4 오일 스프레이를 뿌린 실리콘 판 몰드(높이 0.5cm)를 테프론시트 위에 두고 가장자리부터 가운데까지 파이핑한다.

5 L자 스패출러로 평평하게 정리한다.

6 몰드를 제거한 후 120℃로 예열된 오븐에서 뎀퍼를 100%로 열고 30~35분간 굽는다.

7 오븐에서 나오자마자 라즈베리 스프레이를 분사하고 굳혀 밀폐용기에 보관한다.

1 Put egg whites and isomalt powder in a mixing bowl. Place it over hot water and whisk until it reaches 50°C.

2 Whip until it becomes 30°C.

3 Mix with sifted sugar powder and smoked paprika powder.

4 Place a greased flat silicon mold (0.5 cm in height) on a Teflon sheet and pipe from the edge to the center.

5 Flatten with an offset spatula.

6 Remove the mold. Bake in an oven preheated to 120°C with the damper opened at 100% for 30~35 minutes.

7 As soon as it's out of the oven, spray with raspberry spray mixture and store in an airtight container after it's set.

1

2

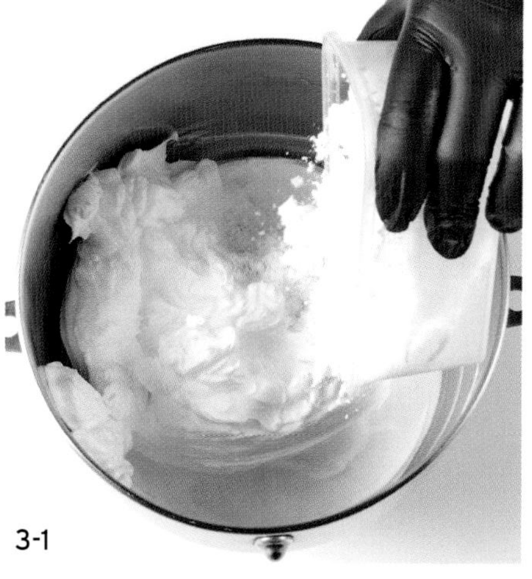
3-1

3-2

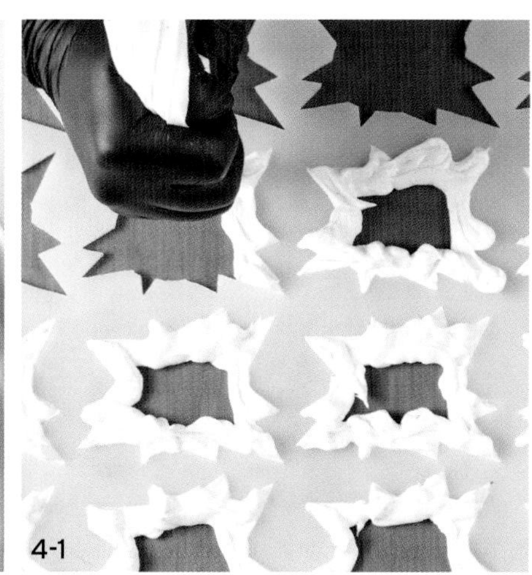
4-1

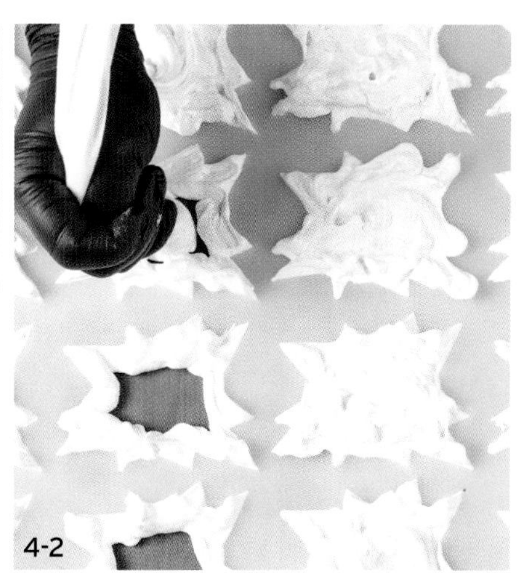
4-2

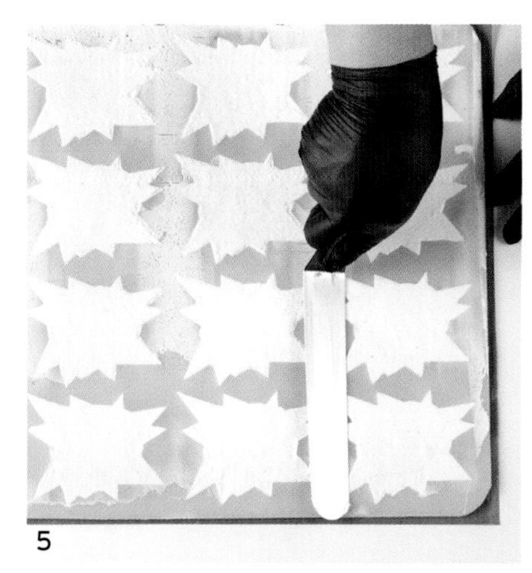
5

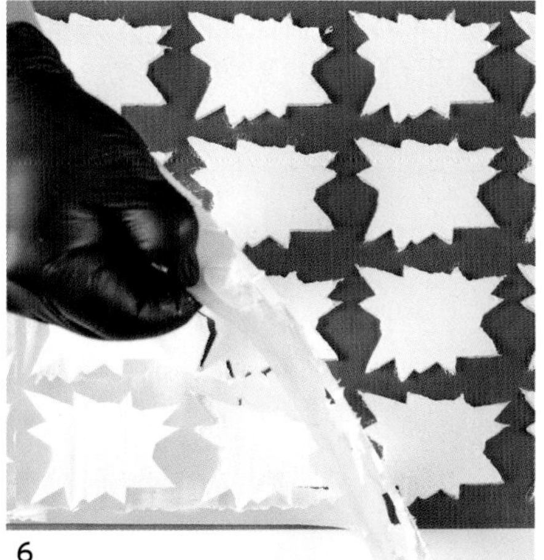
6

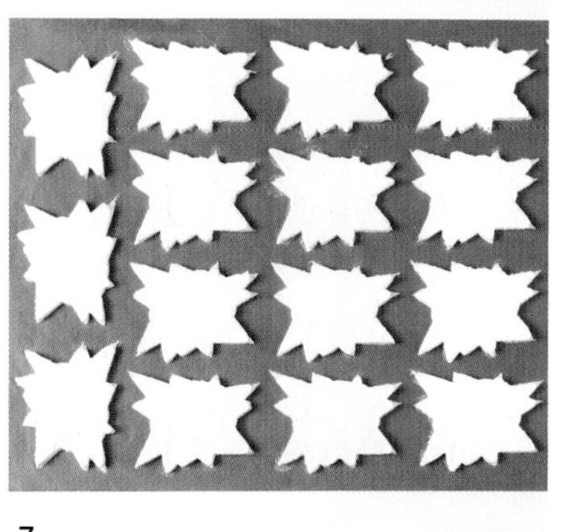
7

✤ ✤

BASIL SOUR CREAM MOUSSE

바질 사워크림 무스

1 냄비에 우유, 생크림A, 바질, 젤라틴매스를 넣고 젤라틴매스가 녹을 때까지 가열한다. (65~70℃)

2 불에서 내려 냄비 입구를 랩핑해 10분 동안 인퓨징한다.

3 화이트초콜릿이 담긴 비커에 담는다.

TIP. 바질이 걸러질 수 있도록 체에 내린다.

4 바믹서로 블렌딩한다.

5 마스카르포네, 사워크림을 넣고 블렌딩한다.

6 볼에 생크림B를 넣고 손으로 가볍게 60~70% 정도로 휘핑한다.

7 차가운 상태의 **6**에 **5**를 두세 번 나눠 넣어가며 섞는다.

8 소스건을 이용해 몰드에 25g씩 채운다.

9 L자 스패출러로 윗면을 평평하게 정리하고 급속 냉동(-35 ~ -40℃)한 후 몰드에서 빼내 냉동(-18℃) 보관한다.

1 Heat milk, heavy cream A, basil, and gelatin mass until the gelatin melts (65~70°C).

2 Remove from the heat, cover the pot, and infuse for 10 minutes.

3 Pour into a beaker with white chocolate.

TIP. Pour over a sieve to strain the basil.

4 Combine with an immersion blender.

5 Blend with mascarpone and sour cream.

6 In a separate bowl, lightly whip heavy cream B with a whisk.

7 Add (**5**) in two or three batches into the cold (**6**) and mix.

8 Use a sauce gun to fill 25 grams in the mold.

9 Organize the top with an offset spatula, blast freeze (-35 ~ -40°C), remove from the mold, and freeze (-18°C).

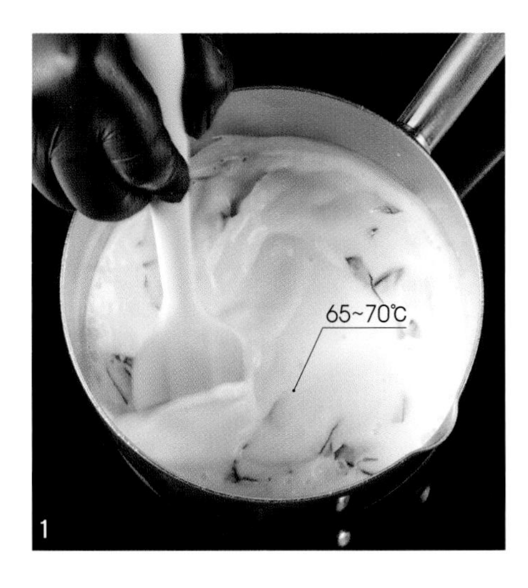

1

2

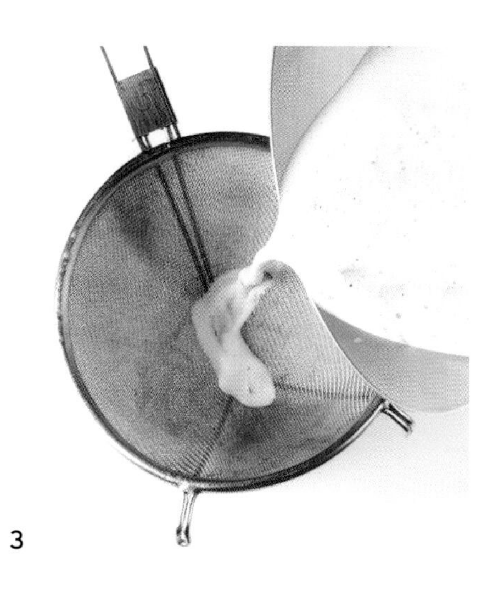
3

4

5

6

7

8

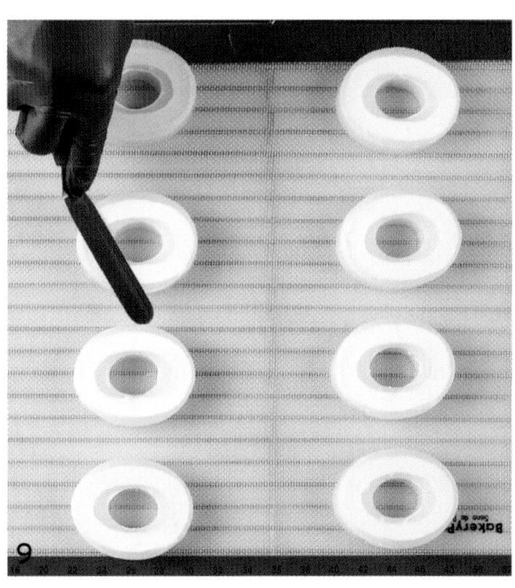
9

STRAWBERRY JAM

딸기잼

1 냄비에 딸기 퓌레, 라즈베리 퓌레를 넣고 가열하다가 45℃가 되기 전에 미리 섞어둔 설탕과 NH펙틴을 넣고 휘퍼로 섞어가며 가열한다.

2 펙틴 반응이 일어나면 젤라틴매스가 녹을 때까지 주걱으로 저어가며 가열한다.

3 불에서 내려 라임즙과 카이엔페퍼파우더를 넣고 섞는다.

4 점도를 확인한 후 바트에 담아 밀착 랩핑해 냉장고에 보관한다.

1 Heat strawberry purée and raspberry purée. Mix with previously mixed sugar and pectin NH before it reaches 45°C, stirring with a whisk.

2 When the pectin activates, add gelatin mass and stir to cook until it melts.

3 Remove from the heat and mix with lime juice and cayenne pepper.

4 Check the viscosity, pour onto a stainless-steel tray, adhere the plastic wrap to the preparation to cover, and refrigerate.

FINISH

마무리

1 바질 사워크림 무스 위에 카이엔페퍼파우더를 살짝 체 친다.

2 바질 사워크림 무스 안쪽에 인서트 가나슈 12g을 동그랗게 파이핑하고 가운데에 딸기잼을 10g 채운다.

3 라즈베리 5개를 올리고 딸기잼을 소량 파이핑한 후 올리브오일과 발사믹 식초를 스포이드로 소량 뿌린다.

4 라즈베리 시럽을 붓으로 바른 후 허브를 올려 마무리한다.

1 Lightly sift cayenne pepper on the basil sour cream mousse.

2 Pipe 12 grams of insert ganache in a circle inside the basil sour cream mousse and fill the center with 10 grams of strawberry jam.

3 Put five raspberries, pipe a small amount of strawberry jam, and drizzle some olive oil and balsamic vinegar using the spoids.

4 Brush with raspberry syrup and top with herbs to finish.

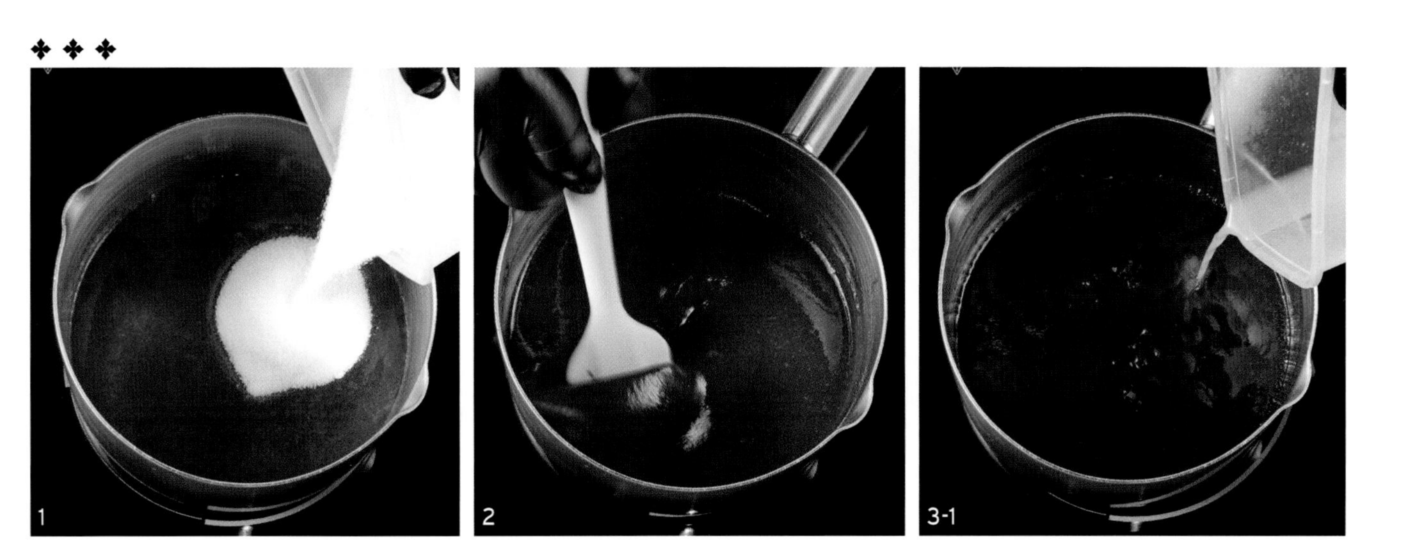
1
2
3-1

3-2
4

FOR OUR TOMORROW

1

SOYBEAN PULP FINANCIER

콩비지 피낭시에

- 2022 WCM 대회의 주제였던 'Tomorrow(내일, 미래)'와 관련된 디저트를 개발하던 중 아이디어가 떠올라 만들게 된 메뉴이다.
- 리사이클링(재활용)이 가능한 재료 중 콩비지를 선택했고, 요리의 재료로 사용되고 남은 콩비지 찌꺼기를 건조시켜 가루 형태로 만들어 피낭시에에 사용했다.
- 아몬드파우더를 콩비지 가루로 100% 대체해 만든 메뉴이다.

- The idea for this menu came up while we were developing a dessert related to the theme of the 2022 WCM (World Chocolate Masters) competition, 'Tomorrow (future).'
- We chose soybeans as a recyclable ingredient, so we dried and powdered the remaining soybean pulp residue to use in the financier.
- Almond powder is replaced 100% with the soybean pulp powder in this menu.

업사이클링 VS 리사이클링

업사이클링은 사용된 제품을 새로운 제품 또는 기존 대비 더 높은 가치를 가진 제품으로 재가공하거나 재활용하는 과정을 의미한다.

→ 창의성이 요구되며, 자원의 효율적인 활용이 강조된다.

리사이클링은 사용된 제품 또는 재료를 수집하여 이를 다시 가공하고, 이러한 과정을 통해 새로운 제품이나 재료를 재생산하는 과정을 의미한다.

→ 자원의 절약과 환경 보호가 강조된다.

UPCYCLING VS RECYCLING

Upcycling is a process of reprocessing or recycling used products into new or higher-value products.

→ It requires creativity and emphasizes the efficient utilization of resources.

Recycling is a process of collecting used products or materials, reprocessing them, and using them to create new products or materials.

→ Conservation of resources and environmental protection are emphasized.

✦ HOW TO COMPOSE THIS RECIPE ✦

STEP 1.	메뉴 정하기	넥스트 피낭시에
	Decide on the menu	NXT(Next) Financier

STEP 2.	메인 맛 정하기	콩비지
	Choose the primary flavor	Soybean Pulp

STEP 3. **메인 맛(콩비지)과의 페어링 선택하기**

Select a pairing flavor (Soybean Pulp)

- ☑ 아몬드 Almond
- ☑ 조청 Grain Syrup
- ☑ 누아제트 버터 Noisette Butter

STEP 4. **구성하기**

Assemble

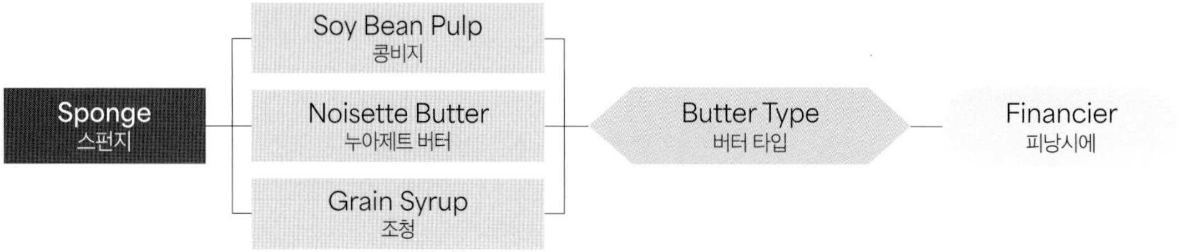

Sponge 스펀지
- Soy Bean Pulp 콩비지
- Noisette Butter 누아제트 버터
- Grain Syrup 조청

→ Butter Type 버터 타입 → Financier 피낭시에

피낭시에를 만들 때 일반적으로 사용하는 아몬드파우더 대신 콩비지가루로 만든 피낭시에.
아몬드에서는 느낄 수 없는 콩 특유의 구수한 맛을 씹을수록 더 진하게 느낄 수 있다.

Instead of almond flour, which is commonly used to make financiers, this financier is made with soybean pulp powder. The more you chew, the more you can taste the savory flavor of the beans, which is different from almonds.

INGREDIENTS 37개 분량/ 37 financiers

콩비지 피낭시에

Soybean Pulp Financier

INGREDIENTS		g
콩비지 가루	Soybean pulp powder	150g
흰자	Egg whites	386g
조청	Grain syrup	180g
T55밀가루	T55 flour	156g
황설탕	Brown sugar	412g
누아제트 버터	Noisette butter	390~400g
소금	Salt	5g
베이킹파우더	Baking powder	4g
TOTAL		**1683~1693g**

캐러멜라이즈 너츠

Caramelized Nuts

INGREDIENTS		g
물	Water	450g
물엿	Corn syrup	650g
설탕	Sugar	550g
견과류	Nuts	500g
TOTAL		**2150g**

*** 만드는 법**

❶ 냄비에 물, 물엿, 설탕을 넣고 강불에서 가열한다.

❷ 확실하게 끓어오르는 게 보이면 견과류를 넣고 중불로 낮춰 주걱으로 저어가며 가열한다.

❸ 시럽이 견과류 표면에 고르게 코팅되고 다시 한번 확실하게 끓어오르면, 이때부터 2분간 저어가며 추가로 가열한다.

❹ 테프론시트를 깐 철판에 고르게 펼친 후, 150℃로 예열된 오븐에서 17분간 굽는다.

- 견과류의 색을 확인하면서 굽는 시간을 조절한다. (견과류 표면이 하얗게 되지 않도록 주의한다.)
- 재사용할 경우 남은 양의 10% 물을 추가한 후 재가열해 사용한다.

*** Procedure**

❶ Heat water, corn syrup, and sugar in a saucepan over high heat.

❷ When it comes to a solid boil, add the nuts, reduce to medium-low heat, and cook while continuing to stir.

❸ Once the syrup has evenly coated the nuts and is once again at a solid boil, cook for another 2 minutes while stirring.

❹ Spread evenly on a baking tray lined with a Teflon sheet and bake in an oven preheated to 150°C for 17 minutes.

- Check the color of the nuts and adjust the baking time. (Be careful the surface of the nuts doesn't turn white.)
- To reuse, add 10% water to the remaining amount and reheat before using.

NOTE.

✤

SOYBEAN PULP FINANCIER

콩비지 피낭시에

1 콩비지는 유산지를 깐 철판에 펼쳐 100℃로 예열된 오븐에서 뎀퍼를 100% 열고 옅은 갈색이 날 때까지 중간중간 섞어가며 구운 후, 믹서에 곱게 갈고 체에 내려 가루 상태로 만들어 사용한다.

TIP. 30분간 굽고 섞은 후, 다시 30분간 구워 구움색을 고르게 낸다.
콩비지의 양에 따라 굽는 시간을 조절한다. 수분감이 없는 완벽한 가루 상태가 되도록 굽는다.

2 비커에 흰자, 조청을 넣고 바믹서로 블렌딩한 후 냉장고에서 하루 동안 숙성시킨다.

3 믹싱볼에 체 친 콩비지 가루와 T55밀가루, 황설탕을 넣고 고르게 섞이도록 믹싱한 후 30℃로 온도를 맞춘 **2**를 흘려 넣으면서 믹싱한다.

TIP. 콩비지 가루와 T55밀가루는 작업 전 50℃ 오븐에 넣어두고 온도가 50℃가 되면 사용한다.

4 누아제트 버터(50℃)를 흘려 넣으면서 저속으로 믹싱한다.

TIP. 버터는 미리 태워 390~400g으로 계량해 사용한다.

5 소금, 베이킹파우더를 넣고 믹싱한다. (최종 반죽 온도 30~32℃)

6 버터를 칠한 보트 모양 틀에 45g씩 반죽을 채운다.

TIP. 여기에서는 가로 11cm, 세로 5.5cm, 높이 2cm 보트 모양 틀을 사용했다.

7 반으로 자른 캐러멜라이즈 넛츠를 5개씩 올린다.

TIP. 여기에서는 아몬드를 사용했다. 취향에 따라 다른 견과류로 대체해도 좋다.

8 180℃로 예열된 오븐에서 17분간 굽는다. (박텔 오븐 기준 바람 세기 3)

1 Spread the soybean pulp on a parchment-lined baking sheet and bake in an oven preheated to 100°C with the damper open at 100% until lightly browned, stirring occasionally. Then, grind them into a food processor to a fine powder and sieve.

TIP. Bake for 30 minutes, mix around, and bake for another 30 minutes to even out the color. Adjust baking time according to the amount of soybean pulp. You want the powder to be dehydrated with no moisture.

2 In a beaker, blend egg whites and grain syrup with an immersion blender and refrigerate overnight.

3 In a mixing bowl, beat to combine the sifted soybean pulp powder, T55 flour, and brown sugar thoroughly. Slowly pour in (**2**), set to 30°C, while mixing.

TIP. Place the soybean pulp powder and T55 flour in a 50°C oven ahead of time and use when the powders are 50°C.

4 Mix on low speed while drizzling in the noisette butter (50°C).

TIP. Cook the butter ahead of time and measure out 390~400 grams.

5 Mix in salt and baking powder. (Final batter temperature: 30~32°C)

6 Fill the butter-coated boat-shaped mold with 45 grams of batter.

TIP. I used boat-shaped molds of W 11 cm × H 5.5 cm × D 2 cm in size.

7 Place five halved caramelized nuts on top.

TIP. I used almonds. You can substitute different nuts to your taste.

8 Bake in an oven preheated to 180°C for 17 minutes. (Fan speed 3 for Bechtel oven)

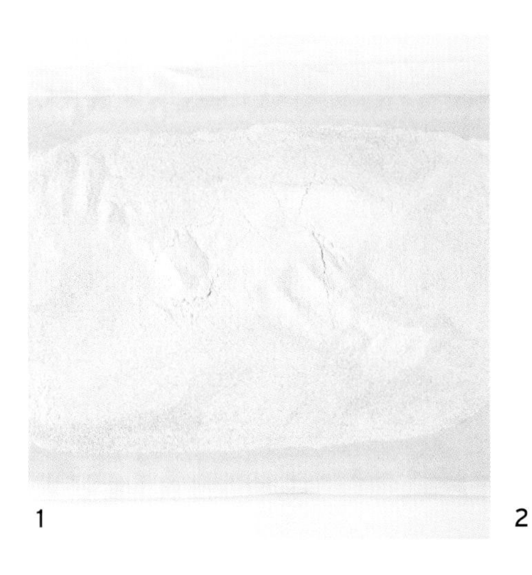
1

2

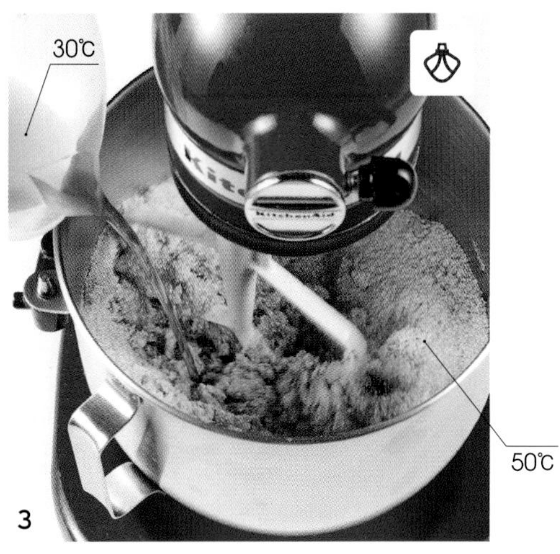
30℃
50℃
3

50℃
4

5-1

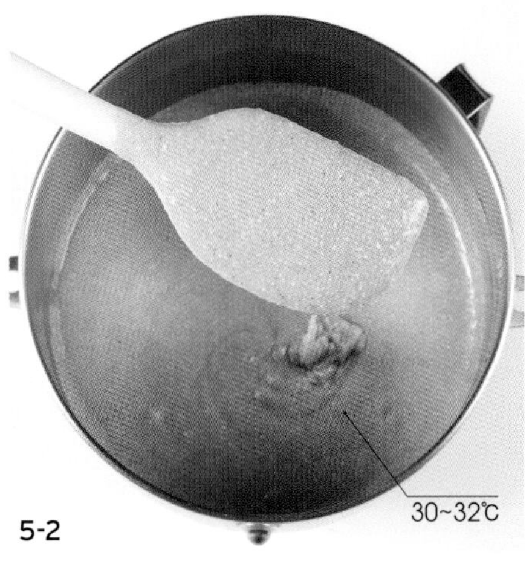
30~32℃
5-2

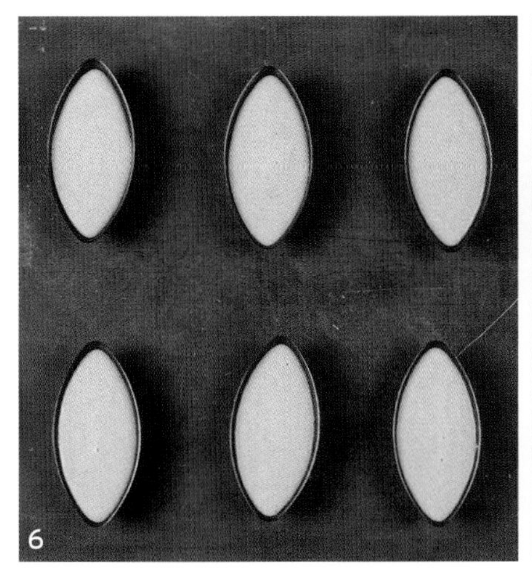
6

7

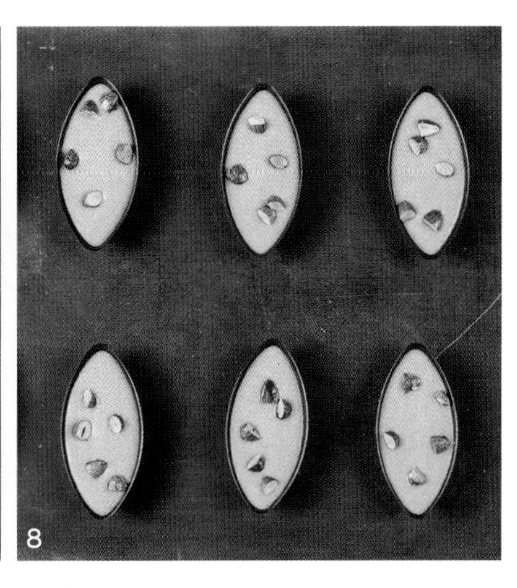
8

2 SOYBEAN PULP MADELEINE

콩비지 마들렌

- 콩비지 피낭시에와 마찬가지로 2022 WCM 대회의 주제였던 'Tomorrow(내일, 미래)'와 관련된 디저트를 개발하던 중 아이디어가 떠올라 만들게 된 메뉴이다.

- 콩비지 피낭시에와 마찬가지로 아몬드파우더를 콩비지 가루로 100% 대체해 만든 메뉴이다.

- 마들렌 윗면에 토피를 사용해 더욱 먹음직스러운 비주얼과 바삭한 식감을 더했다.

- As with the Soybean Pulp Financier, the idea for this menu came up while we were developing a dessert related to the theme of the 2022 WCM (World Chocolate Masters) competition, 'Tomorrow (future).'

- As with the Soybean Pulp Financier, the almond powder is replaced 100% with the soybean pulp powder.

- I used toffee on top of the madeleines to give it a more appetizing look and crunchy texture.

✦ HOW TO COMPOSE THIS RECIPE ✦

STEP 1.	메뉴 정하기	넥스트 마들렌
	Decide on the menu	NXT(Next) Madeleine

STEP 2.	메인 맛 정하기	콩비지
	Choose the primary flavor	Soybean Pulp

STEP 3. **메인 맛(콩비지)과의 페어링 선택하기**

Select a pairing flavor (Soybean Pulp)

☑	캐러멜초콜릿	Caramel Chocolate
☑	피칸	Pecan
☑	정제버터	Clarified Butter

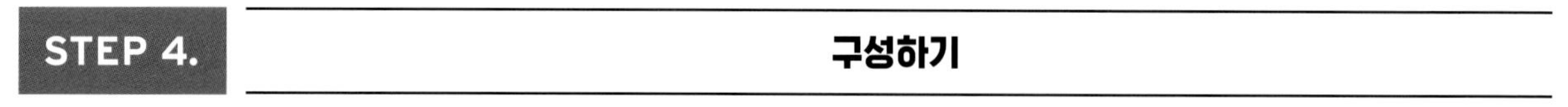

STEP 4. **구성하기**

Assemble

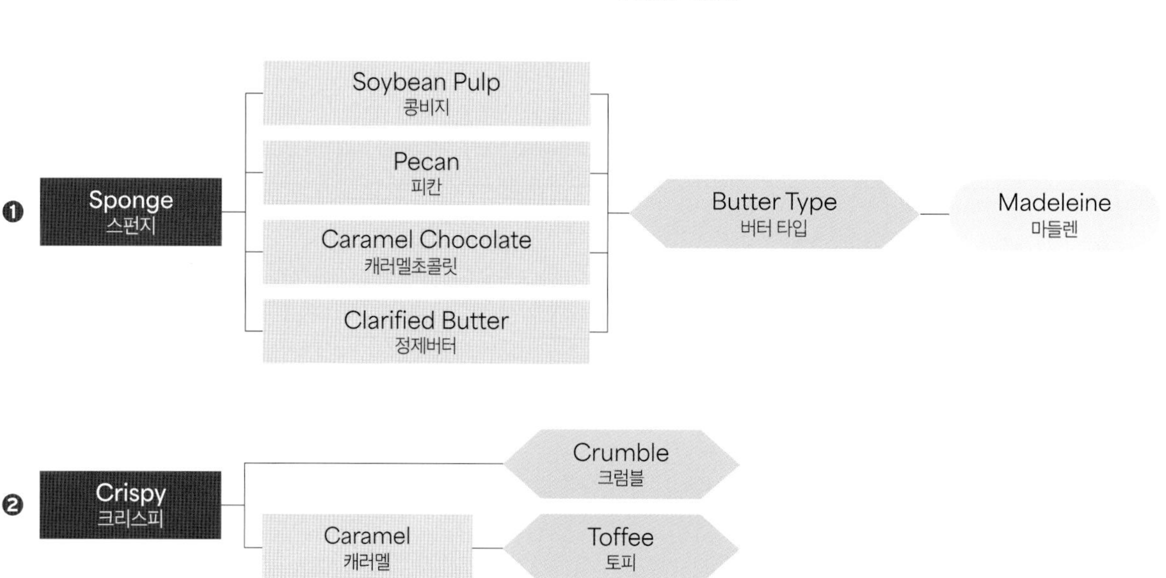

토피
Toffee
마들렌
Madeleine
플레인 크럼블
Plain Crumble

INGREDIENTS

87개 분량/ 87 madeleines

토피

Toffee

INGREDIENTS		g
설탕	Sugar	300g
버터 (Bridel)	Butter	30g
TOTAL		**330g**

콩비지 마들렌

Soybean Pulp Madeleine

INGREDIENTS		g
콩비지 가루	Soybean pulp powder	150g
달걀	Eggs	376g
설탕	Sugar	270g
꿀	Honey	50g
전화당	Inverted sugar	80g
캐러멜초콜릿 (Zephyr Caramel 35%)	Caramel chocolate	290g
우유	Milk	80g
피칸 페이스트	Pecan paste	50g
바닐라에센스 (Aroma Piu)	Vanilla essence	10g
중력분	All-purpose flour	360g
식용유	Vegetable oil	100g
정제버터	Clarified butter	430g
베이킹파우더	Baking powder	13g
소금	Salt	3g
TOTAL		**2262g**

플레인 크럼블

Plain Crumble

• p.388

INGREDIENTS		g
버터 (Bridel)	Butter	360g
코코넛슈거	Coconut sugar	200g
설탕	Sugar	100g
소금	Salt	8g
박력분	Cake flour	600g
옥수수전분	Cornstarch	60g
물	Water	36g
TOTAL		**1364g**

NOTE.

✤

TOFFEE

토피

1 냄비를 불에 올려 예열한 후 설탕을 나눠 넣어가며 캐러멜화시킨다.

2 잔거품이 생기기 시작하면 포마드 상태의 버터를 넣고 휘퍼로 저어가며 녹인다.

3 버터가 녹으면 테프론시트를 깐 차가운 대리석 바닥에 붓는다.

4 토피가 식고 완전히 굳으면 사용하기 적당한 크기로 자르거나 믹서에 곱게 갈아 사용한다.

TIP. 식으면서 점점 투명한 상태로 변한다.

1 Preheat a saucepan and add sugar in several batches to caramelize.

2 When small bubbles begin to form, add softened butter and stir with a whisk to melt.

3 Once the butter melts, pour the toffee onto a cold marble plate lined with a Teflon sheet.

4 When the toffee has cooled and completely hardened, cut it into desired-sized pieces or finely grind it in a food processor to use.

TIP. It will become transparent as it cools.

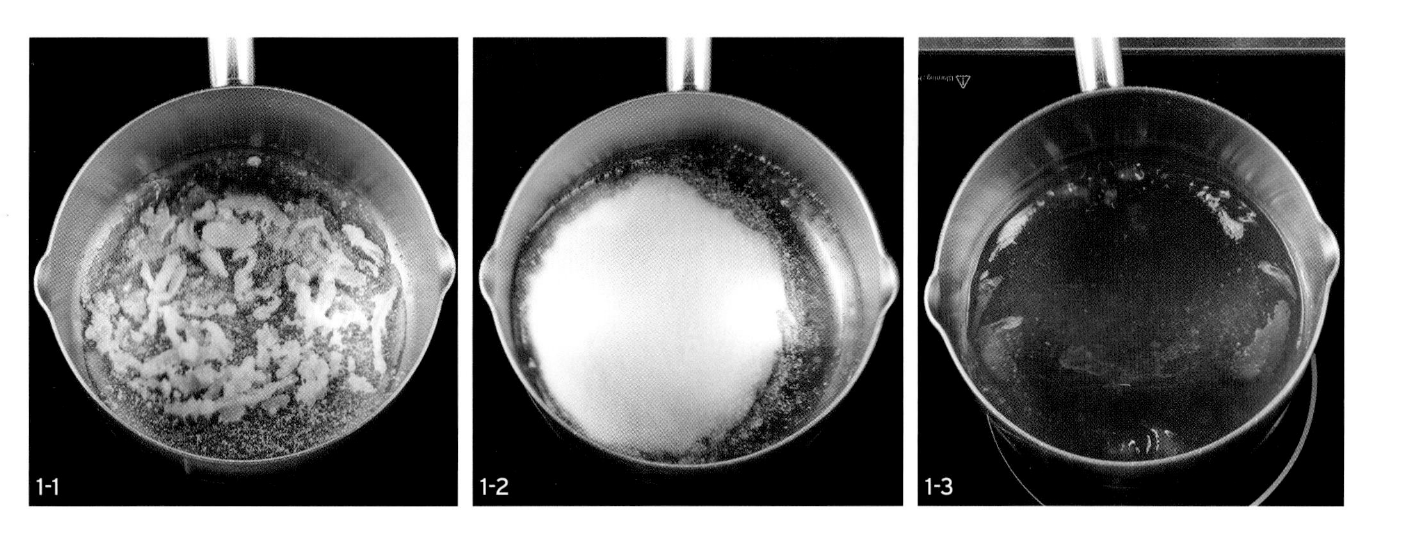
1-1
1-2
1-3

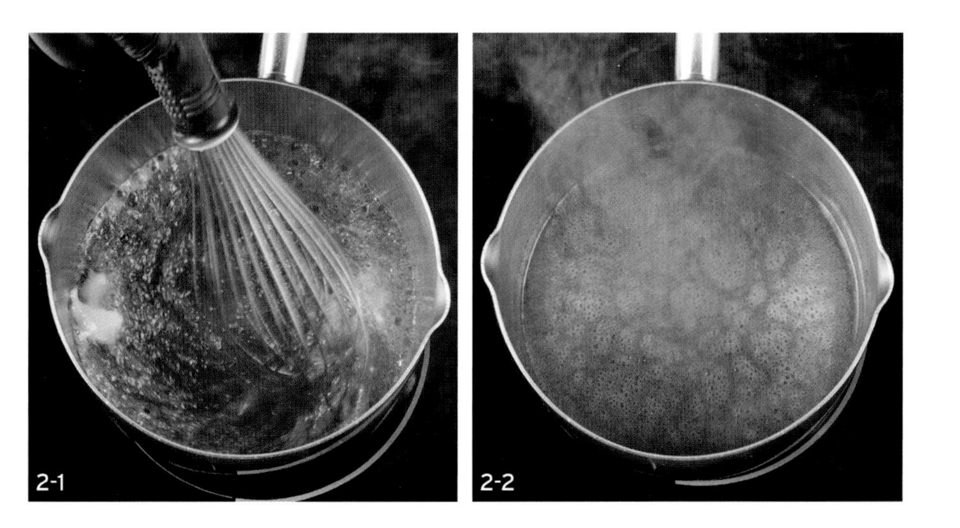
2-1
2-2

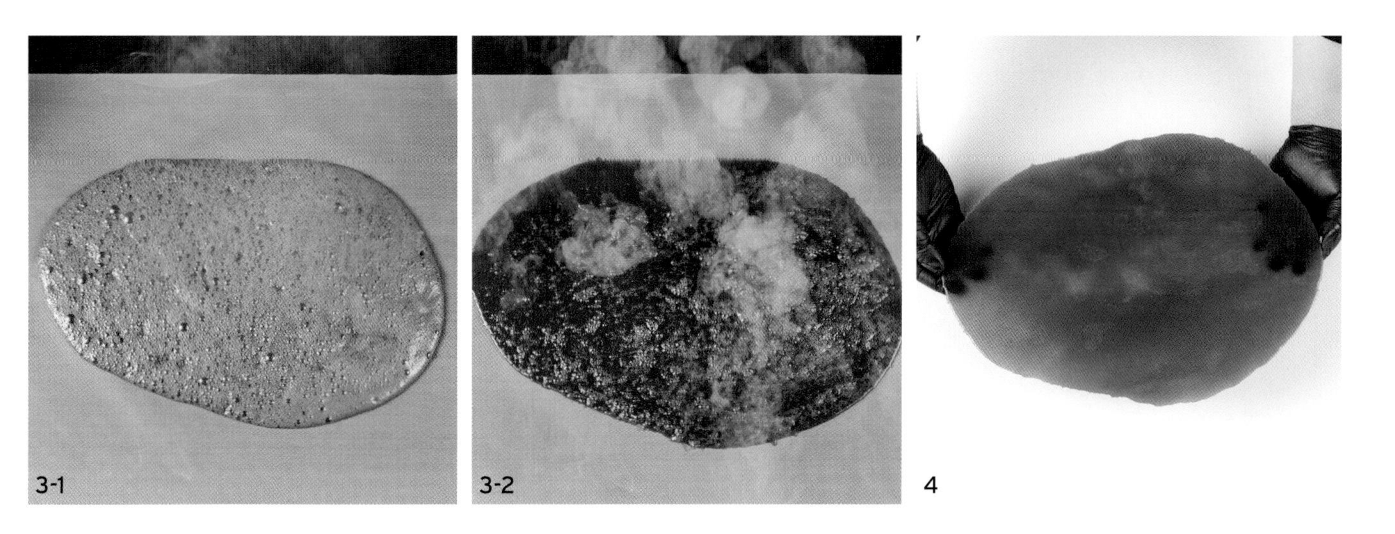
3-1
3-2
4

✤ ✤

SOYBEAN PULP MADELEINE

콩비지 마들렌

1 콩비지는 유산지를 깐 철판에 펼쳐 100℃로 예열된 오븐(데크 오븐의 경우 뎀퍼를 열고)에서 옅은 갈색이 날 때까지 중간중간 섞어가며 구운 후, 믹서에 곱게 갈고 체에 내려 가루 상태로 만들어 사용한다.

TIP. 30분간 굽고 섞은 후, 다시 30분간 구워 구움색을 고르게 낸다.
콩비지의 양에 따라 굽는 시간을 조절한다. 수분감이 없는 완벽한 가루 상태가 되도록 굽는다.

2 비커에 달걀, 설탕, 꿀, 전화당을 넣고 바믹서로 블렌딩한 후 냉장고에서 하루 동안 숙성시킨다.

3 30℃로 맞춘 **2**에 녹인 캐러멜초콜릿을 넣고 바믹서로 블렌딩한다.

4 우유(30℃)를 넣고 블렌딩한다.

5 피칸 페이스트, 바닐라에센스를 넣고 블렌딩한다.

6 믹싱볼에 체 친 콩비지 가루와 중력분을 넣고 **5**를 흘려 넣어가며 믹싱한다.

TIP. 콩비지 가루와 중력분은 작업 전 30℃ 오븐에 넣어두고 온도가 30℃가 되면 사용한다.

7 식용유와 정제버터(45℃)를 넣고 믹싱한다.

8 베이킹파우더, 소금을 넣고 믹싱해 반죽을 완성한다.

1 Spread the soybean pulp on a parchment-lined baking sheet and bake in an oven preheated to 100°C (with the vents open for a deck oven) until lightly browned, stirring occasionally. Then, grind them in a food processor into a fine powder and sieve.

TIP. Bake for 30 minutes, mix around, and bake for another 30 minutes to even out the color. Adjust baking time according to the amount of soybean pulp. You want the powder to be dehydrated with no moisture.

2 In a beaker, blend eggs, sugar, honey, and inverted sugar with an immersion blender; refrigerate overnight.

3 Add melted caramel chocolate into (**2**) at 30°C and continue to blend.

4 Blend in milk (30°C).

5 Blend in pecan paste and vanilla essence.

6 Sift in soybean pulp powder and all-purpose flour into a mixing bowl. Combine while drizzling (**5**) into the mixture.

TIP. Place the soybean pulp powder and all-purpose flour in a 30°C oven ahead of time and use when the powders are 30°C.

7 Mix in vegetable oil and clarified butter (45°C).

8 Mix with baking powder and salt to finish.

1

2

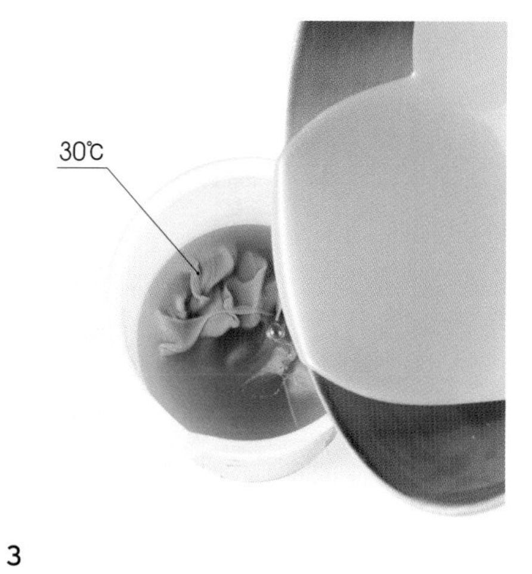

3

4

5

6

7

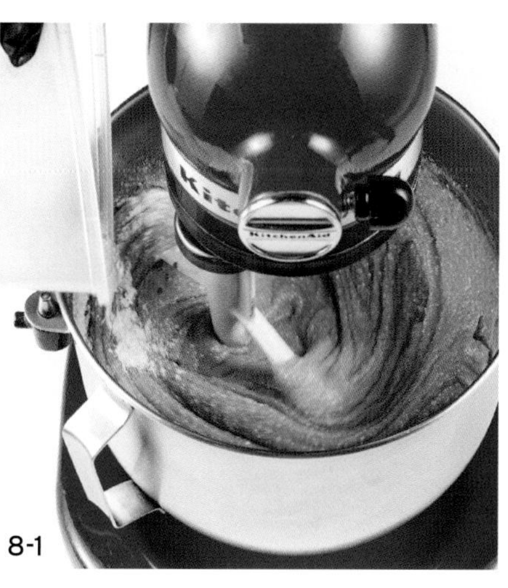
8-1

8-2

9 마들렌 모양 실리콘 몰드(플렉시판 FP2511)에 반죽을 26g씩 채운다.

10 적당한 크기로 부순 토피를 1g씩 올린다.

11 코코넛 크럼블을 4g씩 올린다.

12 190℃로 예열된 오븐에서 12분간 굽는다.

TIP. 오븐에서 나온 직후 30분간 냉장고에 둔다. 완전히 식으면 공기가 들어가지 않게 포장해 24시간 동안 냉장고에서 숙성시킨 후 바로 섭취/판매하거나 냉동고에 보관한다.

9 Fill the madeleine-shaped silicon molds (Flexipan FP2511) with 26 grams of batter.

10 Place 1 gram of toffee cut into moderate size.

11 Top with 4 grams of coconut crumble.

12 Bake in an oven preheated to 190°C.

TIP. Immediately after baking, place in a refrigerator for 30 minutes. Once completely cooled, wrap them airtight and let rest for 24 hours before consuming/selling or storing in a freezer.

9

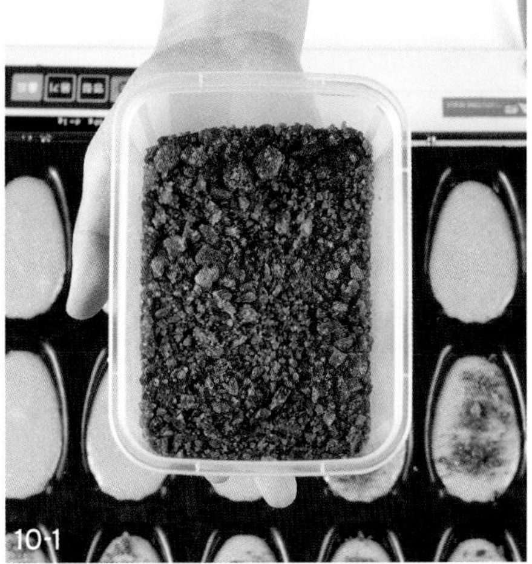
10-1

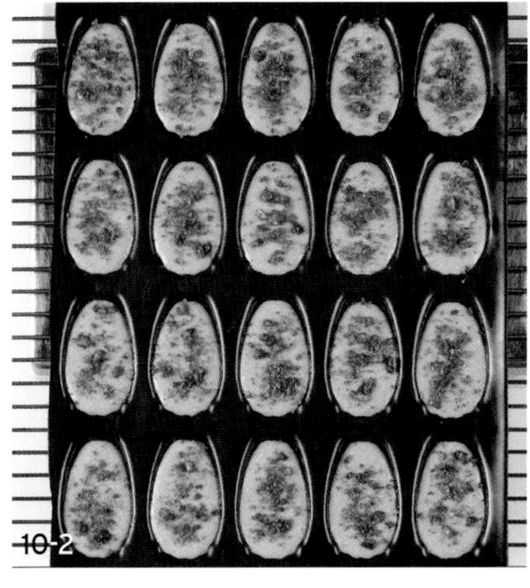
10-2

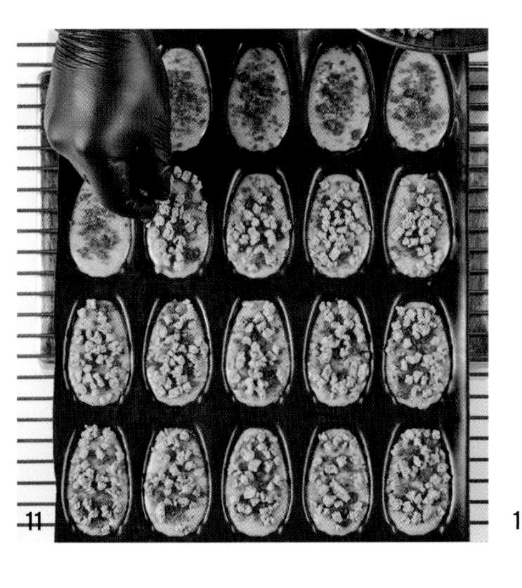
11

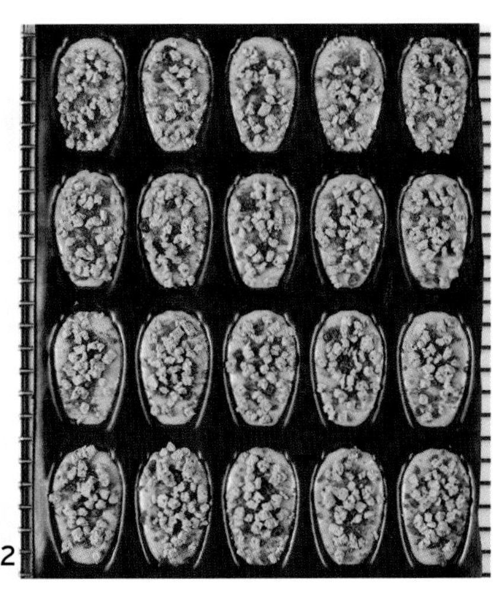
12

3

COFFEE GROUNDS POUND CAKE

커피박 파운드 케이크

- 콩비지 피낭시에, 콩비지 마들렌과 마찬가지로 2022 WCM 대회의 주제였던 'Tomorrow(내일, 미래)'와 관련된 디저트를 개발하던 중에 아이디어가 떠올라 만들게 된 메뉴이다.

- 리사이클링(재활용)이 가능한 재료 중 커피박(에스프레소 추출 후 남은 커피 원두 찌꺼기)을 사용했다.
  → 커피박은 커피와 관련된 디저트에 사용하면 인위적이지 않으면서 커피 본연의 자연스러운 맛과 향을 연출할 수 있다.

- 룸 템퍼라처 글레이즈에도 커피박을 첨가해 맛과 향을 더했다.

- 제품 윗면에 카카오 크럼블을 올려 식감과 시각적인 포인트를 주었다.

- As with the Soybean Pulp Financier and Soybean Pulp Madeleine, the idea for this menu came up while we were developing a dessert related to the theme of the 2022 WCM (World Chocolate Masters) competition, 'Tomorrow (future).'

- We used coffee grounds (the leftover after brewing espresso) as a recyclable ingredient.
  → When the coffee grounds are used in coffee-related desserts, it can bring out the natural flavor and aroma of coffee without being artificial.

- I used the coffee grounds in the room temperature glaze to add flavor and aroma.

- Cacao crumbles are used for texture and visually appealing design.

✦ HOW TO COMPOSE THIS RECIPE ✦

STEP 1.

메뉴 정하기	파운드 케이크 (공립법)
Decide on the menu	Pound Cake (Sponge method)

STEP 2.

메인 맛 정하기	커피박
Choose the primary flavor	Coffee Grounds

STEP 3.

메인 맛(커피박)과의 페어링 선택하기

Select a pairing flavor (Coffee Grounds)

- ☑ 다크초콜릿 Dark Chocolate
- ☑ 밀크초콜릿 Milk Chocolate
- ☑ 헤이즐넛 Hazelnut
- ☑ 카카오파우더 Cacao powder

STEP 4.

구성하기

Assemble

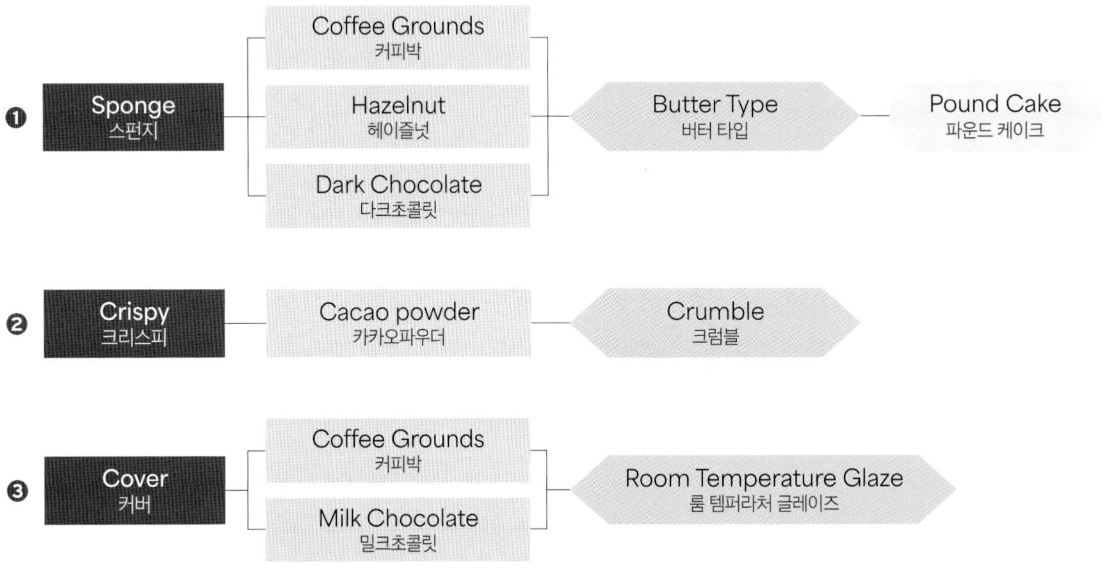

❶ Sponge 스펀지 — Coffee Grounds 커피박 / Hazelnut 헤이즐넛 / Dark Chocolate 다크초콜릿 — Butter Type 버터 타입 — Pound Cake 파운드 케이크

❷ Crispy 크리스피 — Cacao powder 카카오파우더 — Crumble 크럼블

❸ Cover 커버 — Coffee Grounds 커피박 / Milk Chocolate 밀크초콜릿 — Room Temperature Glaze 룸 템퍼라처 글레이즈

카카오 크럼블
Cacao Crumble
커피박 파운드 케이크
Coffee Grounds Pound Cake
가나 커피박 글레이즈
Ghana Coffee Grounds Glaze

INGREDIENTS 6개 분량/ 6 pound cakes

커피박 파운드 케이크

Coffee Grounds Pound Cake

INGREDIENTS		g
달걀	Eggs	350g
전화당	Inverted sugar	30g
설탕	Sugar	200g
다크초콜릿 (NXT Dark 55.7%)	Dark chocolate	300g
버터 (Bridel)	Butter	250g
헤이즐넛파우더	Hazelnut powder	70g
박력분	Cake flour	25g
커피박	Coffee grounds	90g
베이킹파우더	Baking powder	5g
소금	Salt	3g
바닐라에센스 (Aroma Piu)	Vanilla essence	20g
TOTAL		**1343g**

카카오 크럼블

Cacao Crumble

INGREDIENTS		g
버터 (Bridel)	Butter	360g
코코넛슈거	Coconut sugar	200g
설탕	Sugar	100g
소금	Salt	8g
박력분	Cake flour	600g
카카오파우더 (Extra Brute)	Cacao powder	120g
옥수수전분	Cornstarch	60g
물	Water	100g
바닐라에센스 (Aroma Piu)	Vanilla essence	적당량/ QS
TOTAL		**1548g**

*** 만드는 법**

❶ 믹싱볼에 차가운 상태의 버터, 코코넛슈거, 설탕, 소금을 넣고 부드럽게 풀어준다.

❷ 체 친 [박력분, 카카오파우더, 옥수수전분]을 넣고 믹싱하면서 물과 바닐라에센스를 흘려 넣어 반죽 상태로 만든다.

❸ 테프론시트를 깐 철판 위에서 체(간격 약 0.5cm)에 반죽을 내려 크럼블 상태로 만들어 냉동한다.

❹ 냉동시킨 크럼블은 밀폐용기에 담아 냉동 보관하면서 사용하고, 사용하기 전 손으로 가볍게 풀어준다.

*** Procedure**

❶ In a mixing bowl, cream the cold butter, coconut sugar, sugar, and salt until soft.

❷ Add sifted [cake flour, cacao powder, and cornstarch]. Drizzle water and vanilla essence while mixing until it forms a dough.

❸ Pass through a coarse sieve (0.5 cm mesh) onto a baking sheet lined with a Teflon sheet to make it into crumbles; freeze.

❹ Keep the frozen crumbles stored in an airtight container and use as needed. Gently break them apart with your hands before using them.

가나 커피박 글레이즈

Ghana Coffee Grounds Glaze

INGREDIENTS		g
밀크초콜릿 (Ghana 40%)	Milk chocolate	1000g
식용유	Vegetable Oil	200g
커피박	Coffee grounds	50g
TOTAL		**1250g**

＊만드는 법

❶ 비커에 녹인 밀크초콜릿(45~50℃), 식용유, 커피박을 넣고 바믹서로 블렌딩한다.

❷ 30℃로 맞춰 사용한다.

＊Procedure

❶ In a beaker, blend melted milk chocolate (45~50°C), vegetable oil, and coffee grounds using an immersion blender.

❷ Use at 30°C.

✤

COFFEE GROUNDS POUND CAKE (SPONGE METHOD)

커피박 파운드 케이크 (공립법)

1 믹싱볼에 달걀, 전화당, 설탕을 넣고 뜨거운 물이 담긴 볼에 받쳐 중탕으로 45℃로 온도를 맞춘 후 휘핑한다.

2 기공이 조밀하고 단단한 상태(26~27℃ 정도)가 되면 마무리한다.

3 녹인 다크초콜릿(45~50℃)과 녹인 버터(45℃)를 넣고 가볍게 섞는다.

4 체 친 [헤이즐넛파우더, 박력분, 커피박, 베이킹파우더], 소금, 바닐라에센스를 넣고 고르게 섞는다.

5 가로 15cm, 세로 6cm, 높이 6cm 직사각형 틀에 반죽을 200g씩 채운다.

6 카카오 크럼블을 30g씩 올린 후 170℃로 예열된 오븐에서 30분간 굽는다. 굽고 난 후에는 틀에서 빼 식힌 다음 개별로 랩핑해 냉장 보관한다.

7 완전히 식은 파운드 케이크는 가나 커피박 글레이즈로 2/3 정도 높이까지 코팅한다.

1 Warm eggs, inverted sugar, and sugar over a double boiler to 45°C.

2 Whip until dense and stiff (26~27°C).

3 Add melted dark chocolate (45~50°C) and melted butter (45°C) and mix lightly.

4 Mix evenly with sifted [hazelnut powder, cake flour, coffee grounds, baking powder], salt, and vanilla essence.

5 Fill in 15 cm wide, 6 cm long, and 6 cm high rectangular molds with 200 grams of batter.

6 Top each with 30 grams of cacao crumble and bake in an oven preheated to 170°C for 30 minutes.
After baking, let them cool, wrap them individually, and refrigerate.

7 When completely cooled, coat the pound cakes with the Ghana coffee grounds glaze to about two-thirds from the bottom.

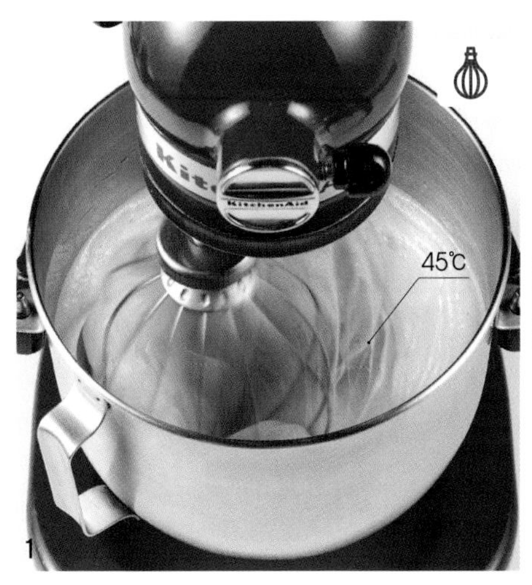

1

2-1

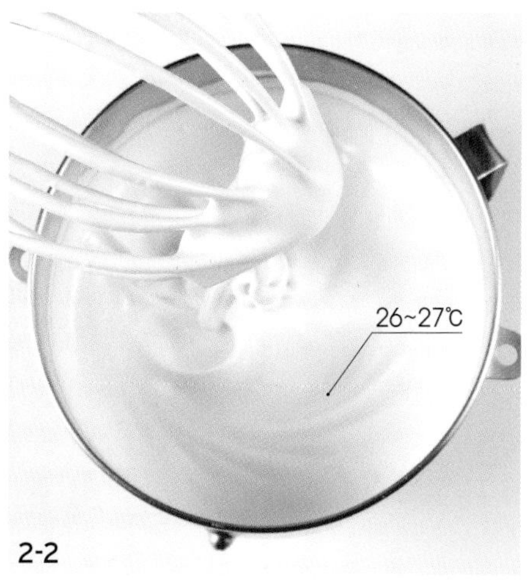

2-2

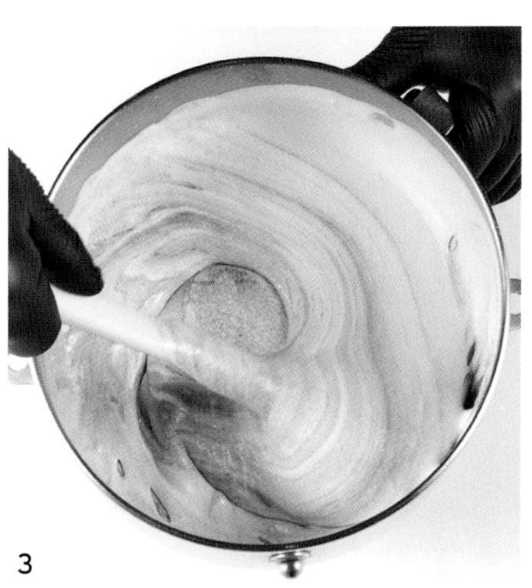
3

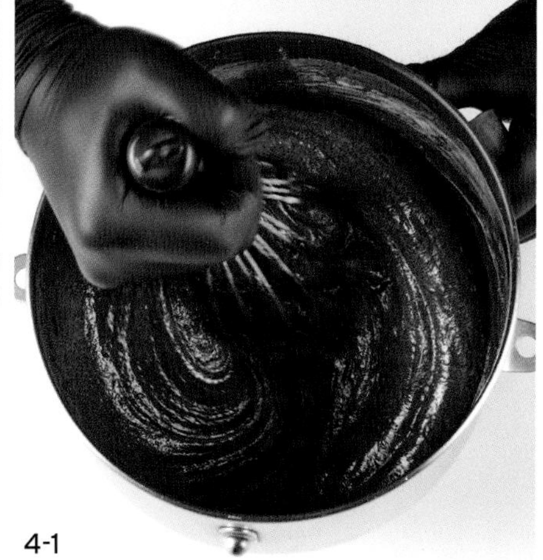
4-1

4-2

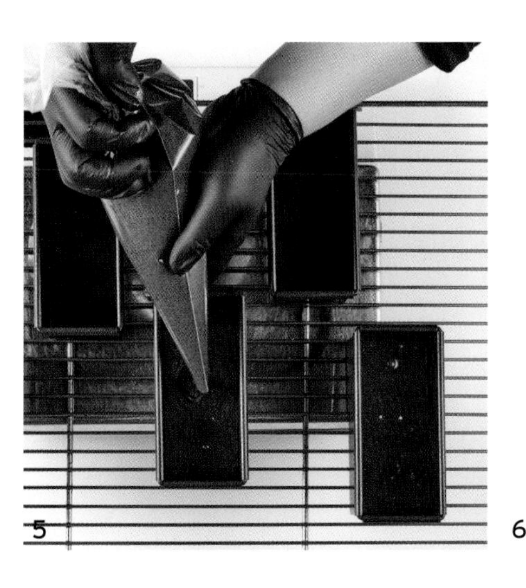
5

6

7

BRANDS AND DISTRIBUTORS USED IN THS BOOK

이 책에서 사용한 브랜드 & 유통사

제원 인터내셔널
Zewon International

이 책에서 사용한 초콜릿, 퓌레 브랜드인 카카오바리와 브와롱의 유통사입니다.
The distributor of Cacao Barry and Boiron, the chocolate and purée brands used in this book.

—

홈페이지 www.jewon1986.com

카카오바리
Cacaobarry

브와롱
Boiron

성우테크
Sungwoo Tech

robot coupe®

로보쿱
Robot Coupe

이 책에서 사용한 로보쿱의 유통사입니다.
The distributor of Robot Coupe used in this book.

—

홈페이지 www.ichef.co.kr

솜 인터내셔널
SOM International

이 책에서 사용한 동결건조 과일, 건과일을 제조하는 브랜드입니다.
The brand that manufactures the freeze-dried and dried fruits used in this book.

—

홈페이지 www.sominternational.com
쇼핑몰 딜라잇가든 delightgarden.co.kr

비앤씨마켓
B&C Market

이 책에서 사용한 실리코마트사의 몰드를 판매하는
제과제빵 도구 및 재료 쇼핑몰입니다.
An online shopping mall for bakery tools and ingredients that
sells the Silikomart molds used in this book.

—

홈페이지 www.bncmarket.com

실리코마트
Silikomart

아이푸드넷
I-Food Net

이 책에서 사용한 소사(Sosa) 제품, 에센스류, 알부민파우더의 유통사입니다.
The distributor of Sosa ingredients, essence ranges, and
albumin powder used in this book.

—

홈페이지 www.ifoodnet.co.kr
쇼핑몰 www.ifoodnetmall.com

올리커
Allliquor

이 책에서 사용한 리큐르의 유통사입니다.
The distributor of the liqueurs used in this book.

—

홈페이지 www.all-liquor.co.kr

malang company

말랑컴퍼니
Malang Company

이 책에서 사용한 자체 제작 몰드를 만든 회사입니다. 책을 위해 자체 제작한 몰드를 독자 분들도
구매하고 사용하실 수 있도록 말랑컴퍼니 사이트에서도 판매하고 있습니다.
The company that created the custom-made molds used in this book.
The molds we made for the book are also available for sale on the Malang Company site so
readers can purchase and use them.

—

홈페이지 www.malangstore.com

Eric Kim

The first Korean judge in the Baking & Pastry category of the World Association of Chefs' Societies (WACS) accredited international competition, recognized by top chefs around the world. In 2022, he achieved a remarkable result in the World Chocolate Masters (WCM) competition, placing 6th in the world and 1st in Asia. With over 20 years of experience and knowledge in the field, he travels domestically and internationally to conduct seminars, lectures, and menu consulting. He has been teaching at Seoul Hoseo Occupational Training College in Seoul for 12 years, where he has been training students in the Baking & Pastry Arts department.

Awards & Honors

2022
1st place in Asia, World Chocolate Masters

2018
World Association of Chefs Societies (WACS) Continental Judge in Confectionery

National Judge for Confectionery Division, Korea Association of Chefs Societies

CACAO BARRY Brand Ambassador

CACAO BARRY Collective Lab Head Chef

2016
World Culinary Olympics (IKA Culinary Olympics), Germany

Gold Medal in Cold Display Dessert

2014
National Team at the Expogast Culinary World Cup in Luxembourg

2012
National Team at the IKA Culinary Olympics, Germany

HOW TO MAKE *My own* RECIPE

First edition printed May 1, 2024
First edition published May 17, 2024

Author Kim Dongseok
Translated by Kim Eunice
Publisher Bak Yunseon
Published by THETABLE Inc.

Plan & Edit Bak Yunseon
Proofreading Kim Youngran
Design Kim Bora
Photograph Park Sungyoung
Sales/Marketing Kim Namkwon, Cho Yonghoon, Moon Seongbin
Management support Kim Hyoseon, Lee Jungmin

Address 122, Jomaru-ro 385beon-gil, Bucheon-si, Gyeonggi-do, Republic of Korea
Website www.icoxpublish.com
Instagram @thetable_book
E-mail thetable_book@naver.com
Phone 82-32-674-5685
Registration date August 4, 2022
Registration number 386-2022-000050
ISBN 979-11-92855-10-3 (14590)
979-11-92855-09-7 (14590) [SET]